Springer Monographs in Mathematics

This series publishes advanced monographs giving well-written presentations of the "state-of-the-art" in fields of mathematical research that have acquired the maturity needed for such a treatment. They are sufficiently self-contained to be accessible to more than just the intimate specialists of the subject, and sufficiently comprehensive to remain valuable references for many years. Besides the current state of knowledge in its field, an SMM volume should ideally describe its relevance to and interaction with neighbouring fields of mathematics, and give pointers to future directions of research.

Tsuyoshi Kato • Nobuhiro Nakamura

Monopole Maps and L^2 -cohomology

Springer

Tsuyoshi Kato
Graduate School of Science, Division of Mathematics and Mathematical Sciences
Kyoto University
Kyoto, Japan

Nobuhiro Nakamura
Center for Integrated Sciences and Humanities
Fukushima Medical University
Fukushima, Japan

ISSN 1439-7382 ISSN 2196-9922 (electronic)
Springer Monographs in Mathematics
ISBN 978-981-96-8471-7 ISBN 978-981-96-8472-4 (eBook)
https://doi.org/10.1007/978-981-96-8472-4

Mathematics Subject Classification: 19K56, 19L47, 20E26, 43A05, 55P20, 55P42, 57R20, 57R91, 58D27, 32Q15, 32Q45, 43A07, 47A53, 53C27, 55Q10, 55Q91, 55R40, 57-03, 57R55, 58A12, 58B15, 58D17, 58J20, 19E20, 19K14, 19L64, 43A15, 46E35, 53C05, 55N15, 55P91, 55Q55, 57K41, 57N75, 57R15, 57R19, 57R22, 58A10, 58D19, 58J10

This Springer imprint is published by the registered company Springer Nature Singapore Pte Ltd.
The registered company address is: 152 Beach Road, #21-01/04 Gateway East, Singapore 189721, Singapore

To Mikio Furuta

Preface

The theme of this book is to introduce two theories that have not often been discussed together: gauge theory and L^2-cohomology theory. We will explain basic materials on these theories and related subjects in a unified way, and propose new differential-topological conjectures with various evidence that supports them. This book is intended to be almost self-contained so that it is accessible not only for graduate students in geometry and topology but also for those who are non-experts in one of the two theories.

The style of this book is a hybrid of different branches in differential topology and related topics. In general, attempts to gain knowledge of different fields by referring to individual references can be quite confusing, for example, due to differences in notational styles or unconscious assumptions. To make the book accessible to readers, we have tried to keep the style of the descriptions as consistent as possible.

Kyoto, Japan — Tsuyoshi Kato
Fukushima, Japan — Nobuhiro Nakamura
April, 2025

Contents

Symbols

$b_i(M)$	The i-th Betti number of M	Sect. 2.1
$\chi(M)$	Euler characteristic of M	Sect. 2.1
$\sigma(M)$	Signature of M	Sect. 2.1
$b^{\pm}(M)$	Dimension of a maximal positive (resp. negative) definite subspace of $H_{2k}(M;\mathbb{Q})$	Sect. 2.1
$e(M)$	Euler class of M	Sect. 2.2
$p_i(M)$	Pontryagin class of M	Sect. 2.2
$w_i(M)$	Stiefel–Whitney class of M	Sect. 2.2
$c_i(M)$	Chern class of M	Sect. 2.2
$\mathrm{ind}(D)$	Index of the elliptic operator D	Sect. 2.3
TM	Tangent bundle of M	Sect. 2.3
$\hat{\mathcal{A}}(TM)$	$\hat{\mathcal{A}}$-genus of TM	Sect. 2.3
$ch(S)$	Chern character of the vector bundle S	Sect. 2.3
$\mathcal{L}(M)$	$\mathcal{L}$-genus of M	Sect. 2.3
$\mathbb{C}P^m$	Complex projective space	Sect. 2.3
Σ_g	Riemann surface of genus g	Sect. 2.3
$\mathbf{H}_{2k}$	$2k$-dimensional hyperbolic space	Sect. 2.3
$\tilde{M}$	Universal covering space of M	Sect. 2.3
$L^2(X;\Lambda^p)$	Space of L^2-forms of degree p on X	Sect. 2.3
$\mathcal{H}^p(X)$	Space of L^2-harmonic p-forms of X	Sect. 2.3
P_Γ	The L^2-projection to $\mathcal{H}^p(X)$ in $L^2(X;\Lambda^p)$	Sect. 2.3
tr_Γ	Γ-trace	Sect. 2.3
$b_\Gamma^p(X)$	L^2-Betti number of X	Sect. 2.3
$\pi_1(M)$	Fundamental group of M	Sect. 2.3
T^n	n-dimensional torus	Sect. 2.4
S^n	n-dimensional sphere	Sect. 2.5
$M \hookrightarrow W$	Inclusion of M into W	Sect. 2.5
$\tau(f)$	Whitehead torsion of f	Sect. 2.5
$Wh(\Gamma)$	Whitehead group of Γ	Sect. 2.5
$\mathbf{S}(X)$	Structure set of X	Sect. 2.5

G/O	Classifying space for stable bundles with fiber homotopy trivialization Sect. 2.5
$L_*(A)$	Wall's L group of A Sect. 2.5
∂M	Boundary of the manifold M Sect. 2.6.1
E_8	Cartan matrix of E_8 Sect. 2.8.1
H	Hyperbolic form Sect. 2.8.1
Q_X	Intersection form of X Sect. 2.8.2
(1)	Bilinear form specified by the matrix (1) Sect. 2.8.2
$E(n)$	Elliptic fibration Sect. 2.8.4
$\twoheadrightarrow$	Surjective map Sect. 2.9
$\sqcup$	Disjoint union Sect. 2.9
C_p	Cyclic group of order p Sect. 2.9
$\pi_m(X)$	m-th homotopy group of X Problem 2.55
$\mathrm{Cl}(V)$	Clifford algebra of V Sect. 3.1.1
$\mathbb{H}$	Space of quaternions Sect. 3.1.1
$\mathbb{H}^{\pm}$	Decomposition of $\mathrm{Cl}(\mathbb{R}^3)$ Sect. 3.1.1
$\mathrm{Pin}(V)$	Pin group of V Sect. 3.1.2
$\mathrm{Spin}(V)$	Spin group of V Sect. 3.1.2
$\mathrm{Pin}(n) = \mathrm{Pin}(\mathbb{R}^n)$	Sect. 3.1.2
$\mathrm{Spin}(n) = \mathrm{Spin}(\mathbb{R}^n)$	Sect. 3.1.2
$\mathrm{Sp}(n)$	Group of unit quaternions Sect. 3.1.2
$\mathrm{Aut}(\mathrm{Cl}(V), V)$	Group of automorphisms on $\mathrm{Cl}(V)$ that preserves V Sect. 3.1.2
$\omega_{\mathbb{C}}$	Involutive element in $\mathrm{Cl}(V) \otimes \mathbb{C}$ Sect. 3.1.3
$(\mathrm{Cl}(V) \otimes_{\mathbb{R}} \mathbb{C})^{\pm}$	Decomposition of $\mathrm{Cl}(V) \otimes_{\mathbb{R}} \mathbb{C}$ with respect to $\omega_{\mathbb{C}}$ Sect. 3.1.3
$\mathrm{M}_k(\mathbb{C})$	$(k \times k)$ matrix algebra over $\mathbb{C}$ Sect. 3.1.3
$S_{\mathbb{C}}(V)$	Complex irreducible representation $S_{\mathbb{C}}(V)$ of $\mathrm{Cl}(V) \otimes_{\mathbb{R}} \mathbb{C}$ Sect. 3.1.3
$\mathrm{End}_{\mathbb{C}}(S_{\mathbb{C}}(V))$	Algebra of endomorphisms of $S_{\mathbb{C}}(V)$ Sect. 3.1.3
$S^{\pm}_{\mathbb{C}}(V)$	Decomposition of $S_{\mathbb{C}}(V)$ with respect to $\omega_{\mathbb{C}}$ Sect. 3.1.3
$\Delta_{\mathbb{C}}$	Complex spin representation of $\mathrm{Spin}(V)$ Sect. 3.1.3
$\mathbb{H}_T$	Space of quaternions identified with $V = \mathbb{R}^4$ Sect. 3.1.3
$\Lambda^* V$	Exterior algebra of V Sect. 3.1.3.1
$\mathrm{Spin}^c(V)$	Spin^c group of V Sect. 3.1.3.2
P_X	Oriented frame bundle of X Sect. 3.2
$\mathrm{Or}(E)$	Bundle of orientations Sect. 3.2
BG	Classifying space of G Sect. 3.2
EG	Universal G bundle over BG Sect. 3.2
$K(G, n)$	Eilenberg–MacLane space Sect. 3.2
$W_{k+1}(P)$	Image of $w_k(P)$ by the Bockstein homomorphism Sect. 3.2
$\mathrm{Ext}(A, B)$	Ext group Sect. 3.2
$\mathrm{Cl}(P)$	Clifford algebra bundle of P Sect. 3.2
∇	Connection Sect. 3.3.1

$\Omega^p(E)$	Space of E-valued p-forms Sect. 3.3.1
$\Omega^0(E) = \Gamma(E)$	Space of smooth sections of E Sect. 3.3.1
R_∇	Curvature of ∇ Sect. 3.3.1
$(\omega_i^j)_{0\le i,j\le n}$	Connection matrix Sect. 3.3.1
$T^\vee P$	Bundle of vertical vectors of P Sect. 3.3.3
$\mathfrak{g}$	Lie algebra of G Sect. 3.3.3
X^H	Horizontal projection and $\pi : P \to X$. Sect. 3.3.3.1
$\mathrm{Aut}(P)$	Gauge transformation group of P Sect. 3.3.5
$\mathcal{G}$	Gauge transformation group of spin^c structure Sect. 3.3.5
D_A	Dirac operator Sect. 3.5.2
$\mathrm{div}\, V$	Divergence of the vector field V Sect. 3.5.2
$\mathrm{ind}\, D$	Index of D Sect. 3.5.2
$\det E$	Determinant line bundle of E Sect. 3.5.2
κ	Scalar curvature Sect. 3.5.2
∇^*	L^2-adjoint of ∇ Sect. 3.5.2
$V^{p,q}$	Space of (p,q)-exterior elements in $\Lambda^*_{\mathbb{C}}(V\otimes\mathbb{C})$ Sect. 3.6
$\pi^{p,q}$	Projection to $V^{p,q}$ Sect. 3.6
$\Omega^{p,q}(Z)$	Space of (p,q)-forms of the complex manifold Z Sect. 3.7
∂	Holomorphic differential Sect. 3.7
$\bar\partial$	Anti-holomorphic differential Sect. 3.7
ρ	Clifford multiplication Sect. 4.1.1
$\mathfrak{s}$	Spin^c structure Sect. 4.1.2
$\Lambda_T^\pm$	Some Spin$^c(4)$ modules Sect. 4.1.3
q	Quadratic map Sect. 4.1.3
$\Omega^+(X)$	Space of self-dual 2-forms Sect. 4.1.3
F_A^+	Self-dual part the curvature F_A Sect. 4.1.3
$\mathcal{A}(L)$	Space of U(1) connections on the determinant line bundle L Sect. 4.1.3
$\mathcal{M}$	Moduli space Sect. 4.1.3
$\|\cdot\|_{L_k^p}$	Sobolev norm Sect. 4.2
P^*	Formal adjoint of the operator P Sect. 4.2
d^+	Composition of the exterior differentiation with projection to the self-dual part Sect. 4.2
$R(d)$	Range of d Sect. 4.2
H^1	Space of harmonic 1-forms Sect. 4.2
$\mathcal{A}_k$	Space of L_k^2-connections Sect. 4.3
C_k	Space of L_k^2-configurations Sect. 4.3
$\mathcal{G}_k$	Group of L_k^2-gauge transformations Sect. 4.3
$\mathcal{B}_k$	Quotient space Sect. 4.3.1
$\Gamma_{(A,\Phi)}$	Stabilizer of $\mathcal{G}$ at (A,Φ) Sect. 4.3.1
C_k^*	Irreducible part of C_k Sect. 4.3.2
$\mathcal{B}_k^*$	Irreducible part of $\mathcal{B}_k$ Sect. 4.3.2
$L_{(A,\Phi)}$	Linearization of $\mathcal{G}$-action Sect. 4.3.2

$\mathrm{Lie}(G)$	Lie algebra of the Lie group G Sect. 4.3.2
$\mathbb{L}$	Universal line bundle Sect. 4.3.3
μ	μ-map Sect. 4.3.3
$/$	Slant product Sect. 4.3.3
$\times$	Cross product Sect. 4.3.3
$\mathrm{Hol}_{\tilde{\gamma}}(A)$	Holonomy of U(1) connection A around the loop $\tilde{\gamma}$ Sect. 4.3.3
deg	Degree of map Sect. 4.3.3
Re	Real part Sect. 4.3.3
$\nabla_{L.C.}$	Levi-Civita connection Sect. 4.3.3
$c_1(\mathfrak{s})_{\mathbb{R}}$	First Chern class in $\mathbb{R}$-coefficient Sect. 4.4.1
$\mathcal{P}_{k-1}$	Space of perturbations Sect. 4.5
$\mathcal{M}_\eta = \mathcal{M}(\mathfrak{s}, \eta)$	Moduli space perturbed with η Sect. 4.5
$\tilde{m}$	Map defined by the Seiberg–Witten equations Sect. 4.5
$\mathbb{M}^*$	Parametrized moduli space Sect. 4.5
$d(\mathfrak{s})$	Virtual dimension of the moduli space $\mathcal{M}(\mathfrak{s}, \eta)$ Sect. 4.5
$h_g^+(\eta)$	g-self-dual harmonic part of the 2-form η Sect. 4.5
$\mathcal{W}$	Wall Sect. 4.5
$L^{sw}_{(A,\Phi)}$	Linearization of $\tilde{m}$ Sect. 4.6
Λ	Direct sum of operators $L^*_{(A,\Phi)}$ and $L^{sw}_{(A,\Phi)}$ Sect. 4.6
$\tilde{\mathcal{M}}^*(\mathfrak{s}, \eta)$	Irreducible solutions to the perturbed Seiberg–Witten equations Sect. 4.6
$\det(\Lambda)$	Determinant line bundle of a family of Fredholm operators Sect. 4.6, Sect. 4.9
$H^+(X)$	Space of self-dual harmonic 2-forms Sect. 4.6
$\mathrm{Spin}^c(X)$	Set of equivalence classes of spin^c structures on X Sect. 4.7
$\mathrm{SW}_{X,\mathfrak{s},O}$	Seiberg–Witten invariant Sect. 4.7
O	Orientation of the moduli space Sect. 4.7
$\mathrm{SW}^{\pm}_{X,\mathfrak{s}}$	Seiberg–Witten invariants when $b^+ = 1$ Sect. 4.9
$\mathbf{H}$	Hyperboloid Sect. 4.9
$\omega_g(\mathfrak{h})$	g-self-dual harmonic 2-form Sect. 4.9
$\mathfrak{S}(g, \eta, \mathfrak{s}, \mathfrak{h})$	Cohomology class used in the characterization of $\mathcal{W}$ Sect. 4.9
$\mathscr{C}^{\pm}(\mathfrak{h})$	Chambers Sect. 4.9
$\mathscr{Met}(X)$	Space of Riemannian metrics on X Sect. 4.9
$\mathcal{F}(X, Y)$	Space of Fredholm operators Sect. 4.9
J	torus $H^1(X; \mathbb{R})/H^1(X; \mathbb{Z})$ Sect. 5.1
$\mathcal{E}, \mathcal{F}$	Hilbert vector bundles over J Sect. 5.1
$\boldsymbol{\mu}_{SW}$	Monopole map Sect. 5.1
ℓ	Linear part of the monopole map Sect. 5.1
c	Quadratic part of the monopole map Sect. 5.1
$\boldsymbol{\mu}$	U(1)-equivariant map Sect. 5.2
D_R	Disk bundle of $\mathcal{E}$ Sect. 5.2

S_R	Sphere bundle of $\mathcal{E}$ Sect. 5.2
$\mathcal{U}$	Hilbert space such that $\mathcal{F} = J \times \mathcal{U}$ Sect. 5.2
E_U, F_U	finite-dimensional vector bundles associated with a linear subspace $U \subset \mathcal{U}$ Sect. 5.2
P_U	Projection Sect. 5.2
f_U	Finite-dimensional approximation Sect. 5.2
$\hat{U}$	Vector space Sect. 5.2
D_U	Disk bundle in E_U Sect. 5.2
S_U	Sphere bundle in E_U Sect. 5.2
$T(E_U)$	Thom space of E_U Sect. 5.2
S^U	1-point compactification of U Sect. 5.2
$D_r(U)$	Disk with radius r in U Sect. 5.2
$\wedge$	Smash product Sect. 5.2
$\{T(\mathrm{ind}\,\ell), S^0\}^{\mathrm{U}(1)}_{\mathcal{U}}$	U(1)-equivariant stable cohomotopy group Sect. 5.2
$\widetilde{\mathrm{SW}}_{X,\mathfrak{s}}$	Stable cohomotopy Seiberg–Witten invariant or Bauer–Furuta invariant Sect. 5.2
$\mathcal{E}^*$	Irreducible part of $\mathcal{E}$ Sect. 5.3.1
$\mathcal{V}$	Hilbert vector bundle over $\mathcal{E}^* / \mathrm{U}(1)$ Sect. subsec:approx
E_U^*	Irreducible part of E_U Sect. 5.3.1
V_U	Vector bundle over $E_U^* / \mathrm{U}(1)$ Sect. 5.3.1
$M(v)$	Approximated moduli space Sect. 5.3.1
F_U	U(1)-equivariant map Sect. 5.3.1
μ_f	μ-map for a finite-dimensional approximation Sect. subsec:approx
$\tilde{\gamma}$	Circle-valued map from J Sect. subsec:approx
τ	Homomorphism from $\{T(\mathrm{ind}\,\ell), S^0\}^{\mathrm{U}(1)}_{\mathcal{U}}$ Sect. 5.3.2
$\pi^{b-1}(X)$	Cohomotopy group of X Sect. 5.3.2
Ci	Cone of the map i Sect. 5.3.2
$PD(\alpha)$	Poincaré dual of the cohomology class α Sect. 5.4
$H_G^*(M)$	Equivariant cohomology Sect. 5.4.1
$\tau_G(E)$	Equivariant Thom class Sect. 5.4.1
$e_G(E)$	Equivariant Euler class Sect. 5.4.1
$f^{\mathrm{U}(1)}$	The restriction of f to the U(1) fixed point set Sect. 5.4.2
$\bar{f}$	Map between disks induced from f Sect. 5.4.2
$K_G(X)$	Equivariant K-group Sect. 5.5.3
$R(G)$	Complex representation ring of the group G Sect. 5.5.3
$\lambda_{-1}(V)$	Alternative sum of exterior algebra in $R(G)$ Sect. 5.5.3
ε	Homomorphism from Pin(2) to $\{\pm 1\}$ Sect. 5.5.3
$\tilde{\mathbb{R}}$	Real 1-dimensional representation of Pin(2) Sect. 5.5.3
$\tilde{\mathbb{C}}$	Complex 1-dimensional representation of Pin(2) Sect. 5.5.3
$\Omega_c^m(X)$	Space of smooth m-forms on X with compact support Sect. 6.1

$H^m_{(2)}(X)$	Unreduced L^2-cohomology group of X Sect. 6.1
$\bar{H}^m_{(2)}(X)$	Reduced L^2-cohomology group of X Sect. 6.1
Δ	Laplace–Beltrami operator Sect. 6.1
$\mathcal{H}^m(X)$	Space of L^2-harmonic forms of degree m Sect. 6.1
Δ_m	Laplace–Beltrami operator of degree m Sect. 6.1.1
E^m_λ	Spectral density function of the Laplace operator Sect. 6.1.1
$\mathrm{spec}(\Delta_m)$	Spectral distribution of the Laplace operator Sect. 6.1.1
F_n	Free group of rank n Sect. 6.2.1
$SL_n(\mathbb{Z})$	Special linear group of degree n over $\mathbb{Z}$ Sect. 6.2.2
$A \triangle B$	Symmetric difference Sect. 6.2.3
$\delta_y(x)$	Delta function Sect. 6.2.3
$l^p(\Gamma)$	l^p-space over Γ Sect. 6.2.3
$\mathcal{P}(\Gamma)$	Power set of Γ Sect. 6.2.3
$\mathrm{vol}(H)$	Volume of H Sect. 6.3
$\mathrm{area}(\partial H)$	Area of ∂H Sect. 6.3
χ_H	Characteristic function of H Sect. 6.3
$h(S)$	Cheeger's constant Sect. 6.3.1
$\{f \geq t\}$	Subset $\{x \in Y : f(x) \geq t\}$ in Y Sect. 6.3.1
$R_1(f)$	Normalized L^1-norm of ∇f Sect. 6.3.1
$R(f)$	Normalized L^2-norm of ∇f Sect. 6.3.1
∂Q	Boundary of the finite set Q Sect. 6.3.2
$h(X), h(Q), h(P)$	Variants of Cheeger's constant Sect. 6.3.2
$\mathrm{inj}(X)$	Injective radius Sect. 6.3.2
$V_K(r)$	Volume of the ball of radius r in the simply connected complete Riemannian manifold of constant curvature $-K^2$ Sect. 6.3.2
$A_K(r)$	Area of the boundary of the ball of radius r Sect. 6.3.2
$\mathbb{Z}(i)$	$2^i \cdot \mathbb{Z}$ Sect. 6.2.4
$\bar{H}^m_{(2)}(L)$	L^2-cohomology group of simplicial complex L Sect. 6.4
$C^i_{(2)}(\tilde{K})$	Space of real oriented L^2-cochains of degree i Sect. 6.4.1
d_c	Coboundary operator Sect. 6.4.1
$\hat{\sigma}$	Barycenter of the simplex σ Sect. 6.4.2
U_v	Star closure of v Sect. 6.4.3
$c_{ij}(A)$	$c_{ij}(A) = \langle Ae_i, e_i\rangle$ for a bounded operator A on a Hilbert space with orthonormal basis $\{e_i\}$ Sect. 7.1.1
$\lVert \cdot \rVert_{HS}$	Hilbert–Schmidt norm Sect. 7.1.1
$\langle \cdot \rangle_{HS}$	Hilbert–Schmidt inner product Sect. 7.1.1
$\mathrm{tr}(T)$	Trace for a trace class operator T Sect. 7.1.1
$\mathbb{C}\Gamma$	Group ring of a discrete group Γ Sect. 7.1.2
$*$	Convolution on $\mathbb{C}\Gamma$ Sect. 7.1.2
$\mathcal{N}(\Gamma)$	Group von Neumann algebra of Γ Sect. 7.1.2
$\mathrm{tr}_\Gamma(f)$	von Neumann Γ-trace Sect. 7.1.2
$\widehat{\ }$	Fourier transform Sect. 7.1.2

$\dim_\Gamma$	von Neumann Γ-dimension Sect. 7.1.2
$k_A(x, y)$	Kernel function of the operator A Sect. 7.1.3
$C_b^m(X)$	Space of functions with uniformly bounded C^m-norm Sect. 7.1.3
$D = d + d^*$	de Rham operator Sect. 7.1.3
$f(D)$	Functional calculus Sect. 7.1.3
$b_{(2)}^m(X)$	L^2-Betti number of X Sect. 7.1.4
$[\Gamma : \Gamma_m]$	Index of the subgroup $\Gamma_m \subset \Gamma$ Sect. 7.1.5
$K(i)$	Simplicial decomposition on X_i Sect. 7.1.5
$C_{(2)}^m(\tilde{K})$	Space of L^2-cochains on $\tilde{K}$ Sect. 7.1.6
$P(\lambda)$	Spectral decomposition Sect. 7.1.6
$E_i(\lambda)$	Ordered set of eigenvalues μ with $\mu \leq \lambda$ Sect. 7.1.6
$F_i(\lambda)$	Normalized cardinality of $E_i(\lambda)$ Sect. 7.1.6
$\overline{F_\infty}(\lambda)$, $\underline{F_\infty}(\lambda)$	Limits of $F_i(\lambda)$ Sect. 7.1.6
$\overline{F_\infty}^+(\lambda)$, $\underline{F_\infty}^+(\lambda)$	Limits from the right of $\overline{F_\infty}(\lambda)$ and $\underline{F_\infty}(\lambda)$ Sect. 7.1.6
$C^*(K; V)$	cochain group with coefficient V Sect. 7.1.6
$C_{(2)}^*(K; V)$	L^2-cochain group with coefficient V Sect. 7.1.6
d_*^V	Differential on $C_{(2)}^*(K; V)$ Sect. 7.1.6
$\mathrm{tr}_{\mathbb{C}\Gamma}\left(\sum_g \lambda_g \cdot g\right)$	Linear functional on $\mathbb{C}\Gamma$ Sect. 7.1.7
$\lvert G \rvert$	Number of elements of G Sect. 7.2
$\mathrm{ind}_\Gamma(\tilde{D})$	Γ-index of the operator $\tilde{D}$ Sect. 7.3
Q	Parametrix of the operator D Sect. 7.3.1
$\tilde{U} A$	Polar decomposition of $\tilde{D}$ Sect. 7.3.2
$\exp_x$	Exponential map Sect. 7.4
$\log_x$	Inverse of the exponential map $\exp_x$ Sect. 7.4
$ch(\nabla_1, \nabla_2)$	Relative Chern form Sect. 7.4.1
$ch(E_1, E_2)$	Relative Chern character Sect. 7.4.1
σ_Γ	Γ-signature Sect. 8.1
χ_{AHS}	Euler characteristic of the Atiyah–Hitchin–Singer complex Sect. 8.1
Z_g	Complex algebraic surface with non-zero signature due to Atiyah Sect. 8.2.1
Ric	Ricci curvature Sect. 8.4
CK	Geodesic cone of the subset K in a complete Riemannian manifold X Sect. 8.5
K_g	Sectional curvature Sect. 8.5
L_0^k	Linear map on exterior algebra Sect. 8.6
Λ_0^k	Algebraically dual operator of L_0^k Sect. 8.6
L^k	Linear map on differential forms Sect. 8.6
Λ^k	Algebraically dual operator of L_K Sect. 8.6

Chapter 1
Introduction

John Milnor, who is one of the greatest pioneers of differential topology, gave us a natural definition: "Differential topology may be defined as the study of those properties of differentiable manifolds which are invariant under diffeomorphism" [125]. Although over sixty-five years have passed since Milnor's words, and as is well known, differential topology has made remarkable progress in many directions and continues to evolve further, the main theme remains the same.

Surgery theory is the standard method for the classification of high-dimensional (≥ 5) smooth manifolds. In fact, the classification problem up to diffeomorphisms is reduced to the machinery of the surgery exact sequence (Theorem 2.19) which computes the structure set (the set of h-cobordism classes) in terms of a kind of topological K-group (the normal set) and the algebraic L-group over the fundamental group ring $\mathbb{Z}[\pi_1]$. This reveals, among other things, that the fundamental group of an underlying manifold is strongly associated with the diversity of the smooth structures. In general, finitely presented groups are very difficult to study as algebraic objects. Such a situation has led us to the analysis of the fundamental groups from the view-point of large-scale metric geometry on the universal covering spaces. Both L^2-cohomology theory and Atiyah's Γ-index theory are among the main subjects of such metric geometry, and have been extensively developed.

Unfortunately, surgery theory does not work well in the smooth category of dimension 4. Instead, gauge theory has made a spectacular appearance, and gave a number of revolutionary applications to the topology of smooth 4-manifolds. It has been revealed that smooth structures of 4-manifolds show quite different aspects from the cases of the other dimensions. For instance, it is known that there are uncountably many mutually different smooth structures on $\mathbb{R}^4$, while the Euclidean space of any other dimension has a unique one.

At least at the present time, it would not be so common to study both L^2-cohomology theory and gauge theory at the same time. One of the main reasons for this is the different characteristics of both theories. Roughly speaking, the former is

T. Kato, N. Nakamura, *Monopole Maps and L^2-cohomology*, Springer Monographs in Mathematics, https://doi.org/10.1007/978-981-96-8472-4_1

based on the linear analysis on non-compact manifolds, and the latter on the non-linear analysis mainly on compact manifolds.

On the other hand, both of these theories have in common that they are closely related to the Atiyah–Singer index theory! In fact, L^2-cohomology theory and gauge theory can be thought of as lying on two branches toward which the index theory has developed. One branch is the moduli theories of non-linear elliptic PDEs, such as gauge theory and Gromov–Witten theory, where the index theory is applied to analyze local linearizations of the equations. The other one is index theories on non-compact manifolds, such as L^2-cohomology theory and Atiyah's Γ-index theory.

Combining L^2-cohomology theory and gauge theory, one encounters quite interesting situations. From the gauge theory part, the following theorem is one of the core themes of this book.

Theorem 1.1 (Furuta [59]) *Let M be a compact spin smooth 4-manifold with indefinite form.*

Then, we have the inequality

$$b_2(M) \geq \frac{10}{8} \cdot |\sigma(M)| + 2, \tag{1.1}$$

where $\sigma(M)$ is the signature and $b_2(M)$ is the second Betti number of M.

The inequality (1.1) is often called the $10/8$-*inequality*.

Let us combine the inequality with Lück's convergence theorem in L^2-cohomology theory (Theorem 7.22). Then, we immediately obtain the following variation of the Furuta's $\frac{10}{8}$-inequality (Proposition 8.10). The extra term " + 2" is eliminated in a canonical way, during taking the limiting process. (See also Problem 2.61.)

Proposition 1.2 *Let M be a compact spin smooth 4-manifold with the residually finite fundamental group. Then, the inequality*

$$b_\Gamma^2(\tilde{M}) \geq \frac{10}{8} \cdot |\sigma(M)|$$

holds, where $\tilde{M}$ is the universal covering space, and $b_\Gamma^2(\tilde{M})$ is the L^2-Betti number of $(\tilde{M}, \Gamma)$ with $\Gamma := \pi_1(M)$.

Even though $\pi_1(M)$ can be infinite and hence $\tilde{M}$ can be non-compact, the argument of the proof essentially reduces to the compact case. This leads us to propose the following covering version of the $\frac{10}{8}$-inequality (Conjecture 8.1):

Conjecture 1.3 Let M be a compact smooth spin 4-manifold. Then, the inequality

$$b_\Gamma^2(\tilde{M}) \geq \frac{10}{8} \cdot |\sigma(M)|$$

holds, where $\Gamma = \pi_1(M)$ and $\tilde{M}$ is the universal covering space.

From the L^2-cohomology theory part, the following Singer conjecture is one of the core themes of this book (Conjecture 2.9):

Conjecture 1.4 Let M be a compact aspherical smooth manifold of even dimension $\dim M = 2k$. Then, the L^2-Betti numbers $b^i_\Gamma(\tilde{M}) = 0$ vanish for $i \neq k$, where $\Gamma = \pi_1(M)$ and $\tilde{M}$ is the universal covering space.

Now we arrive at the aspherical $\frac{10}{8}$-inequality conjecture, combining the covering $\frac{10}{8}$-inequality conjecture with the Singer conjecture (Lemma 8.4 and Conjecture 8.2):

Conjecture 1.5 Let M be a compact aspherical smooth spin 4-manifold. Then, the inequality

$$\chi(M) \geq \frac{10}{8} \cdot |\sigma(M)|$$

holds.

This inequality is stronger than Furuta's $\frac{10}{8}$-inequality, and the applicable class of smooth 4-manifolds is smaller.

This book is organized as follows.

Chapter 2 gives a quick overview of historical progress of differential topology.

Chapter 3 contains the basic subjects of spin geometry, vector bundles and connections. We give detailed definitions and proofs for almost all the statements.

Chapter 4 treats with the foundations of the Seiberg–Witten theory, including the analytical framework of the theory, the detailed analysis of the Seiberg–Witten moduli spaces, and the construction and basic properties of the Seiberg–Witten invariants.

In Chap. 5, the Bauer–Furuta theory, which is one of main themes of this book, is described. After introducing the monopole map, we discuss the finite-dimensional approximation method and the construction of the Bauer–Furuta invariant which is a stable cohomotopy refinement of the Seiberg–Witten invariant. As an application, we reprove Donaldson's diagonalization theorem for the definite 4-manifolds. Finally, the proof of Furuta's $\frac{10}{8}$-inequality (Theorem 1.1) is given.

In Chap. 6, we present a basic theory of L^2-cohomology theory. It is deeply related to the spectral behavior near zero of the Laplace operators on L^2-forms. A fundamental subject is a relation of amenability with the spectra at zero called Brook's theorem. We describe basic subjects of amenability and residual finiteness of discrete groups. The basic de Rham theorem states that the de Rham cohomology group is isomorphic to the singular cohomology group over a compact manifold. The de Rham cohomology group is also isomorphic to simplicial cohomology group over a compact manifold. We present a detailed proof of the extension to the universal covering space of a compact manifold.

In Chap. 7, we present basic theory of L^2-Betti numbers. In the case when a manifold X is the universal covering space of a manifold M, then the L^2-cohomology group admits a linear action by the fundamental group $\pi_1(M)$.

When X is non-compact, it would be impossible in the usual sense to define Betti numbers of the L^2-cohomology group of X, because the rank can be infinite. On the other hand, when M is compact, then one can define the "average" of the rank. It turns out that the average is a finite number that is called the L^2-Betti number. We present definitions and basic properties of the von Neumann traces and the L^2-Betti numbers.

When $\Gamma := \pi_1(M)$ is residually finite, then there exists infinite tower of normal and finite coverings of M that "converges to X" in a sense. A finite covering of a compact manifold is still compact, and hence one can obtain the well-defined Betti number. Lück's convergence theorem states that the Betti numbers divided by the indices of the covering converge to the L^2-Betti number of (X, Γ). We present a detailed proof of this property. Atiyah's Γ-index theorem is formulated by using the L^2-Betti numbers. It states that the Γ-index over the universal covering space coincides with the Atiyah–Singer index over the underlying manifold. We present a proof of Atiyah's Γ-index theorem.

In Chap. 8, we treat the Singer conjecture. In particular, we describe Gromov's solution to the Singer conjecture for Kähler hyperbolic manifolds. Then, as already stated, we describe various variations of Furuta's $\frac{10}{8}$-inequalities and how the aspherical $\frac{10}{8}$-inequalities conjecture is induced. We present the evidence by examining various classes of 4-manifolds, such as aspherical surface bundles and complex surfaces.

We assume some basic knowledge in geometry and topology on the undergraduate level. This is a quite unique and basic book, containing plenty of unsolved problems, suggesting future directions of topology of smooth 4-manifolds by using various analytic methods.

Chapter 2
A Glimpse of Progress of Differential Topology

In this chapter, we give a quick overview of advances in differential topology for the positioning of the theme of this book. Most of the proofs of the results in this chapter are omitted.

2.1 Topological Invariants

Differential topology explores the interplay between smooth structures and underlying topological structures. It can happen that there are more than two distinct smooth structures within the same topological type. Namely, for two smooth manifolds M and N that admits a homeomorphism $f : M \overset{\cong}{\to} N$, it can happen that there exist no diffeomorphisms between them.

The following topological invariants are fundamental:

- Betti numbers $b_i(M) := \dim H_i(M; \mathbb{Q})$;
- Euler characteristic $\chi(M) := \sum_{i=0}^{\dim M} (-1)^i b_i(M)$;
- Signature $\sigma(M) := b^+(M) - b^-(M)$ when $\dim M$ is a multiple of 4, where $b^+(M)$ (resp. $b^-(M)$) is the dimension of a maximal positive (resp. negative) definite subspace of $H_{2k}(M; \mathbb{Q})$.

In fact, these invariants are all homotopy invariants, that is, two homotopy equivalent manifolds have the same values.

Why study such homotopically stable objects?

There are strong constraints on the values these invariants can take. Let us list some of basic and remarkable properties of these invariants.

- The equality $b_i(M) = b_{m-i}(M)$ holds for a closed oriented manifold for $0 \leq i \leq m := \dim M$. (Poincaré duality.)
- The Euler characteristic $\chi(M)$ vanishes when $\dim M$ is odd.

T. Kato, N. Nakamura, *Monopole Maps and L^2-cohomology*, Springer Monographs in Mathematics, https://doi.org/10.1007/978-981-96-8472-4_2

- If the orientation on M is reversed, then the signature $\sigma(M)$ changes its sign. In particular, if M admits an orientation-reversing homotopy equivalence, then $\sigma(M) = 0$.

2.2 Characteristic Classes

Let M be a smooth manifold. Characteristic classes of the tangent bundle TM are more refined than the invariants in the previous section:

- The Euler class $e(M) \in H^m(M; \mathbb{Z})$, where $m := \dim M$, when M is oriented.
- The Pontryagin classes $p_i(M) \in H^{4i}(M; \mathbb{Z})$.
- The Stiefel–Whitney classes $w_i(M) \in H^i(M; \mathbb{Z}/2)$.
- The Chern classes $c_i(M) \in H^{2i}(M; \mathbb{Z})$, when M is a complex manifold.

A priori, these invariants are diffeomorphism invariants of M, because they are determined by the tangent bundle that is defined by using a smooth structure of M.

To what extent do these invariants depend on the smooth structure?

Theorem 2.1 (Poincaré–Hopf) *The Euler characteristic coincides with the number of zeros with counted sign of a generic vector field.*

In particular:

- The number of zeros counted with sign of a generic vector field V is independent of the choice of V.
- If the Euler characteristic $\chi(M) \neq 0$ is non-zero, then any vector field on M must have zero at some points. Conversely, if there exists a nowhere vanishing vector field on M, then $\chi(M)$ must vanish.
- Hence, the Euler characteristic of any Lie group must vanish.

Theorem 2.2 *The Euler characteristic is given by the integral of the Euler class.*

In particular, the Euler class is an oriented homotopy invariant of M. It is also known that the Stiefel–Whitney classes of closed manifolds are homotopy invariant by the Wu formula (see [123, Chapter 23]).

The following is a deep result:

Theorem 2.3 (Novikov [141]) *The rational Pontryagin classes are topological invariants.*

Examples

- The first Stiefel–Whitney class $w_1(M)$ vanishes, if and only if M is orientable (Lemma 3.20).
- The second Stiefel–Whitney class $w_2(M)$ vanishes, if and only if M admits a spin structure (Lemma 3.22).
- The top Chern class $c_m(M)$ coincides with the Euler class $e(M)$, when M is a complex manifold of $\dim_{\mathbb{C}} M = m$.

2.3 Atiyah–Singer Index Theorem

Let D be an elliptic differential operator from E to F over a compact oriented smooth manifold M. The following is well-known:

Proposition 2.4 *Both the spaces* $\ker D$ *and* $\operatorname{coker} D$ *are finite dimensional. In particular, the bounded operator*

$$D : L_1^2(M; E) \to L^2(M; F)$$

is of Fredholm.

Some special cases are treated in Sect. 4.2.

Recall the definition of the index by

$$\operatorname{ind}(D) := \dim\ker D - \dim\operatorname{coker} D \ \in \ \mathbb{Z}.$$

A formulation of the Atiyah–Singer index theorem expresses the index of D by the integration of some form determined by both D and M.

Theorem 2.5 (The Atiyah–Singer Index Theorem [9])

$$\operatorname{ind}(D) = \int_M \hat{\mathcal{A}}(TM) \wedge ch(S').$$

Here, S' is the "coefficient bundle" (see [147], page 164), and $ch(S')$ is the Chern character of S'. The cohomology class $\hat{\mathcal{A}}(M)$ is a universal polynomial with coefficients by the Pontryagin classes of TM, and it is called the $\hat{\mathcal{A}}$-genus.

In particular, if M is spin and D is the Dirac operator acting on the spinor bundle, then $\mathrm{ind}(D) = \int_M \hat{\mathcal{A}}(TM)$ holds.

The Atiyah–Singer index theorem lies at the crossroads of several areas, including algebraic topology, differential geometry, and linear analysis. From the viewpoint of differential topology, it is important that several topological invariants have analytic expression. Let us endow M with a Riemannian metric g.

- **The signature operator**: Suppose the dimension of M is a multiple of 4, say $\dim M = 4k$. Then, the Hodge star operator acting on the space of differential forms satisfies the equality $*^2 = 1$ on $2k$-forms. Then, $\Lambda^{2k}(M)$ decomposes as the eigenbundles as $\Lambda^{2k}(M) = \Lambda^{2k}_+(M) \oplus \Lambda^{2k}_-(M)$. The signature operator is given by

$$D := d \oplus d^* : L^2(M; \Lambda^{2k}_+(M)) \to L^2(M; \Lambda^{2k}_-(M)).$$

Then, the index of the signature operator gives the signature, and the formula is given by

$$\mathrm{ind}(D) = \sigma(M) = \int_M \mathcal{L}(M),$$

where $\mathcal{L}(M)$ is a universal polynomial with coefficients by the Pontryagin classes of TM, and it is called the L-genus. The second equality is called the *Hirzebruch signature theorem*.
- **The de Rham operator**: The de Rham operator is given by

$$D := d \oplus d^* : L^2(M; \Lambda^{ev}(M)) \to L^2(M; \Lambda^{odd}(M)).$$

Then, the index of the de Rham operator gives the Euler characteristic, and the formula is given by

$$\mathrm{ind}(D) = \int_M e(M) = \chi(M) = \int_M \mathrm{Pf}(R),$$

where $e(M)$ is the Euler class of TM and $\mathrm{Pf}(R)$ is the Pfaffian of the curvature form. The last equality is called the *Chern–Gauss–Bonnet theorem*.

The following property is important:

Both the signature and the Euler characteristic are multiplicative under taking a finite covering.

This is because both the signature and Euler characteristic are given by the integrations of the rational Pontryagin classes and the Euler class respectively that can be given by differential forms constructed locally from the Riemannian curvature tensors. In particular, if the index is non-zero and is not divisible by n, then there is no free action of order n on M.

Remark 2.6 On the other hand, the Betti numbers are, in general, far from being multiplicative under taking finite coverings. (See Problem 2.52.)

Examples

- The signature of $\mathbb{C}P^m$ is equal to 1. Hence, there is no free action of any finite groups with order ≥ 2 on $\mathbb{C}P^m$. Note that $\mathbb{C}P^m$ admits non trivial group actions with any order (but have fixed points). See Problem 2.53.
- Let us consider Riemann surfaces Σ_g of genus $g \geq 2$. The Euler characteristics are given by $2 - 2g < 0$. Suppose it admits a k-fold covering. Then the Euler characteristics are given by $k \cdot (2 - 2g) = 2 - 2((g - 1) \cdot k + 1)$. One can read from this formula that say, if g is odd, then the genus of the covering space must be odd.

Let us consider Riemann surfaces Σ_g of genus $g \geq 2$. Then the first Betti numbers $b_1(\tilde{\Sigma}_g)$ of k-fold coverings $\tilde{\Sigma}_g$ diverge as $k \to \infty$. In particular, they are all non-zero. By the Hodge theory, the dimension of the space of L^2-harmonic 1-forms diverges to infinity on $\tilde{\Sigma}_g$, as $k \to \infty$. One predicts that there may exist non-zero L^2-harmonic 1-forms on the universal covering space. Equip Σ_g with the Riemannian metric with -1 curvature. Then the universal covering space is isometric to the 2-dimensional hyperbolic space $\mathbf{H}_2$. Then, we have the following (see Problem 2.57):

- The space of L^2-harmonic 1-forms is infinite dimensional on $\mathbf{H}_2$.
- The space of L^2-harmonic 0 or 2-forms is zero on $\mathbf{H}_2$.

In general, let $X := \tilde{M}$ be the universal covering space of a compact oriented Riemannian manifold (M, g). Then X is also a Riemannian manifold with the metric $\tilde{g}$, which is the lift of g. The lift $\tilde{g}$ is Γ-invariant with $\Gamma := \pi_1(M)$.

Let $\mathcal{H}^p(X)$ be the space of L^2-harmonic p-forms on X. This space is Γ-invariant. Let P_Γ be the L^2-projection to $\mathcal{H}^p(X)$ in $L^2(X; \Lambda^p)$. The operator commutes with the Γ-action. Then, one can take the trace with respect to the Γ-action, called the Γ-trace, denoted by tr_Γ. The L^2-Betti number $b_\Gamma^p(X)$ is given by

$$b_\Gamma^p(X) := \mathrm{tr}_\Gamma(P_\Gamma) \in \mathbb{R}.$$

See Definition 7.12.

Theorem 2.7 *The Euler characteristic is equal to the alternative sum of the L^2-Betti numbers:*

$$\chi(M) = \sum_{i=0}^{\dim M} (-1)^i \cdot b_\Gamma^i(X).$$

This is a special case of the Atiyah's Γ-index theorem. (See Sect. 7.3.)

Hence, we have:

$$b^0_\Gamma(\mathbf{H}_2) = b^2_\Gamma(\mathbf{H}_2) = 0, \quad b^1_\Gamma(\mathbf{H}_2) = 2g - 2,$$

where $\Gamma := \pi_1(\Sigma_g)$.

Theorem 2.8 ([41], see also [34]) *Let M be a compact manifold that admits a Riemannian metric of constant negative curvature. Suppose M is even dimensional, say* $\dim M = 2k$.

Then, we have:

$$b^i_\Gamma(\tilde{M}) \begin{cases} = 0 & i \neq k, \\ > 0 & i = k, \end{cases}$$

where $\tilde{M}$ is the universal covering space that is isometric to the hyperbolic space $\mathbf{H}_{2k}$.

2.4 Euler Characteristics and Aspherical Manifolds

Let M be a compact Riemannian manifold.

We say that M is aspherical, if the universal covering space $\tilde{M}$ of M is contractible. It is known that the homotopy type of an aspherical manifold is uniquely determined by its fundamental group $\pi_1(M)$. By definition, M is a classifying space of $\pi_1(M)$.

If a compact manifold admits a Riemannian metric of non-positive sectional curvature, then it is aspherical by the Cartan–Hadamard theorem. In particular, a hyperbolic manifold is aspherical.

Examples

- The torus T^n is aspherical, and its universal covering space is diffeomorphic to the Euclidean space $\mathbb{R}^n$.
- The Riemann surface of genus $g \geq 2$ is hyperbolic, and therefore aspherical.

For compact aspherical manifolds with even dimension, the following conjectures are proposed and investigated.

Conjecture 2.9 (Singer Conjecture, See [117]) Let M be a compact aspherical manifold of even dimension $\dim M = 2k$, and let $\tilde{M}$ be the universal covering space. Then, the L^2-Betti numbers vanish except for the middle degree:

$$b^i_\Gamma(\tilde{M}) = 0 \qquad \text{for } i \neq k.$$

Let M be as in Conjecture 2.9. Assume moreover that M admits a Riemannian metric of negative sectional curvature. The *strong Singer conjecture* states the following properties:

$$b^i_\Gamma(\tilde{M}) \begin{cases} = 0 & i \neq k, \\ > 0 & i = k. \end{cases}$$

The strong Singer conjecture is true for hyperbolic spaces by Theorem 2.8.

Remark 2.10 In his PhD thesis, Chern verified that for a 4-dimensional compact hyperbolic manifold M, the Euler characteristic $\chi(M)$ is positive [29].

Conjecture 2.11 (Hopf Conjecture, See [117]) For an even-dimensional compact aspherical manifold M with $\dim M = 2k$, the Euler characteristic satisfies that $(-1)^k \cdot \chi(M) \geq 0$ holds.

The Hopf conjecture follows from the Singer conjecture. We study how to approach the Singer conjecture in Chap. 8.

Lemma 2.12 *Vanishing of the L^2-cohomology group at a specified degree on the universal covering space $\tilde{M}$ is an invariant property under change of the Riemannian metrics on a compact manifold M.*

This follows from Theorem 6.46.

The signature and the Euler characteristic are closely related to each other. By combining Theorem 8.29 and Atiyah's Γ-index theorem, we obtain the following qualitative topological result:

Theorem 2.13 *Let M be an oriented compact aspherical smooth manifold with the amenable fundamental group. Then, both the signature and the Euler characteristic $\sigma(M) = \chi(M) = 0$ vanish.*

Compare Theorem 2.13 with Conjecture 1.5.

2.5 Smooth Structures and Fundamental Groups

Let us consider how the fundamental groups affect the topology of manifolds.

Aspherical manifolds can be considered as an extreme case. As mentioned before, two aspherical manifolds with isomorphic fundamental groups are homotopy equivalent. In particular, if they are closed, then their dimensions must coincide! (Problem 2.58.) Naturally all their homotopy invariants coincide, such as:

- Betti numbers,
- intersection forms,
- signature, and so on.

Another extreme case is the case of simply-connected manifolds. Even in this case, it is observed the existence of homotopy equivalent or homeomorphic but non-diffeomorphic manifolds. In fact, it was first discovered by Milnor [124] that a sphere may admit two or more distinct smooth structures:

Theorem 2.14 ([93, 124]) *There exist smooth manifolds of dimension* ≥ 7 *such that they are homeomorphic but non-diffeomorphic to the spheres.*

Example
There is an 11-dimensional smooth manifold M that is homeomorphic but not diffeomorphic to S^{11}. It can be realized as the boundary of a smooth 12-dimensional manifold $\bar{W}$ whose intersection form is isomorphic to the definite form E_8 (see Sect. 2.8.1). One can argue as follows; if M were diffeomorphic to S^{11}, then it would give a 12-dimensional closed manifold W, and a direct computation gives a non-integer value of the integral of $\mathcal{L}(W)$. This cannot happen by the signature theorem.

It is said that Theorem 2.14 marked the beginning of the differential topology. Now the primary concern in differential topology is the following question.

What is the difference between smooth and topological categories?

In the simply-connected case, the condition for two homotopy equivalent manifolds to be diffeomorphic is given by the *h-cobordism theorem* (Theorem 2.15) below. A cobordism (W, M, M') is called h-cobordism if the inclusions $M \hookrightarrow W$ and $M' \hookrightarrow W$ are homotopy equivalences.

Theorem 2.15 (Smale [155], see also [126]) *Suppose* $\dim M \geq 5$ *and* $\pi_1(M) = 1$. *Then, an h-cobordism* (W, M, M') *is diffeomorphic to the product* $(M \times [0, 1], M, M)$.

The condition $\dim M \geq 5$ is necessary for the Whitney trick to work, as described below (Sect. 2.6.1).

Note that there is an h-cobordism (W, M, M') diffeomorphic to the product if and only if M and M' are mutually diffeomorphic. Hence, two simply-connected manifolds of dim ≥ 5 are mutually diffeomorphic, if and only if they admit an h-cobordism.

From this, the famous solution to the high-dimensional Poincaré conjecture is obtained:

Corollary 2.16 (Smale [155], See Also [126]) *If a compact manifold of* $\dim M \geq 5$ *is homotopy equivalent to the sphere, then it is also homeomorphic to the sphere.*

Theorem 2.14 and Corollary 2.16 tell us that smooth and topological categories show completely different aspects. Let us present a recent discovery:

Theorem 2.17 ([88]) *There exists a closed topological fiber bundle such that its total space is smoothable but is non-smoothable as a fiber bundle.*

The h-cobordism theorem is generalized to non-simply-connected manifolds. A cobordism (W, M, M') is called s-cobordism if the inclusions $M \hookrightarrow W$ and $M' \hookrightarrow W$ are simple homotopy equivalences. The following theorem is called the *s-cobordism theorem* (see [146]):

Theorem 2.18 (Barden–Mazur–Stallings) *Suppose* $\dim M \geq 5$. *Then, an h-cobordism (W, M, M') is diffeomorphic to the product, if and only if it is an s-cobordism.*

Let M and X be compact oriented smooth manifolds. Generally speaking, surgery theory attempts to answer the following question.

When Is a Homotopy Equivalence $f : M \to X$ Homotopic to a Diffeomorphism?

The above question is considered through the surgery exact sequence which involves three components: the structure set $\mathbf{S}(X)$, the normal set $[X, G/O]$, and the L-group $L(\mathbb{Z}[\pi_1(X)])$. (See [146] for a comprehensive introduction to surgery theory.)

For a compact oriented smooth manifold X, the *structure set* $\mathbf{S}(X)$ is defined as the set of equivalence classes of pairs (M, f), where M is a smooth manifold and $f : M \to X$ is a homotopy equivalence. The pairs (M, f) and (M', f') are equivalent, if there exists an h-cobordism W between M and M' that extends the homotopy equivalences $(F, f, f') : (W, M, M') \to (X \times [0, 1], X \times \{0\}, X \times \{1\})$.

If X is simply-connected, then the structure set $\mathbf{S}(X)$ is identified with the set of diffeomorphism classes of smooth manifolds homotopy equivalent to X by the h-cobordism theorem (Theorem 2.15).

For a homotopy equivalence $f: M \to X$, if TM and $f^*(TX)$ are not isomorphic as the vector bundles over M, then M and X cannot be diffeomorphic through f. (See Problem 2.59.) One can say that the tangent space of a manifold is the first approximation of a smooth structure. In surgery theory, the information on the tangent bundles is encoded in the normal set, which is a kind of topological K-theory identified with a homotopy set $[X, G/O]$ that classifies the stable vector bundle reductions of a stable spherical fibration over X.

Roughly speaking, the *L-group* $L_*(\mathbb{Z}[\pi_1])$ is defined as the set of equivalence classes of unimodular forms over $\mathbb{Z}[\pi_1]$, that is, π_1-equivariant intersection forms of manifolds with fundamental group π_1.

Now the *surgery exact sequence* due to Browder, Novikov, Sullivan, and Wall is as follows.

Theorem 2.19 *Let X be a smooth manifold of dimension* ≥ 5. *Then, there exists an exact sequence*

$$\cdots \to L_{n+1}(\mathbb{Z}[\pi_1(X)]) \to \mathbf{S}(X) \to [X, G/O] \to L_n(\mathbb{Z}[\pi_1(X)]). \tag{2.1}$$

Remark 2.20 There exists the s-version of the sequence (2.1). For instance, the s-version of the structure set classifies the simple homotopy equivalences via the equivalence relation by s-cobordisms.

We can say that the structure set $\mathbf{S}(X)$ heavily depends on:

- the tangent bundle,
- the intersection form of X, and
- the fundamental group $\pi_1(X)$.

It is known to be impossible to classify finitely presented groups. More precisely, the word problem is undecidable for some finitely presented groups [140].

How do we understand the fundamental groups?

In general, this is a very hard problem. One of the reasons for this is that most of the fundamental groups are non-commutative. This has motivated the creation of the fields of noncommutative geometry [31] and geometric group theory [67].

There are various conjectures related to the fundamental groups. These topics are outside scope of this book. We just mention one of them, which is called the *zero-in-the-spectrum conjecture* ([67] for the aspherical case, [115]). (See Sect. 7.4.)

Conjecture 2.21 The unreduced L^2-cohomology group is non-zero at some degree over the universal covering space of a compact aspherical manifold.

Note that this includes the case of finite fundamental groups. The conjecture trivially holds for the case, because constant functions are in L^2.

There exist counterexamples for the non-aspherical case.

Theorem 2.22 (Farber–Weinberger [44]) *There exist compact manifolds of any dim ≥ 6 that do not satisfy the zero-in-the-spectrum conjecture.*

2.6 Distinctions by Dimension and Simply-Connectedness

There are at least two viewpoints from which to distinguish manifolds in differential topology that have led to the different developments:

1. High-dimensional (≥ 5) or low-dimensional (≤ 4);
2. Simply-connected or multiply-connected.

Viewpoint 1 concerns whether the h- and s-cobordism theorems hold. The key point is the Whitney trick explained in the next subsection (Sect. 2.6.1).

As for the smooth structures of low-dimensional manifolds, it is known that any manifold of dim ≤ 3 admits a unique smooth structure. (The case when dim ≤ 2 is classical. The case of dimension 3 is due to E. Moise [129].)

The situation is significantly different in dimension 4. Even in the simply-connected case, we have the following amazing facts:

Theorem 2.23 (Donaldson [37]) *There are h-cobordant non-diffeomorphic simply-connected smooth* 4*-manifolds. Hence, the h-cobordism theorem is not valid in dimension* 4.

Theorem 2.24 (Taubes [161]) *There exist uncountably many distinct smooth structures on $\mathbb{R}^4$.*

Remark 2.25 For $n \neq 4$, $\mathbb{R}^n$ has a unique smooth structure.

For viewpoint 2, we have seen that the fundamental group of a manifold deeply affects the smooth structures in the previous section (Sect. 2.5).

The case when $\pi_1 \neq 1$ and $\dim = 4$ is the most mysterious in smooth manifold theory.

2.6.1 *Whitney's Trick*

Let M, N and X be connected smooth manifolds such that $\dim M = m$, $\dim N = n$ and $\dim X = m+n$, and assume M and N are embedded in X. Suppose $M \cap N \neq \phi$ and they intersect transversely. For each intersect point $x \in M \cap N$, the orientations of M, N and X determine the sign of the point x. Take two points $x, y \in M \cap N$ with the opposite signs. Choose two segments $I \subset N$ with $\partial I = \{x, y\}$, $I \cap M = \partial I$, and $J \subset M$ with $\partial J = \{x, y\}$, $I \cap N = \partial J$ such that their union $I \cup J$ consists of a contractible loop. Suppose $\dim X \geq 5$ and $\dim N \geq 3$. Then, one can span a disk $D \subset X$ with $\partial D = I \cup J$ and $\{M \cup N\} \cap D = \partial D$. Hence along D, one can push out the segment I, and finally can cancel the pair of the intersections. Hence, we have the following lemma called the Whitney trick (see [126]):

Lemma 2.26 *Under the above conditions, two points $x, y \in M \cap N$ can be cancelled by an isotopy of X.*

The Whitney trick (Lemma 2.26) is essential to the proof of the h-cobordism theorem (Theorem 2.15), and the surgery exact sequence works well with this trick.

This is not the case if $\dim X = 4$. Actually, whether the Whitney trick works or not makes the essential difference between dimension ≥ 5 or 4.

To what extent, does the strategy in high-dimensional differential topology work for 4-manifold theory?

2.7 Topological 4-Manifolds: Casson–Freedman Theory

In general, the Whitney trick does not work on 4-dimensional spaces. However, Casson discovered that one can iterate infinitely many times of the process of handle-cancelling on a simply-connected 4-manifold. As a result, one can obtain an open subset of the standard 2-handle, which is called the *Casson handle*. If it were the standard 2-handle, then the Whitney trick might work as well as the high-dimensional case.

Theorem 2.27 ([48, 49]) *A Casson handle is homeomorphic to an open 2-handle.*

There are many Casson handles that are parametrized by the set of rooted infinite trees. Some of them are known to be non-diffeomorphic to the standard open 2-handle.

By Theorem 2.27, some of the mechanisms of the high-dimensional differential topology work in the 4-dimensional topological category. Freedman proved the following classification theorem for simply-connected topological 4-manifolds.

Theorem 2.28 (Freedman [48]) *Let Q be a non-degenerate symmetric bilinear form over $\mathbb{Z}$. If Q is an even (resp. odd) form, then there exists exactly one (resp. two) simply-connected closed topological manifold(s) representing Q.*

Recall that a symmetric bilinear form $Q(\cdot, \cdot)$ is even when $Q(x, x)$ is even for every x, and is odd if it is not even. In the odd case, the two manifolds in Theorem 2.28 are distinguished by their Kirby–Siebenmann obstructions. One of them has a vanishing Kirby–Siebenmann obstruction, and the other one has a nontrivial one. As a corollary, we obtain the following.

Corollary 2.29 (Freedman [48]) *The* 4*-dimensional Poincaré conjecture holds.*

The classical Poincaré conjecture concerns the homeomorphism types of manifolds. There is a smooth version of the 4-dimensional Poincaré conjecture.

Conjecture 2.30 A smooth manifold homotopy equivalent to S^4 is diffeomorphic to the sphere.

Namely, it is a question whether an exotic smooth structure could exists on S^4. This is a very hard problem, and is still open.

2.8 Smooth 4-Manifolds

2.8.1 Spin 4-Manifolds and Rokhlin's Theorem

For spin 4-manifolds, the following facts are well-known (see [94]).

- A closed spin 4-manifold has an even intersection form.
- If $H_1(X;\mathbb{Z}) = 0$ for a closed oriented 4-manifold X, then X is spin if and only if the intersection form of X is even.

These facts are consequences of the Wu formula.

Let E_8 be the symmetric bilinear form over $\mathbb{Z}$ which is represented by the Cartan matrix of E_8:

$$E_8 = \begin{pmatrix} 2 & -1 & 0 & 0 & 0 & 0 & 0 & 0 \\ -1 & 2 & -1 & 0 & 0 & 0 & 0 & 0 \\ 0 & -1 & 2 & -1 & 0 & 0 & 0 & 0 \\ 0 & 0 & -1 & 2 & -1 & 0 & 0 & 0 \\ 0 & 0 & 0 & -1 & 2 & -1 & 0 & -1 \\ 0 & 0 & 0 & 0 & -1 & 2 & -1 & 0 \\ 0 & 0 & 0 & 0 & 0 & -1 & 2 & 0 \\ 0 & 0 & 0 & 0 & -1 & 0 & 0 & 2 \end{pmatrix}.$$

Let k be a nonnegative integer. The direct sum $\underbrace{E_8 \oplus \cdots \oplus E_8}_{k}$ is denoted by kE_8, and let $-kE_8$ be $\underbrace{(-E_8) \oplus \cdots \oplus (-E_8)}_{k}$. Let H be the hyperbolic form:

$$H = \begin{pmatrix} 0 & 1 \\ 1 & 0 \end{pmatrix}.$$

It is known that indefinite forms are classified by ranks, signatures and types as follows (see [127]).

Proposition 2.31 *An even indefinite form is isomorphic to $kE_8 \oplus \ell H$ for some integers k and ℓ with $\ell \geq 1$. An odd indefinite form is isomorphic to $p(1) \oplus q(-1)$ for some positive integers p and q.*

Remark 2.32 On the other hand, definite forms are much more complicated. It is known that the number of distinct definite forms increases very rapidly with the rank.

The following theorem due to V. Rokhlin is one of a few classical structure theorems for smooth 4-manifolds.

Theorem 2.33 (Rokhlin [148]) *For a closed spin 4-manifold M, the signature $\sigma(M)$ is divisible by 16.*

A proof of Theorem 2.33 by using the index theorem is given in Sect. 3.5.2 (Theorem 3.79). Theorem 2.33 implies the following.

Corollary 2.34 *If X is a closed spin 4-manifold with intersection form $kE_8 \oplus \ell H$, then k is even.*

Combining with Freedman's theorem (Theorem 2.28), we obtain a sequence of non-smoothable 4-manifolds. By Theorem 2.28, we have a unique simply-connected topological 4-manifold $X_{k,\ell}$ whose intersection form is isomorphic to $kE_8 \oplus \ell H$.

Corollary 2.35 *If k is odd, then $X_{k,\ell}$ is not smoothable.*

2.8.2 *Definite 4-Manifolds*

At about the same time Freedman's theorem appeared, S. K. Donaldson proved his groundbreaking theorem, which is the first application of gauge theory to 4-dimensional differential topology.

Theorem 2.36 (Donaldson [35, 38]) *If the intersection form Q_X of X is positive or negative definite, then Q_X is diagonalizable, that is,*

$$Q_X \cong \pm(\underbrace{(1) \oplus \cdots \oplus (1)}_{n}).$$

Donaldson proved Theorem 2.36 by using the (A)SD equations in the Yang–Mills theory (see [47].) Later, an alternative proof by the Seiberg–Witten theory is given. (See [151, 165].) A proof of Theorem 2.36 by the Seiberg–Witten theory is given in Sect. 5.4.

Note that the intersection $Q_{X_{k,0}}$ of the 4-manifold $X_{k,0}$ in the previous subsection is definite. It cannot be determined by Theorem 2.33 whether the topological manifold $X_{2m,0}$ is smoothable or not. But it turns out non-smoothable by Theorem 2.36.

Corollary 2.37 *For a non-zero integer m, $X_{2m,0}$ is not smoothable.*

2.8.3 *$\frac{11}{8}$-Conjecture and $\frac{10}{8}$-Inequality*

Y. Matsumoto [122] stated the $\frac{11}{8}$-conjecture in relation to the "bounding genera" of homology 3-spheres defined and investigated by him.

Conjecture 2.38 (The $\frac{11}{8}$-Conjecture, Y. Matsumoto [122]) For a closed smooth spin 4-manifold M, the following inequality holds:

$$b_2(M) \geq \frac{11}{8}|\sigma(M)|. \tag{2.2}$$

By Corollary 2.34, a closed spin 4-manifold has an intersection form isomorphic to $(2k)E_8 \oplus \ell H$. Then (2.2) is equivalent to

$$\ell \geq 3|k|. \tag{2.3}$$

Remark 2.39 The equality in (2.3) is realized by the connected sums of $K3$ surfaces. Every combination of (k, l) satisfying (2.3) is realized by the intersection form of a smooth 4-manifold such as a connected sum of several $K3$ surfaces and $S^2 \times S^2$.

On the other hand, Furuta's theorem (Theorem 1.1) is rewritten as follows.

Theorem 2.40 *For a closed spin 4-manifold with the intersection form isomorphic to $(2k)E_8 \oplus \ell H$ $(\ell \geq 1)$, the following inequality holds:*

$$\ell \geq 2|k| + 1.$$

Hence, the $\frac{11}{8}$-conjecture is still open in the range between $3|k|$ and $2|k| + 1$.

Corollary 2.41 *If $\ell \leq 2|k|$, then $X_{2k,\ell}$ is not smoothable.*

Compare this corollary with Corollaries 2.35 and 2.37.

2.8.4 Donaldson and Seiberg–Witten Invariants

In contrast to higher dimensions, the classification of the diffeomorphism types of smooth 4-manifolds is very far from resolved. In general, it is difficult to distinguish smooth structures on an underlying topological 4-manifold. Gauge theory provides diffeomorphism invariants of smooth 4-manifolds, Donaldson and Seiberg–Witten invariants, which enable us to distinguish smooth structures. The basic strategy of application of gauge theory to topology is as follows: If the values of a gauge-theoretic invariant of two manifolds are different, then they are not diffeomorphic to each other. The construction of the Seiberg–Witten invariants is explained in Chap. 4. We refer the readers to [40] for Donaldson invariants.

For instance, elliptic surfaces with two multiple fibers are classified up to diffeomorphisms by using these invariants. (See [64].) Let $E(1)$ be the elliptic fibration $\mathbb{C}P^2\#9\overline{\mathbb{C}P}^2 \to \mathbb{C}P^1$, and $E(n)$ be the fiber sum of n $E(1)$'s. Then $E(2)$ is a $K3$ surface. Performing logarithmic transformations on two generic fibers of $E(n)$ with multiplicities p and q, we obtain an elliptic surface $E(n)_{p,q}$ with two multiple fibers.

Theorem 2.42 *If $n \geq 2$, then $E(n)_{p,q}$ is diffeomorphic to $E(n)_{p',q'}$ if and only if $\{p, q\} = \{p', q'\}$.*

The diffeomorphism types of the elliptic surfaces $E(n)_{p,q}$ have been studied first by using Donaldson invariants, and the final form of Theorem 2.42 is proved by using Seiberg–Witten invariants.

Historically, Donaldson [39] introduced the Donaldson invariant in late 1980s, which is defined by using the moduli spaces of the solutions to the ASD equation in the Yang–Mills theory. In 1994, N. Seiberg and E. Witten introduced the Seiberg–Witten equations in their works of quantum field theory. The Seiberg–Witten invariant is introduced in Witten's paper [169], where the equivalence of Donaldson and Seiberg–Witten invariants is conjectured (Witten's conjecture) through the structure theorem for Donaldson invariants due to Kronheimer and Mrowka [100]. The structure theorem by Kronheimer–Mrowka is proved under the condition called *simple type*. The Seiberg–Witten version of the notion of simple type first appeared in Witten's paper [169], and a mathematical proof of Witten's conjecture is given by Feehan and Leness [45] for a class of 4-manifolds of Seiberg–Witten simple type.

Roughly speaking, a closed oriented 4-manifold with $b^+ \geq 2$ is said to be of Seiberg–Witten simple type if the Seiberg–Witten invariant can be non-zero only when the dimension of the moduli space is zero. (The precise definition is given in Sect. 5.3.2.) The next is known as the *simple type conjecture* (see [102, Conjecture 1.6.2]):

Conjecture 2.43 (Simple Type Conjecture) All closed oriented smooth 4-manifolds with $b^+ \geq 2$ are of Seiberg–Witten simple type.

There is a result on the mod 2 version of the simple type, supporting Conjecture 2.43:

Theorem 2.44 ([89]) *For a closed, connected, oriented smooth 4-manifold with $b^+ \geq 2$ is of mod 2 Seiberg–Witten simple type under some cohomological conditions.*

For example, $b^+ - b_1 \equiv 3 \bmod 2$ and $b_1 \leq 1$ suffice. See [89] for details.

Recently, Manolescu disproved the longstanding problem of the so-called "triangulation conjecture" by using the Seiberg–Witten Floer theory:

Theorem 2.45 ([121]) *There exist non-triangulable n-dimensional topological n-manifolds for any $n \geq 5$.*

Even though Gauge theory mainly concerns 4- or 3-dimensional manifolds, it is applied to verify existence of remarkable high-dimensional topological manifolds.

2.8.5 Metrics of Positive Scalar Curvature

Let us briefly state a recent development on the existence of positive scalar curvature metrics.

Conjecture 2.46 ([68]) There exist no metrics of positive scalar curvature on a compact aspherical spin manifold.

There are partial solutions as the consequence of the strong Novikov conjecture and Baum–Connes conjecture for various classes of fundamental groups. (see [31].)

Here is a recent development:

Theorem 2.47 ([30]) *There exist no metrics of positive scalar curvature on a compact aspherical manifold of dimensions* 4 *and* 5.

Compare this with Theorem 4.70.

2.9 Toy Models of Gauge Theories

Let X and Y be finite-dimensional compact smooth manifolds. Consider a smooth map

$$f : X \to Y.$$

Recall that $y \in Y$ is a regular value, if the differential $df_x : T_xX \twoheadrightarrow T_yY$ is surjective for any $x \in f^{-1}(y)$.

Lemma 2.48 *If* $y \in Y$ *is a regular value, then* $M_y := f^{-1}(y) \subset X$ *is empty or a smooth submanifold of* $\dim M_y = \dim X - \dim Y$.

The following properties are well-known:

- The set of regular values are dense in Y (Sard's theorem).
- For two regular values $y_1, y_2 \in Y$, there is a smooth cobordism W with $\partial W = M_{y_1} \sqcup M_{y_2}$.

Corollary 2.49 *Suppose both* X *and* Y *are oriented. For a regular value* y*, the submanifold* M_y *has a fundamental class* $[M_y] \in H_*(X; \mathbb{Z})$*, which is independent of the choice of regular value* y*, and uniquely determined by the homotopy class of* $f : X \to Y$.

The framework of gauge-theoretic invariants is an infinite-dimensional version of Corollary 2.49. In the case of the Seiberg–Witten theory, we have a non-linear map between Hilbert spaces (See (4.20)):

$$\mathcal{F} : C_k \to L^2_{k-1}(X; S^- \oplus \sqrt{-1}\Lambda^+)$$

given by $(A, \Phi) \to (D_A(\Phi), F_A^+ - q(\Phi))$. (The equation $\mathcal{F}(A, \Phi) = 0$ is the Seiberg–Witten equations. The map $\mathcal{F}$ is essentially the *monopole map* which is one of the main topics of this book.) The map $\mathcal{F}$ is equivariant under the action of an infinite-dimensional Lie group $\mathcal{G}_{k+1}$ of gauge symmetry. Then the Seiberg–Witten invariant is defined from the fundamental class of the moduli space $\mathcal{M}_y = \mathcal{F}^{-1}(y)/\ \mathcal{G}_{k+1}$, which turns out to be a finite-dimensional compact manifold for a regular value y. By an analogue of Corollary 2.49, the fundamental class $[\mathcal{M}_y]$ is independent of the auxiliary data such as the choice of regular value y and Riemannian metric on the underlying 4-manifold. From this, it is proved that the Seiberg–Witten invariant is a diffeomorphism invariant.

Let us return to our toy model. Assume moreover that both X and Y are closed and $\dim X = \dim Y = n$.

- The fundamental class $[M_y] \in H_0(X; \mathbb{Z})$ is identified with an integer $\sharp M_y \in \mathbb{Z}$.
- The homotopy class of $f : X \to Y$ is classified by the degree of $f_* : H_n(X; \mathbb{Z}) \to H_n(Y; \mathbb{Z})$, which is denoted by $\deg(f)$.

The following is a basic fact in the differential topology (Problem 2.60):

Lemma 2.50 *The equality holds:*

$$\sharp M_y = \deg(f) \in \mathbb{Z}.$$

This relation is a toy model that bridges the Seiberg–Witten theory and Bauer–Furuta theory.

Furuta's paper [59] is a turning point of the development of gauge theory in that, instead of the inverse image $\mathcal{M}_y$, the map $\mathcal{F}$ itself is the subject of consideration. Since it is not easy to define directly the degree between infinite-dimensional spaces,

Furuta constructed finite-dimensional approximations of the map $\mathcal{F}$. This is made possible by the Fredholm property of the map $\mathcal{F}$ and the compactness of the moduli space $\mathcal{M}_y$. Finally, Bauer and Furuta [18] defines a refinement of the Seiberg–Witten invariant called the Bauer–Furuta invariant as a stable homotopy class of the finite-dimensional approximation. Then, when $\dim \mathcal{M}_y = 0$, the stable homotopy class (the Bauer–Furuta invariant) recovers the ordinary Seiberg–Witten invariant by an analogue of Lemma 2.50. In general, the homotopy class of a map has more information than the fundamental class of its inverse image. Bauer [15] showed that the Bauer–Furuta invariant is actually more refined than the Seiberg–Witten invariant by showing a class of 4-manifolds, e.g., connected sums of several $K3$ surfaces, has the property that the Seiberg–Witten invariant is trivial but the Bauer–Furuta invariant is not (Theorem 5.27).

The Borsuk–Ulam theorem gives a constraint on equivariant maps between spheres with antipodal actions. (See e.g. [71], and Problem 5.58.)

Proposition 2.51 (The Borsuk–Ulam Theorem) *Let C_2 be the cyclic group of order 2 that acts on both S^n and S^m antipodally. If there exists a continuous C_2-equivariant map $f : S^m \to S^n$, then $m \leq n$ holds.*

Furuta's $\frac{10}{8}$-inequality (Theorem 1.1) is proved by applying a Pin(2)-equivariant analogue of Proposition 2.51 to the finite-dimensional approximation of the monopole map (Sect. 5.5.4).

Problems

2.52 Give examples of compact manifolds whose Betti numbers are not multiplicative under taking finite coverings.

2.53 (Finite Group Actions on $\mathbb{C}P^m$)

(a) Construct nontrivial finite group actions of order $n \geq 1$ on $\mathbb{C}P^m$.
(b) Verify that there is no free action of any finite groups with order ≥ 2 on $\mathbb{C}P^{2m}$.

2.54 (Complex Structures on Even-Dimensional Spheres)

(a) Verify that there does not exist any complex structure on S^4. (Hint: Use a formula on the characteristic classes in [76, §4.6].)
(b) Verify that there do not exist any complex structures on S^{2m} for all $m \geq 4$. (Hint: Use the Atiyah–Singer index theorem.)

2.55 (Aspherical Space) Verify that a finite complex X is aspherical, if and only if its homotopy groups $\pi_k(X)$ are trivial except for $k = 1$.

2.56 (Poincaré–Hopf Theorem) Verify that if a compact complex surface M with $c_1(M)^2 = 0$ admits a nowhere vanishing vector field, then the signature must vanish. The 4-torus, bi-elliptic surfaces, and Kodaira surfaces satisfy $c_1^2 = 0$. (See [14].)

2.57 Let X be the universal covering space of a manifold M with infinite $\pi_1(M)$. Let us equip it with a complete Riemannian metric on M and lift it on X.

Verify that if there exists a non-zero L^2-harmonic form on X, then the space of L^2-harmonic forms on X is infinite dimensional.

2.58 Verify that if a pair of closed manifolds are homotopy equivalent, then their dimensions must coincide.

2.59 Verify that if a homotopy equivalence $f : M \to M'$ is homotopic to a diffeomorphism between compact oriented smooth manifolds, then the equalities $f^*(p_i(M')) = p_i(M)$ hold for any $i \geq 0$.

2.60 Verify Lemma 2.50.

2.61 For a compact spin smooth 4-manifold, verify

$$b_2(M) \geq \frac{10}{8} \cdot |\sigma(M)|$$

from Theorem 1.1 and Theorem 2.36.

Chapter 3
Spin Geometry

Chapter 3 gives a short introduction to the spin geometry (See [106].) After introducing the Clifford algebra and spin groups, spin and spin^c structures are formulated, on which the Seiberg–Witten equations will be defined. Next, connections, gauge transformations and Chern–Weil theory are explained, and then Dirac operator is introduced. This chapter is concluded with a short discussion on holomorphic vector bundles.

3.1 Clifford Algebras and Spin Groups

3.1.1 Clifford Algebras

Let V be a finite-dimensional Euclidean space with the inner product $\langle\ ,\ \rangle$. We write $\|v\|^2 := \langle v, v\rangle$. Let $T(V) := \oplus_{n\geq 0} V \otimes \cdots \otimes V$ (n times) be the tensor algebra of V. The algebra $T(V) = T_0(V) \oplus T_1(V)$ splits with respect to even or odd gradings. Let $I(V)$ be the ideal in $T(V)$ generated by elements of the form $v \otimes v + \|v\|^2 \cdot 1$ for $v \in V$.

Definition 3.1 The *Clifford algebra* of V is defined as

$$\mathrm{Cl}(V) := T(V) \,/\, I(V).$$

The element represented by $v_1 \otimes \cdots \otimes v_k \in T(V)$ is written as $v_1 \cdots v_k$.

Let us list several properties of the Clifford algebras.

T. Kato, N. Nakamura, *Monopole Maps and L^2-cohomology*, Springer Monographs in Mathematics, https://doi.org/10.1007/978-981-96-8472-4_3

- The action of $\mathrm{SO}(V)$ on V extends to the action on $\mathrm{Cl}(V)$ as an automorphism by

$$\rho(v_1 \cdots v_l) := \rho(v_1) \cdots \rho(v_l) \tag{3.1}$$

for any $\rho \in \mathrm{SO}(V)$. Note that the action of $\mathrm{SO}(V)$ on V extends to the one on $T(V)$, that preserves $I(V)$.

- $\mathrm{Cl}(V)$ admits the $\mathbb{Z}_2$-graded structure

$$\mathrm{Cl}(V) \;=\; \mathrm{Cl}_0(V) \oplus \mathrm{Cl}_1(V)$$

induced by the grading on $T(V) = T_0(V) \oplus T_1(V)$. In particular, $\mathrm{Cl}(V)$ admits an involution

$$\epsilon : \; \mathrm{Cl}_0(V) \oplus \mathrm{Cl}_1(V) \;\cong\; \mathrm{Cl}_0(V) \oplus \mathrm{Cl}_1(V)$$

given by the matrix form

$$\epsilon = \begin{pmatrix} 1 & 0 \\ 0 & -1 \end{pmatrix}.$$

Moreover, $\mathrm{Cl}_0(V)$ is an algebra, but $\mathrm{Cl}_1(V)$ is not.

- Let $\{e_1, \cdots, e_n\}$ be an orthonormal basis of V. Then, the following relations hold:

$$\begin{cases} e_i \cdot e_j + e_j \cdot e_i = 0 \quad (i \neq j), \\ e_i^2 = -1. \end{cases}$$

- $\mathrm{Cl}(V)$ is spanned by $\{ e_{i_1} \cdots e_{i_l} \}_{i_1, \ldots, i_l}$ as a vector space. In particular,

$$\dim_{\mathbb{R}} \mathrm{Cl}(V) = 2^n.$$

- Let $e := 1 \in \mathbb{R}$ be the unit vector. Then, the algebra isomorphism

$$\mathrm{Cl}(V) \cong \mathrm{Cl}_0(V \oplus \mathbb{R})$$

is given by $v_0 + v_1 \to v_0 + v_1 \cdot e$.

Examples

- $V = \mathbb{R}$ case: Let $x = e_1$.

$$\mathrm{Cl}(\mathbb{R}) = \; \mathbb{R}[x] \,/\, (x^2 + 1) = \mathbb{R} \cdot 1 \oplus \mathbb{R} \cdot x \;\cong \mathbb{C} = \mathbb{R} \oplus i\mathbb{R} \cong \mathrm{Cl}_0(\mathbb{R}) \oplus \mathrm{Cl}_1(\mathbb{R}).$$

- $V = \mathbb{R}^2$ case: Let $x = e_1$ and $y = e_2$.

$$\mathrm{Cl}(\mathbb{R}^2) = \mathbb{R}\langle x, y\rangle \,/\, \langle xy + yx, x^2 + 1, y^2 + 1\rangle,$$

where $\mathbb{R}\langle x, y\rangle$ is the free algebra over $\mathbb{R}$ generated by two elements x and y with unit 1. $\langle xy + yx, x^2 + 1, y^2 + 1\rangle$ is the ideal generated by three elements $xy + yx$, $x^2 + 1$ and $y^2 + 1$.

We have the isomorphisms

$$\begin{aligned}\mathrm{Cl}(\mathbb{R}^2) &= \mathrm{Cl}_0(\mathbb{R}^2) \oplus \mathrm{Cl}_1(\mathbb{R}^2) = (\mathbb{R}\cdot 1 \oplus \mathbb{R}\cdot xy) \oplus (\mathbb{R}\cdot x \oplus \mathbb{R}\cdot y)\\ &\cong \mathbb{H} = \mathbb{C} \oplus \mathbb{C}j\end{aligned}$$

via the correspondence $xy \mapsto i$, $x \mapsto j$, $y \mapsto k$.

- $V = \mathbb{R}^3$ case:

$$\begin{aligned}&\mathrm{Cl}(\mathbb{R}^3)\\ &\quad = \mathbb{R}\langle e_1, e_2, e_3\rangle \,/\\ &\qquad \langle e_1e_2 + e_2e_1, e_2e_3 + e_3e_2, e_3e_1 + e_1e_3, e_1^2 + 1, e_2^2 + 1, e_3^2 + 1\rangle\end{aligned}$$

Let $\mathbb{H}^{\pm}$ be the subspaces with bases

$$\begin{cases}(\mathbf{1}, \boldsymbol{i}, \boldsymbol{j}, \boldsymbol{k}) := \left(\frac{1-e_1e_2e_3}{2}, \frac{e_2e_3+e_1}{2}, \frac{e_3e_1+e_2}{2}, \frac{e_1e_2+e_3}{2}\right) & \text{for } \mathbb{H}^+,\\ (\mathbf{1}', \boldsymbol{i}', \boldsymbol{j}', \boldsymbol{k}') := \left(\frac{1+e_1e_2e_3}{2}, \frac{e_2e_3-e_1}{2}, \frac{e_3e_1-e_2}{2}, \frac{e_1e_2-e_3}{2}\right) & \text{for } \mathbb{H}^-.\end{cases} \tag{3.2}$$

Then $(\mathbf{1}, \boldsymbol{i}, \boldsymbol{j}, \boldsymbol{k})$ and $(\mathbf{1}', \boldsymbol{i}', \boldsymbol{j}', \boldsymbol{k}')$ satisfy the relations of quaternions $1, i, j, k$. Hence $\mathrm{Cl}(\mathbb{R}^3) = \mathbb{H}^+ \oplus \mathbb{H}^-$ is isomorphic to $\mathbb{H} \oplus \mathbb{H}$. Note that, for $x \in \mathbb{H}^+$ and $x' \in \mathbb{H}^-$, $xx' = x'x = 0$.

The subalgebra $\mathrm{Cl}_0(\mathbb{R}^3)$ embeds diagonally

$$\mathrm{Cl}_0(\mathbb{R}^3) \xrightarrow{\Delta} \mathbb{H}^+ \oplus \mathbb{H}^- = \mathrm{Cl}(\mathbb{R}^3).$$

Note that $\mathrm{Cl}_0(\mathbb{R}^3) \cong \mathbb{H}$.

- $V = \mathbb{R}^4$ case: Note that the isomorphism $\mathrm{Cl}(\mathbb{R}^3) \cong \mathrm{Cl}_0(\mathbb{R}^4)$ is given by the correspondence $v_0 + v_1 \mapsto v_0 + v_1 \cdot e_4$. This induces the splitting $\mathrm{Cl}_0(\mathbb{R}^4) = \mathbb{H}^+ \oplus \mathbb{H}^-$ where $\mathbb{H}^{\pm}$ are the subspaces isomorphic to $\mathbb{H}$. For later purposes, we choose the bases of $\mathbb{H}^{\pm}$ as

$$\begin{cases}(\mathbf{1}, \boldsymbol{i}, \boldsymbol{j}, \boldsymbol{k}) := \left(\frac{1-e_1e_2e_3e_4}{2}, \frac{e_1e_2+e_3e_4}{2}, \frac{e_1e_3+e_4e_2}{2}, \frac{e_1e_4+e_2e_3}{2}\right) & \text{for } \mathbb{H}^+,\\ (\mathbf{1}', \boldsymbol{i}', \boldsymbol{j}', \boldsymbol{k}') := \left(\frac{1+e_1e_2e_3e_4}{2}, \frac{-e_1e_2+e_3e_4}{2}, \frac{-e_1e_3+e_4e_2}{2}, \frac{-e_1e_4+e_2e_3}{2}\right) & \text{for } \mathbb{H}^-.\end{cases} \tag{3.3}$$

Note that $xx' = x'x = 0$ for $x \in \mathbb{H}^+$ and $x' \in \mathbb{H}^-$.

3.1.2 Pin(V) *and* Spin(V)

Let $S \subset G$ be a subset of a group. We denote the group generated by elements of S by $\langle S \rangle$. The groups $\mathrm{Pin}(V)$ and $\mathrm{Spin}(V)$ are defined as follows.

Definition 3.2

- $\mathrm{Pin}(V) := \langle v \in V : \|v\| = 1 \rangle$.
- $\mathrm{Spin}(V) := \mathrm{Pin}(V) \cap \mathrm{Cl}_0(V)$.

Examples

Let us use the notations $\mathrm{Pin}(n) := \mathrm{Pin}(\mathbb{R}^n)$, $\mathrm{Spin}(n) := \mathrm{Spin}(\mathbb{R}^n)$.

- $\mathrm{Pin}(1) = \langle \pm i \rangle \cong \mathbb{Z}_4$.
- $\mathrm{Spin}(1) = \{\pm 1\} \cong \mathbb{Z}_2$.
- $\mathrm{Pin}(2) = \langle \cos\theta \cdot j + \sin\theta \cdot k : \theta \in [0, 2\pi] \rangle \cong \mathrm{U}(1) \sqcup j \cdot \mathrm{U}(1)$.
- $\mathrm{Spin}(2) = \mathrm{U}(1)$.
- $\mathrm{Spin}(3) \cong S^3$. To see this, note the embedding

  $$\mathrm{Cl}(\mathbb{R}^3) \supset \mathbb{R}^3 \cong \mathrm{Im}\,\mathbb{H} \hookrightarrow \mathbb{H} \oplus \mathbb{H}$$

  by $\alpha \to (\alpha, -\alpha)$. Then,

  $$\mathrm{Spin}(3) \cong \langle \alpha \cdot \beta : \alpha, \beta \in \mathrm{Im}\,\mathbb{H}, \|\alpha\| = \|\beta\| = 1 \rangle \cong S^3 \subset \mathbb{H}.$$

- $\mathrm{Spin}(4) \cong S^3 \times S^3$. To see this, note that the embedding $\mathbb{R}^3 \hookrightarrow \mathbb{R}^4$ induces the embedding $\mathrm{Spin}(3) \subset \mathrm{Spin}(4)$.

  For any 3-dimensional linear subspace $V \subset \mathbb{R}^4$, it induces the embedding $\mathrm{Spin}(V) \subset \mathrm{Spin}(4)$. We take all such linear subspaces V. Then it follows that

  $$\mathrm{Spin}(4) \cong S^3 \times S^3 \cong \mathrm{Sp}(1) \times \mathrm{Sp}(1) \subset \mathbb{H} \times \mathbb{H}.$$

Let $\mathrm{Aut}(\mathrm{Cl}(V), V)$ be the group of automorphisms on $\mathrm{Cl}(V)$ that preserves V.

Lemma 3.3 *Let* $\mathrm{Spin}(V)$ *act on* $\mathrm{Cl}(V)$ *by conjugation as*

$$w \mapsto qwq^{-1} \quad \textit{for } w \in \mathrm{Cl}(V) \textit{ and } q \in \mathrm{Spin}(V).$$

Then the conjugation action defines a group homomorphism $\tilde{\rho}\colon \mathrm{Spin}(V) \to \mathrm{Aut}(\mathrm{Cl}(V), V)$. *Moreover the restriction of the action to* V *induces a group homomorphism*

$$\xi_0\colon \mathrm{Spin}(V) \to \mathrm{SO}(V),$$

which is a double covering. If $n = \dim V \geq 3$, *then this is the universal covering.*

Proof Note that the conjugation action of $\mathrm{Spin}(V)$ is the restriction of that of $\mathrm{Pin}(V)$. For $v \in V$, let $R_{v^\perp} : V \to V$ be the reflection with respect to the plane that is orthogonal to v. Then it is easy to see that $qwq^{-1} = -R_{q^\perp}(w)$ for any $w \in V$ and $q \in V$ with $\|q\| = 1$. Thus the conjugation action of $\mathrm{Pin}(V)$, and therefore $\mathrm{Spin}(V)$, preserves V. Then the action of each $w \in \mathrm{Spin}(V)$ on V is given by a product of the even number of reflections. It is well-known that every element of $\mathrm{SO}(V)$ can be written as a product of the even number of reflections. Hence the conjugation action induces a surjective homomorphism $\xi_0\colon \mathrm{Spin}(V) \to \mathrm{SO}(V)$. Note that $\ker \xi_0$ contains ± 1. When $n = 1, 2$, it is easily checked that ξ_0 is a double covering. When $n = 3, 4$, $\pi_1(\mathrm{SO}(V)) = \mathbb{Z}/2$ and $\mathrm{Spin}(V)$ is simply-connected by the above computations. These imply that $\ker \xi_0 = \{\pm 1\}$ and ξ_0 gives the universal covering. For the general cases ($n > 4$), we refer the readers to Morgan's book [131, Lemma 2.2.2]. □

Remark 3.4 Let $\rho : \mathrm{SO}(n) \to \mathrm{Aut}(\mathrm{Cl}(\mathbb{R}^n))$ be the canonical action in (3.1). It is easy to check $\tilde{\rho}$ factors through the double covering ξ_0 and ρ,

$$\tilde{\rho} = \rho \circ \xi_0 : \mathrm{Spin}(n) \to \mathrm{Aut}(\mathrm{Cl}(\mathbb{R}^n)).$$

3.1.3 Representation of Clifford Algebras

Let $\{e_1, \cdots, e_n\}$ be an oriented orthonormal basis of V, and set

$$\omega_{\mathbb{C}} := i^{\lfloor \frac{n+1}{2} \rfloor} \cdot e_1 \cdots e_n \in \mathrm{Cl}(V) \otimes_{\mathbb{R}} \mathbb{C}.$$

The following facts are easy to check:

- $\omega_{\mathbb{C}}^2 = 1$.
- $\omega_{\mathbb{C}}$ is independent of the choice of an oriented orthonormal basis.
- $\omega_{\mathbb{C}}$ is a center if $\dim V$ is odd.
- If $\dim V$ is even, then $\omega_{\mathbb{C}}$ is a center in $\mathrm{Cl}_0(V) \otimes_{\mathbb{R}} \mathbb{C}$, and it anticommutes in $\mathrm{Cl}_1(V) \otimes_{\mathbb{R}} \mathbb{C}$ (see below for the notations).

The $\omega_{\mathbb{C}}$ acts on $\mathrm{Cl}(V) \otimes_{\mathbb{R}} \mathbb{C}$ from the left, and $\mathrm{Cl}(V) \otimes_{\mathbb{R}} \mathbb{C}$ decomposes as

$$\mathrm{Cl}(V) \otimes_{\mathbb{R}} \mathbb{C} \cong (\mathrm{Cl}(V) \otimes_{\mathbb{R}} \mathbb{C})^+ \oplus (\mathrm{Cl}(V) \otimes_{\mathbb{R}} \mathbb{C})^-$$

corresponding to the eigenspaces ± 1 of $\omega_{\mathbb{C}}$.

Let us continue to list some properties of the decomposition.

- It is an algebra isomorphism $\mathrm{Cl}(V) \otimes_{\mathbb{R}} \mathbb{C} \cong (\mathrm{Cl}(V) \otimes_{\mathbb{R}} \mathbb{C})^+ \oplus (\mathrm{Cl}(V) \otimes_{\mathbb{R}} \mathbb{C})^-$, if $\dim V$ is odd.
- Both $(\mathrm{Cl}(V) \otimes_{\mathbb{R}} \mathbb{C})^\pm$ admit algebra isomorphisms to $\mathrm{Cl}_0(V) \otimes_{\mathbb{R}} \mathbb{C}$, if $\dim V$ is odd.

- There is an algebra isomorphism $\mathrm{Cl}_0(V)\otimes_{\mathbb{R}}\mathbb{C} \cong (\mathrm{Cl}_0(V)\otimes_{\mathbb{R}}\mathbb{C})^+ \oplus (\mathrm{Cl}_0(V)\otimes_{\mathbb{R}}\mathbb{C})^-$, if $\dim V$ is even.
- Let $W' \subset V$ be any codimension 1 linear subspace. Then there are algebra isomorphisms

$$(\mathrm{Cl}_0(V) \otimes_{\mathbb{R}} \mathbb{C})^{\pm} \cong (\mathrm{Cl}(W') \otimes_{\mathbb{R}} \mathbb{C})^{\pm}.$$

- Combining the above properties, we have the algebra isomorphisms

$$(\mathrm{Cl}_0(V) \otimes_{\mathbb{R}} \mathbb{C})^+ \cong \mathrm{Cl}_0(W') \otimes_{\mathbb{R}} \mathbb{C} \cong \mathrm{Cl}(W) \otimes_{\mathbb{R}} \mathbb{C}$$

  for any codimension 1 linear subspace $W \subset W'$.

Note that $\mathrm{Cl}(\mathbb{R}^2) \otimes_{\mathbb{R}} \mathbb{C} \cong \mathbb{H} \otimes_{\mathbb{R}} \mathbb{C}$ is isomorphic to the algebra $\mathrm{End}_{\mathbb{C}}(\mathbb{C}^2)$ through the map $\mathbb{H} \to \mathrm{M}_2(\mathbb{C})$ by

$$\alpha + j\beta \to \begin{pmatrix} \alpha & -\bar{\beta} \\ \beta & \bar{\alpha} \end{pmatrix},$$

where $\mathrm{M}_2(\mathbb{C})$ is the space of 2×2 complex matrices. In particular, the algebra isomorphism $\mathbb{H} \otimes_{\mathbb{R}} \mathbb{C} \cong \mathrm{M}_2(\mathbb{C})$ holds.

Lemma 3.5 *The algebra isomorphism holds:*

$$\mathrm{Cl}(V \oplus \mathbb{R}^2) \otimes_{\mathbb{R}} \mathbb{C} \cong (\mathrm{Cl}(V) \otimes_{\mathbb{R}} \mathbb{C}) \otimes_{\mathbb{C}} \left(\mathrm{Cl}(\mathbb{R}^2) \otimes_{\mathbb{R}} \mathbb{C}\right).$$

Proof Let $\{v_1, \cdots, v_n\}$ and $\{e_1, e_2\}$ be orthonormal basis of V and $\mathbb{R}^2$ respectively. Then, consider a linear map

$$V \oplus \mathbb{R}^2 \to (\mathrm{Cl}(V) \otimes_{\mathbb{R}} \mathbb{C}) \otimes_{\mathbb{C}} \left(\mathrm{Cl}(\mathbb{R}^2) \otimes_{\mathbb{R}} \mathbb{C}\right)$$

$$\begin{cases} v_i \to i v_i \otimes e_1 \cdot e_2, \\ e_k \to 1 \otimes e_k. \end{cases}$$

This extends to an algebraic homomorphism $\mathrm{Cl}(V \oplus \mathbb{R}^2) \otimes_{\mathbb{R}} \mathbb{C} \to (\mathrm{Cl}(V) \otimes_{\mathbb{R}} \mathbb{C}) \otimes_{\mathbb{C}} \left(\mathrm{Cl}(\mathbb{R}^2) \otimes_{\mathbb{R}} \mathbb{C}\right)$. Then, it is sufficient to check two things: that both sides have the same dimension as vector spaces and that it is surjective. They are easy to check. □

Corollary 3.6

- *The algebra isomorphism* $\mathrm{Cl}(V) \otimes_{\mathbb{R}} \mathbb{C} \cong \mathrm{M}_{2^n}(\mathbb{C})$ *holds, if* $\dim V = 2n$.
- *The algebra isomorphism* $\mathrm{Cl}(V) \otimes_{\mathbb{R}} \mathbb{C} \cong \mathrm{M}_{2^n}(\mathbb{C}) \oplus \mathrm{M}_{2^n}(\mathbb{C})$ *holds, if* $\dim V = 2n + 1$.

Remark 3.7

- The algebra isomorphism $(\mathrm{Cl}_0(V) \otimes \mathbb{C})^+ \cong \mathrm{M}_{2^{n-1}}(\mathbb{C})$ holds if $\dim V = 2n$.
- The diagonal embedding

$$\Delta : \mathrm{Cl}_0(V) \otimes_{\mathbb{R}} \mathbb{C} \hookrightarrow \mathrm{M}_{2^{n-1}}(\mathbb{C}) \oplus \mathrm{M}_{2^{n-1}}(\mathbb{C})$$

  holds, where the right-hand side sum corresponds to the $\pm$ 1-eigenspace decomposition of $\omega_{\mathbb{C}}$.

Corollary 3.8

- *Suppose* $\dim V = 2n$. *There is a unique complex irreducible* 2^n*-dimensional representation* $S_{\mathbb{C}}(V)$ *of* $\mathrm{Cl}(V) \otimes_{\mathbb{R}} \mathbb{C}$ *such that the algebra isomorphism holds:*

$$\mathrm{Cl}(V) \otimes_{\mathbb{R}} \mathbb{C} \cong \mathrm{End}_{\mathbb{C}}(S_{\mathbb{C}}(V)).$$

- *Suppose* $\dim V = 2n + 1$. *There are two complex irreducible* 2^n*-dimensional representations of* $\mathrm{Cl}(V) \otimes_{\mathbb{R}} \mathbb{C}$.

  Moreover the restrictions of two representations to $\mathrm{Cl}_0(V) \otimes_{\mathbb{R}} \mathbb{C}$ *are isomorphic such that the algebra isomorphism holds:*

$$\mathrm{Cl}_0(V) \otimes_{\mathbb{R}} \mathbb{C} \cong \mathrm{End}_{\mathbb{C}}(S_{\mathbb{C}}(V)).$$

Corollary 3.9 *Suppose* $\dim V = 2n$. *There is a decomposition*

$$S_{\mathbb{C}}(V) = S_{\mathbb{C}}^+(V) \oplus S_{\mathbb{C}}^-(V)$$

where $S_{\mathbb{C}}^{\pm}(V)$ *are* (± 1)*-eigenspaces of* $\Delta(\omega_{\mathbb{C}})$ *which gives the following isomorphisms:*

$$(\mathrm{Cl}_0(V) \otimes_{\mathbb{R}} \mathbb{C})^+ \cong \mathrm{End}(S_{\mathbb{C}}^+(V)),$$

$$(\mathrm{Cl}_0(V) \otimes_{\mathbb{R}} \mathbb{C})^- \cong \mathrm{End}(S_{\mathbb{C}}^-(V)).$$

In particular $S_{\mathbb{C}}^+(V)$ *and* $S_{\mathbb{C}}^-(V)$ *are mutually inequivalent* $\mathrm{Cl}_0(V) \otimes_{\mathbb{R}} \mathbb{C}$ *modules.*

Definition 3.10 We say that the restriction of the above representation $\mathrm{Cl}_0(V) \otimes_{\mathbb{R}} \mathbb{C} \hookrightarrow \mathrm{End}_{\mathbb{C}}(S_{\mathbb{C}}(V))$ to the spin group

$$\Delta_{\mathbb{C}} : \mathrm{Spin}(V) \to \mathrm{Aut}_{\mathbb{C}}(S_{\mathbb{C}}(V))$$

is the complex spin representation.

Let us list several properties of the complex spin representation.

- $\Delta_{\mathbb{C}} = \Delta_{\mathbb{C}}^{+} \oplus \Delta_{\mathbb{C}}^{-}$ splits as $S_{\mathbb{C}}(V) = S_{\mathbb{C}}^{+}(V) \oplus S_{\mathbb{C}}^{-}(V)$ as $\mathrm{Spin}(V)$ representations, if $\dim V = 2n$.
- $\Delta_{\mathbb{C}}$ is the irreducible $\mathrm{Spin}(V)$ representation, if $\dim V = 2n + 1$.

Examples

- The case of $V = \mathbb{R}^2$. Note that $S_{\mathbb{C}}^{\pm}(V) \cong \mathbb{C}$ and $\mathrm{Spin}(2) \cong S^1$. The complex spin representation is given by

$$\Delta_{\mathbb{C}}^{\pm} : \mathrm{Spin}(2) \to \mathrm{Aut}(S_{\mathbb{C}}^{+}(\mathbb{R}^2) \oplus S_{\mathbb{C}}^{-}(\mathbb{R}^2)),$$
$$\alpha \to \begin{pmatrix} \alpha & 0 \\ 0 & \bar{\alpha} \end{pmatrix}$$

Then V is identified with $\mathbb{C}$ by $e_1 \mapsto 1$ and $e_2 \mapsto i$ and the Clifford multiplication is given by

$$\begin{pmatrix} 0 & -\bar{\beta} \\ \beta & 0 \end{pmatrix} \qquad \text{for } \beta \in V = \mathbb{C}.$$

- The case of $V = \mathbb{R}^3$. Note that $\mathbb{H}^+$ and $\mathbb{H}^-$ of (3.2) is the ± 1 eigenspaces of $\omega_{\mathbb{C}} = -e_1e_2e_3$. Two irreducible representations $S_{\mathbb{C}}(V)$ can be taken to be $\mathbb{H}^+$ and $\mathbb{H}^-$. The action of $\mathrm{Cl}(V) \otimes_{\mathbb{R}} \mathbb{C} = (\mathbb{H}^+ \oplus \mathbb{H}^-) \otimes_{\mathbb{R}} \mathbb{C}$ on $\mathbb{H}^{\pm}$ is given by multiplication of quaternions as

$$((v_+ + v_-) \otimes z) \cdot \phi_{\pm} = v_{\pm}\phi_{\pm}z,$$

where $v_{\pm}, \phi_{\pm} \in \mathbb{H}^{\pm}$ and $z \in \mathbb{C}$. (Recall that $x_+x_- = x_-x_+ = 0$ for $x_{\pm} \in \mathbb{H}^{\pm}$.) For the bases of $\mathbb{H}^{\pm}$ in (3.2),

$$e_1\mathbf{1} = \boldsymbol{i}, \qquad e_2\mathbf{1} = \boldsymbol{j}, \qquad e_3\mathbf{1} = \boldsymbol{k},$$
$$e_1\mathbf{1}' = -\boldsymbol{i}', \quad e_2\mathbf{1}' = -\boldsymbol{j}', \quad e_3\mathbf{1}' = -\boldsymbol{k}'.$$

We identify V with $\mathrm{im}\,\mathbb{H}^{\pm}$ by

$$e_1 = \pm i, \quad e_2 = \pm j, \quad e_3 = \pm k.$$

Then the Clifford multiplication of V on $S_{\mathbb{C}}(V) = \mathbb{H}^{\pm}$ is identified with the multiplication of $\mathrm{im}\,\mathbb{H}$ with $\mathbb{H}$. Recall that $\mathrm{Spin}(3)$ is a 3-sphere diagonally embedded in $\mathbb{H}^+ \oplus \mathbb{H}^-$. Since $\mathrm{Spin}(3)$ acts on V by $v \mapsto qvq^{-1}$ for $v \in V$ and $q \in \mathrm{Spin}(3)$, the Clifford multiplication is $\mathrm{Spin}(3)$-equivariant.
- The case of $V = \mathbb{R}^4$. Note that $\mathbb{H}^{\pm}$ in (3.3) are the ± 1 eigenspaces of $\omega_{\mathbb{C}} = -e_1e_2e_3e_4$. We take $S_{\mathbb{C}}^{+}(V)$ to be $\mathbb{H}^+$. Then we can assume $S_{\mathbb{C}}^{-}(V) =$

$e_1 S^+_{\mathbb{C}}(V) = e_1\mathbb{H}^+$. Note that

$$\begin{array}{llll} e_1\mathbf{1} = \mathbf{1}'e_1, & e_1\boldsymbol{i} = \boldsymbol{i}'e_1, & e_1\boldsymbol{j} = \boldsymbol{j}'e_1, & e_1\boldsymbol{k} = \boldsymbol{k}'e_1, \\ \mathbf{1}e_1 = e_1\mathbf{1}', & \boldsymbol{i}e_1 = e_1\boldsymbol{i}', & \boldsymbol{j}e_1 = e_1\boldsymbol{j}', & \boldsymbol{k}e_1 = e_1\boldsymbol{k}'. \end{array}$$

Then we can identify $S^-_{\mathbb{C}}(V) = e_1\mathbb{H}^+$ with $\mathbb{H}^-$ by

$$e_1\mathbf{1} \leftrightarrow \mathbf{1}', \quad e_1\boldsymbol{i} \leftrightarrow \boldsymbol{i}', \quad e_1\boldsymbol{j} \leftrightarrow \boldsymbol{j}', \quad e_1\boldsymbol{k} \leftrightarrow \boldsymbol{k}'.$$

Note that

$$\begin{gathered} e_2\mathbf{1} = -e_1\boldsymbol{i}, \quad e_3\mathbf{1} = -e_1\boldsymbol{j}, \quad e_4\mathbf{1} = -e_1\boldsymbol{k}, \\ e_1e_1\mathbf{1} = -\mathbf{1}, \quad e_2e_1\mathbf{1} = -\boldsymbol{i}, \quad e_3e_1\mathbf{1} = -\boldsymbol{j}, \quad e_4e_1\mathbf{1} = -\boldsymbol{k}. \end{gathered}$$

We identify V with $\mathbb{H}$ by

$$e_1 \leftrightarrow -1, \quad e_2 \leftrightarrow -i, \quad e_3 \leftrightarrow -j, \quad e_4 \leftrightarrow -k.$$

The space $\mathbb{H}$ identified with V is denoted by $\mathbb{H}_T$. Then the Clifford action $V \to \mathrm{End}(\mathbb{H}^+ \oplus \mathbb{H}^-)$ is identified with

$$v \mapsto \begin{pmatrix} 0 & v \\ -\bar{v} & 0 \end{pmatrix}. \tag{3.4}$$

The group $\mathrm{Spin}(4) = \mathrm{Sp}(1) \times \mathrm{Sp}(1) \subset \mathbb{H}^+ \oplus \mathbb{H}^-$ is written as

$$\mathrm{Spin}(4) = \{q_+ + q_- \mid q_\pm \in \mathbb{H}^\pm, |q_\pm| = 1\}.$$

(Note that $(q_+ + q_-)(q'_+ + q'_-) = q_+q'_+ + q_-q'_-$ since $x_+x_- = x_-x_+ = 0$ for any $x_\pm \in \mathbb{H}^\pm$.) The action of $\mathrm{Spin}(4)$ on V is given by

$$v \mapsto (q_+ + q_-)v(q_+^{-1} + q_-^{-1}) = q_+vq_-^{-1} + q_-vq_+^{-1}. \tag{3.5}$$

Then the Clifford action is $\mathrm{Spin}(4)$-equivariant.

Proposition 3.11 *There exists a Hermitian inner product $(\cdot, \cdot)$ on $S_{\mathbb{C}}(V)$ such that the Clifford multiplication by unit vector $e \in \mathbb{R}$ preserves the inner product, i.e.,*

$$(e \cdot u, e \cdot v) = (u, v)$$

for $u, v \in S_{\mathbb{C}}(V)$ and $e \in V$ with $|e| = 1$.

Proof Let O_n be the finite subgroup in $\mathrm{Cl}(V)$ generated by an orthonormal basis $e_1, \ldots, e_n$ of V. Choose a Hermitian inner product on $S_{\mathbb{C}}(V)$ and averaging it over O_n. Then $(e_j u, e_j u) = (u, u)$. For $j \neq k$, $(e_j u, e_k u) = 0$ since

$$(e_j u, e_k u) = (e_k e_j u, -u) = (e_j e_k u, u) = -(e_j u, e_k u).$$

For $e = \sum_j c_j e_j$ with $\sum_j c_j^2 = 1$,

$$(eu, eu) = \sum_j c_j^2 (e_j u, e_j u) + \sum_{j \neq k} c_j c_k (e_j u, e_k u) = (u, u).$$

The proposition follows from $(e(u+v), e(u+v)) = (u+v, u+v)$ and $(e(iu+v), e(iu+v)) = (iu+v, iu+v)$. □

Corollary 3.12 *The Clifford multiplication of a vector $a \in V$ is a skew-symmetric transformation,*

$$(a \cdot u, v) = -(u, a \cdot v).$$

Proof For $a \neq 0$, let $e = a/|a|$. Then $ea = aa/|a| = -|a|^2/|a| = -|a|$.

$$(au, v) = (eau, ev) = -|a|(u, ev) = -(u, |a|ev) = -(u, av).$$

□

3.1.3.1 Clifford Algebra and Exterior Algebra

For a vector space V, let $\Lambda^* V = \sum_i \Lambda^i V$ be the exterior algebra of V.

Lemma 3.13

1. *There is a canonical vector space isomorphism*

$$\mathfrak{K}\colon \Lambda^* V \xrightarrow{\cong} \mathrm{Cl}(V),$$

given by

$$v_1 \wedge \cdots \wedge v_k \mapsto \frac{1}{k!} \sum_{\sigma \in \mathfrak{S}_k} \mathrm{sign}(\sigma) v_{\sigma(1)} \cdots v_{\sigma(k)},$$

where $\mathfrak{S}_k$ is the symmetric group of degree k.

2. *Equip $\Lambda^* V$ with the multiplication of vectors of V by*

$$v \cdot (v_1 \wedge \cdots \wedge v_k) = v \wedge (v_1 \wedge \cdots \wedge v_k) - v \lrcorner (v_1 \wedge \cdots \wedge v_k), \tag{3.6}$$

where

$$v \lrcorner (v_1 \wedge \cdots \wedge v_k) := \sum_{i=1}^{k} (-1)^{i-1} (v, v_i) \cdot v_1 \wedge \cdots \wedge \check{v}_i \wedge \cdots \wedge v_k.$$

Then $\mathfrak{R}$ *is an isomorphism of the modules of* $\mathrm{Cl}(V)$ *with respect to this multiplicative structure.*

Proof Exercise. Problem 3.90. □

Lemma 3.14 *Suppose* $\dim V = 4$. *Then through the linear isomorphism* $\mathrm{Cl}(V) \otimes_{\mathbb{R}} \mathbb{C} \cong \wedge^*(V) \otimes_{\mathbb{R}} \mathbb{C}$ *in Lemma 3.13, the restriction induces the isomorphism*

$$(\mathrm{Cl}_0(V) \otimes_{\mathbb{R}} \mathbb{C})^{\pm} \cong \mathbb{C}\left(\frac{1 \pm \omega_{\mathbb{C}}}{2}\right) \oplus \wedge^2_{\pm}(V) \otimes_{\mathbb{R}} \mathbb{C}$$

as the $\mathrm{Cl}_0(V)$*-modules, where* $\wedge^2_{\pm}(V)$ *is the space of self-dual 2-forms.*

Proof The linear isomorphism restricts as $\mathrm{Cl}_0(V) \cong \wedge^0(V) \oplus \wedge^2(V) \oplus \wedge^4(V)$. Through Lemma 3.13, $\omega_{\mathbb{C}}$ acts on $\wedge^2(V) \otimes_{\mathbb{R}} \mathbb{C}$ to itself, and switches $\wedge^0(V) \otimes_{\mathbb{R}} \mathbb{C}$ with $\wedge^4(V) \otimes_{\mathbb{R}} \mathbb{C}$. For an oriented orthonormal basis $\{e_1, e_2, e_3, e_4\}$, it is easy to check that $\omega_{\mathbb{C}} \cdot e_{i_1} \cdot e_{i_2}$ corresponds to $*(e_{i_1} \wedge e_{i_2})$. □

3.1.3.2 Spin^c Representation

Definition 3.15 The spin^c group is defined by

$$\mathrm{Spin}^c(V) := \langle\, v \in V \otimes_{\mathbb{R}} \mathbb{C} : \|v\| = 1 \,\rangle \cap (\mathrm{Cl}_0(V) \otimes_{\mathbb{R}} \mathbb{C}) \subset \mathrm{Cl}(V) \otimes_{\mathbb{R}} \mathbb{C}.$$

Lemma 3.16

$$\mathrm{Spin}^c(V) \cong \mathrm{Spin}(V) \times_{\{\pm 1\}} \mathrm{U}(1).$$

Proof Because $S^1 \subset \mathrm{Cl}(V) \otimes_{\mathbb{R}} \mathbb{C}$ is a center, there is a surjective homomorphism

$$\mathrm{Spin}^c(V) \twoheadrightarrow \mathrm{Spin}(V).$$

It is immediate to check $\ker = \{(\alpha, \bar{\alpha}) : \alpha \in \mathrm{Spin}(V) \cap S^1\}$. □

Lemma 3.17 *There is an extension of the spin representation as*

$$\hat{\Delta}_{\mathbb{C}} : \mathrm{Spin}^c(V) \to \mathrm{Aut}_{\mathbb{C}}(S_{\mathbb{C}}(V))$$

such that it splits as $\hat{\Delta}^{\pm}_{\mathbb{C}} : \mathrm{Spin}^c(V) \to \mathrm{Aut}_{\mathbb{C}}(S^{\pm}_{\mathbb{C}}(V))$*, if* $\dim V = 2n$.

Proof It is sufficient to check $\Delta_{\mathbb{C}}(-1) = -1$. □

Lemma 3.18 *The conjugation action of* $\mathrm{Spin}^c(V)$ *on* $\mathrm{Cl}(V)$ *is defined by*

$$w \mapsto qwq^{-1}$$

for $w \in \mathrm{Cl}(V)$ *and* $[(q, z)] \in \mathrm{Spin}^c(V) = \mathrm{Spin}(V) \times_{\{\pm 1\}} \mathrm{U}(1)$. *Moreover this action defines a group homomorphism* $\tilde{\rho}^c\colon \mathrm{Spin}^c(V) \to \mathrm{Aut}(\mathrm{Cl}(V), V)$*, and the restriction of the action to* V *induces a group homomorphism*

$$\xi_0^c\colon \mathrm{Spin}^c(V) \to \mathrm{SO}(V).$$

The map $\tilde{\rho}$ *factors through* I^c *and* ρ*,*

$$\tilde{\rho}^c = \rho \circ \xi_0^c : \mathrm{Spin}^c(n) \to \mathrm{Aut}(\mathrm{Cl}(\mathbb{R}^n)).$$

Proof The proof is parallel to that of Lemma 3.3. □

By Proposition 3.11 and Corollary 3.12, we obtain:

Proposition 3.19 *The spin and spin*^{c} *representations are unitary.*

Example

By complexifying the spin representations in the Examples in Sect. 3.1.3, we obtain the spin^c representations for $V = \mathbb{R}^n$ $(n = 2, 3, 4)$. We summarize the case of $V = \mathbb{R}^4$. The spaces V and $S^{\pm}_{\mathbb{C}}(V) = \mathbb{H}^{\pm}$ are identified with the copies of $\mathbb{H}$ as in the Examples in Sect. 3.1.3. We define complex structures on $\mathbb{H}^{\pm}$ by multiplication of complex numbers by right. The $\mathrm{Spin}^c(4)$ module structures on V, $\mathbb{H}^+$ and $\mathbb{H}^-$ are given as follows. For $[q_+, q_-, u] \in \mathrm{Spin}^c(4) = (\mathrm{Sp}(1) \times \mathrm{Sp}(1)) \times_{\{\pm 1\}} \mathrm{U}(1)$, $v \in V, \phi \in \mathbb{H}^+$ and $\psi \in \mathbb{H}^-$,

$$[q_+, q_-, u] \cdot v = q_+ v q_-^{-1},$$

$$[q_+, q_-, u] \cdot \phi = q_+ \phi u^{-1},$$

$$[q_+, q_-, u] \cdot \psi = q_- \phi u^{-1}.$$

Then it is easy to check that the Clifford action (3.4) is $\mathrm{Spin}^c(4)$ equivariant. We identify $\mathbb{H}^{\pm}$ with $\mathbb{C}^2$ by the correspondence

$$\phi = \phi_1 - j\phi_2 \mapsto \begin{pmatrix} \phi_1 \\ \phi_2 \end{pmatrix}. \tag{3.7}$$

Then, for $q = \alpha + \beta j$, $\phi = \phi_1 - j\phi_2 \in \mathbb{H}$ and $z \in \mathbb{C}$, $q\phi z$ is written as

$$\begin{pmatrix} \alpha & \beta \\ -\bar{\beta} & \bar{\alpha} \end{pmatrix} \begin{pmatrix} \phi_1 z^{-1} \\ \phi_2 z^{-1} \end{pmatrix} = z^{-1} \begin{pmatrix} \alpha & \beta \\ -\bar{\beta} & \bar{\alpha} \end{pmatrix} \begin{pmatrix} \phi_1 \\ \phi_2 \end{pmatrix}.$$

3.2 Spin and Spin^c Structures

Let X be an oriented smooth and closed manifold, and $P \to X$ be a principal $\mathrm{SO}(n)$ bundle. Let $\rho : \mathrm{SO}(n) \to \mathrm{Aut}\,(V)$ be a representation. Then $E := P \times_\rho V \to X$ is an oriented vector bundle.

Examples

- Let there be with a Riemannian metric on X with $\dim X = n$. Then the space of oriented orthonormal frames $\{e_1, \cdots, e_n\}$ of $T_x X$ at any $x \in X$ forms a principal $\mathrm{SO}(n)$ bundle P_X. The principal $\mathrm{SO}(n)$ bundle P_X is called the *oriented frame bundle*.
- Let $\rho_0 : \mathrm{SO}(n) \to \mathrm{Aut}(\mathbb{R}^n)$ be the standard representation. Then, the vector bundle isomorphism holds:

$$P_X \times_{\rho_0} \mathbb{R}^n \cong TX.$$

Lemma 3.20 *Let $P \to X$ be a principal $\mathrm{O}(n)$ bundle, and $E := P \times_{\rho_0} \mathbb{R}^n$ be the associated vector bundle. Then, E is orientable, if and only if the first Stiefel–Whitney class $w_1(P) = 0$ vanishes. In the case, P is reduced to a principal $\mathrm{SO}(n)$ bundle.*

Proof 1. The bundle of orientations of E is given by the principal $\mathbb{Z}/2$-bundle $\mathrm{Or}(E) = P/\mathrm{SO}(n)$. Note that E is orientable if and only if $\mathrm{Or}(E)$ is a trivial bundle. The latter condition is equivalent to the condition that P is reduced to a principal $\mathrm{SO}(n)$ bundle. The principal $\mathbb{Z}/2$-bundles are classified by

$$\mathrm{Hom}(\pi_1(X), \mathbb{Z}/2) \cong \mathrm{Hom}(H_1(X), \mathbb{Z}/2) \cong H^1(X; \mathbb{Z}/2).$$

Via the above isomorphisms, $\mathrm{Or}(E)$ determines a class $w_1(E) \in H^1(X; \mathbb{Z}/2)$, and $\mathrm{Or}(E)$ is trivial if and only if $w_1(E) = 0$. We claim this $w_1(E)$ is the first Stiefel–Whitney class. Since $\mathrm{Or}(f^*E) = f^*\mathrm{Or}(E)$ for a continuous map $f : Y \to X$, w_1 satisfies the naturality, $w_1(f^*E) = f^*w_1(E)$. Let $\tilde{E}$ be the universal $\mathbb{R}^n$ bundle over the classifying space $B\,\mathrm{O}(n)$. Then $w_1(\tilde{E})$ is the unique non-zero element in $H^1(B\,\mathrm{O}(n); \mathbb{Z}/2) \cong \mathbb{Z}/2$ because, if otherwise, every $\mathbb{R}^n$ bundle would be orientable. These facts prove that $w_1(E)$ is the first Stiefel–Whitney class. □

Let $\xi_0 :\ \mathrm{Spin}(n) \to \mathrm{SO}(n)$ be the double covering in Lemma 3.3.

Definition 3.21 We say that an SO(n) bundle P over X lifts to a Spin(n)-bundle $\tilde{P}$ over X if there is a fiber-preserving double covering $\xi : \tilde{P} \to P$ such that

$$\xi(pg) = \xi(p)\xi_0(g) \quad \text{for } p \in \tilde{P} \text{ and } g \in \mathrm{Spin}(n).$$

Lemma 3.22 *Let $P \to X$ be a principal* SO(n) *bundle, and $E = P \times_{\rho_0} \mathbb{R}^n$ be the associated vector bundle. Then, P lifts to a* Spin(n) *bundle $\tilde{P}$, if and only if the second Stiefel–Whitney class $w_2(E) = 0$ vanishes. Moreover the set of the isomorphism classes of the lifts has a bijective correspondence with $H^1(X; \mathbb{Z}/2)$.*

Proof If there exists a fiber-preserving double cover $\xi : \tilde{P} \to P$ such that the restriction to each fiber is a nontrivial double cover, then ξ is automatically a lift because the SO(n) action on P lifts to a Spin(n) action on $\tilde{P}$. Therefore the set of isomorphism classes of the lifts has a bijective correspondence with the set of classes in $H^1(P; \mathbb{Z}/2)$ whose restriction to the fiber is non-zero. By using the Leray–Serre spectral sequence for the fibration $\mathrm{SO}(n) \xrightarrow{i} P \xrightarrow{\pi} X$, we obtain the exact sequence

$$0 \to H^1(X; \mathbb{Z}/2) \to H^1(P; \mathbb{Z}/2) \to H^1(\mathrm{SO}(n); \mathbb{Z}/2) \xrightarrow{w_2} H^2(X; \mathbb{Z}/2).$$

Recall that $H^1(\mathrm{SO}(n); \mathbb{Z}/2) \cong \mathbb{Z}/2$. Let g be its generator. By the above exact sequence, there is a class in $H^1(P; \mathbb{Z}/2)$ whose restriction to the fiber is non-zero if and only if $w_2(g) = 0$. Furthermore, the set of such classes is a coset of $H^1(X; \mathbb{Z}/2)$. We shall prove that $w_2(g)$ is the second Stiefel–Whitney class of E. The naturality of w_2 follows from the naturality of the exact sequence. Let $P = E\,\mathrm{SO}(n)$ be the universal SO(n) bundle over $B\,\mathrm{SO}(n)$, and $\tilde{E}$ be the associated $\mathbb{R}^n$ bundle. Since $E\,\mathrm{SO}(n)$ is contractible, the above exact sequence implies that $w_2(g)$ is the non-zero element in $H^2(B\,\mathrm{SO}(n); \mathbb{Z}/2) = \mathbb{Z}/2$, and gives the second Stiefel–Whitney class of $\tilde{E}$. □

Definition 3.23 We say that the lift $\tilde{P}$ of P is a spin structure of P.

We just say that $\tilde{P}$ is a spin structure of X, when P is the oriented frame bundle.

Remark 3.24 Note the following exact sequence:

$$1 \to \{\pm 1\} \to \mathrm{Spin}(n) \xrightarrow{\xi_0} \mathrm{SO}(n) \to 1.$$

If $\tilde{P}$ is a spin structure on P, then we have the isomorphism $P = \tilde{P}/\{\pm 1\}$.

Consider the subgroups of $\mathrm{Spin}^c(n) = \mathrm{Spin}(n) \times_{\{\pm 1\}} \mathrm{U}(1)$,

$$\mathrm{U}(1) = \{[(1, z)] \in \mathrm{Spin}^c(n) \mid 1 \in \mathrm{Spin}(n), z \in \mathrm{U}(1)\},$$

$$\begin{aligned}\mathrm{Spin}(n) &= \{[(s, 1)] \in \mathrm{Spin}^c(n) \mid s \in \mathrm{Spin}(n), 1 \in \mathrm{U}(1)\},\\ \{\pm 1\} &= \mathrm{U}(1) \cap \mathrm{Spin}(n) = \{[(1, 1)], [(1, -1)]\}.\end{aligned}$$

The inclusion maps of these subgroups into $\mathrm{Spin}^c(n)$ induce the following exact sequences:

$$\begin{gathered} 1 \to \mathrm{U}(1) \to \mathrm{Spin}^c(n) \xrightarrow{\xi_0^c} \mathrm{SO}(n) \to 1,\\ 1 \to \mathrm{Spin}(n) \to \mathrm{Spin}^c(n) \to \mathrm{U}(1) \to 1,\\ 1 \to \{\pm 1\} \to \mathrm{Spin}^c(n) \to \mathrm{SO}(n) \times \mathrm{U}(1) \to 1. \end{gathered}$$

Definition 3.25 Let P be an $\mathrm{SO}(n)$-bundle over X. A *spin^c structure on P* is a lift of P to a $\mathrm{Spin}^c(n)$-bundle $\tilde{P}$. A *spin^c structure on X* is a spin^c structure on the oriented frame bundle P_X.

Lemma 3.26 *Let P be an $\mathrm{SO}(n)$-bundle over X. There is a spin^c structure on P if and only if the second Stiefel–Whitney class $w_2(P) \in H^2(X; \mathbb{Z}/2)$ has an integral lift in $H^2(X; \mathbb{Z})$. Moreover the set of the isomorphism classes of spin^c structures has a bijective correspondence with $H^2(X; \mathbb{Z})$.*

Proof The first part of the lemma follows from the following claim:

Claim For a $\mathrm{U}(1)$-bundle $P_{\mathrm{U}(1)}$, the fiber sum $P \times_X P_{\mathrm{U}(1)}$ has a lift to a $\mathrm{Spin}^c(n)$-bundle $\tilde{P}$ if and only if the first Chern class $c_1(P_{\mathrm{U}(1)})$ of $P_{\mathrm{U}(1)}$ is an integral lift of $w_2(P)$.

Assuming the claim, we prove the first part of the lemma. If there is a spin^c structure $\tilde{P}$ on P, then $P = \tilde{P}/\mathrm{U}(1)$ and $P_1 = P/\mathrm{Spin}(n)$ is a $\mathrm{U}(1)$-bundle such that $\tilde{P}/\{\pm 1\} = P \times_X P_1$. Therefore $\tilde{P}$ is a lift of $P \times_X P_1$. Then $w_2(P)$ has an integral lift $c_1(P_1)$ by the claim. Conversely, if $w_2(P)$ has an integral lift α, then there is a $\mathrm{U}(1)$-bundle $P_{\mathrm{U}(1)}$ such that $c_1(P_{\mathrm{U}(1)}) = \alpha$. Then there is a lift of $P \times_X P_{\mathrm{U}(1)}$ to a $\mathrm{Spin}^c(n)$-bundle $\tilde{P}$ by the claim. The bundle $\tilde{P}$ gives a spin^c structure on P.

We prove the claim. Note that $\mathrm{SO}(n) \times \mathrm{U}(1)$ is naturally embedded in $\mathrm{SO}(n+2)$, and there is a commutative diagram

$$\begin{array}{ccccccccc} 1 & \longrightarrow & \{\pm 1\} & \longrightarrow & \mathrm{Spin}^c(n) & \longrightarrow & \mathrm{SO}(n) \times \mathrm{U}(1) & \longrightarrow & 1 \\ & & \| & & \downarrow & & \downarrow & & \\ 1 & \longrightarrow & \{\pm 1\} & \longrightarrow & \mathrm{Spin}(n+2) & \longrightarrow & \mathrm{SO}(n+2) & \longrightarrow & 1 \end{array}$$

Define an $\mathrm{SO}(n+2)$-bundle Q by $Q = (P \times_X P_{\mathrm{U}(1)}) \times_{\mathrm{SO}(n)\times\mathrm{U}(1)} \mathrm{SO}(n+2)$. By Lemma 3.22, there is a lift $\varpi : \tilde{Q} \to Q$ where $\tilde{Q}$ is a $\mathrm{Spin}(n+2)$-bundle if and only if

$$w_2(P') = w_2(P) + w_2(P_{\mathrm{U}(1)}) = 0.$$

Note that $c_1(P_{\mathrm{U}(1)})$ is an integral lift of $w_2(P_{\mathrm{U}(1)})$. If there is a spin^c structure $\tilde{P}$ on P, then the $\mathrm{Spin}(n+2)$-bundle $\tilde{Q} = \tilde{P} \times_{\mathrm{Spin}^c(n)} \mathrm{Spin}(n+2)$ is a lift of Q, and therefore $w_2(P) + w_2(P_{\mathrm{U}(1)}) = 0$, that is, $w_2(P)$ is a mod-2 reduction of $c_1(P_{\mathrm{U}(1)})$. Conversely, if $w_2(P)$ is a mod-2 reduction of $c_1(P_{\mathrm{U}(1)})$, then there is a lift $\varpi : \tilde{Q} \to Q$. Then $\tilde{P} = \varpi^{-1}(P \times_X P_{\mathrm{U}(1)})$ gives a spin^c structure on P.

Let us prove the second part of the lemma. The exact sequence $\mathrm{U}(1) \to \mathrm{Spin}^c(n) \to \mathrm{SO}(n)$ induces the following fibration of classifying spaces:

$$B\,\mathrm{U}(1) \to B\,\mathrm{Spin}^c(n) \xrightarrow{r} B\,\mathrm{SO}(n).$$

For $G = \mathrm{Spin}^c(n)$ or $\mathrm{SO}(n)$, the isomorphism classes of G-bundles over X are classified by the homotopy classes in $[X, BG]$. If the isomorphism class of an $\mathrm{SO}(n)$-bundle P is represented by $f : X \to B\,\mathrm{SO}(n)$, the isomorphism class of a spin^c -structure on P is represented by a lift $\tilde{f} : X \to B\,\mathrm{Spin}^c(n)$ such that $r \circ \tilde{f} = f$. Suppose there are two lifts $\tilde{f}_1$ and $\tilde{f}_2$ of f. Since $\pi_1 B\,\mathrm{SO}(n) = 1$ and $B\,\mathrm{U}(1)$ is the Eilenberg–MacLane space $K(\mathbb{Z}, 2)$, there is a sole difference obstruction of these lifts $\tilde{f}_1$ and $\tilde{f}_2$ in $H^2(X; \pi_2 B\,\mathrm{SO}(2)) = H^2(X, \mathbb{Z})$. This proves the second statement of the lemma. □

Remark 3.27 For a map $f : X \to B\,\mathrm{SO}(n)$ representing a $\mathrm{SO}(n)$-bundle P, there is a sole obstruction to the existence of a lift $\tilde{f} : X \to B\,\mathrm{Spin}^c(n)$ in $H^3(X; \pi_2 B\,\mathrm{SO}(n)) = H^3(X; \mathbb{Z})$. By Lemma 3.26, this obstruction class turns out to be the class $W_3(P)$ which is defined as follows. The coefficient exact sequence $0 \to \mathbb{Z} \xrightarrow{\times 2} \mathbb{Z} \to \mathbb{Z}/2 \to 0$ gives rise to a long exact sequence

$$\to H^k(X; \mathbb{Z}) \xrightarrow{\times 2} H^k(X; \mathbb{Z}) \to H^k(X; \mathbb{Z}/2) \xrightarrow{\beta} H^{k+1}(X; \mathbb{Z}) \to .$$

The map β is called the Bockstein homomorphism. The class $W_{k+1}(P)$ is defined by $W_{k+1}(P) = \beta(w_k(P))$. The long exact sequence shows that $W_{k+1}(P) = 0$ if and only if $w_k(P)$ has an integral lift.

Lemma 3.28 (Hirzebruch and Hopf [76]) *If X is a closed oriented connected smooth 4-manifold, $w_2(X)$ has an integral lift.*

Proof Let T be the subgroup of torsion elements in $H_2(X; \mathbb{Z})$. For $x, y \in H_2(X; \mathbb{Z})/T$, the intersection number $x \cdot y \in \mathbb{Z}$ is defined by using the orientation of X. The correspondence $x \mapsto x \cdot x \bmod 2$ defines a homomorphism $q : H_2(X; \mathbb{Z}) \to \mathbb{Z}/2$. Since q is zero on T, q has a lift $\tilde{q} : H_2(X; \mathbb{Z}) \to \mathbb{Z}$. We have a diagram of the universal coefficient sequences.

$$
\begin{array}{ccccccccc}
0 & \longrightarrow & \mathrm{Ext}(H_1(X),\mathbb{Z}) & \longrightarrow & H^2(X;\mathbb{Z}) & \longrightarrow & \mathrm{Hom}(H_2(X),\mathbb{Z}) & \longrightarrow & 0 \\
 & & \downarrow \rho_0 & & \downarrow \rho_1 & & \downarrow \rho_2 & & \\
0 & \longrightarrow & \mathrm{Ext}(H_1(X),\mathbb{Z}/2) & \longrightarrow & H^2(X;\mathbb{Z}/2) & \xrightarrow{h} & \mathrm{Hom}(H_2(X),\mathbb{Z}/2) & \longrightarrow & 0
\end{array}
$$

Then $h(w_2(X)) = q$ by Wu's formula $\langle w_2(X), x\rangle = x \cdot x \bmod 2$ (see e.g. [123, Chapter 23]). Let $\alpha \in H^2(X;\mathbb{Z})$ be a lift of $\tilde{q} \in \mathrm{Hom}(H_2(X),\mathbb{Z})$. Then $\delta := w_2(X) - \rho_1(\alpha)$ is in $\ker h = \mathrm{Ext}(H_1(X),\mathbb{Z}/2)$. Since ρ_0 is surjective (e.g. [72, Section 3.F]), there is a lift $\tilde{\delta} \in \mathrm{Ext}(H_1(X),\mathbb{Z})$ of δ. Then $\alpha + \tilde{\delta}$ is an integral lift of $w_2(X)$. □

Corollary 3.29 *A closed oriented smooth* 4*-manifold admits a spin*^{c} *structure.*

Remark 3.30 Teichner and Vogt [164] proved that every orientable (possibly non-compact) 4-manifold admits spin^c structures.

Let $\tilde{P}$ be a spin (resp. spin^c) structure of X, and $\Delta_{\mathbb{C}} : \mathrm{Spin}(n) \to \mathrm{Aut}_{\mathbb{C}}(S_{\mathbb{C}}(\mathbb{R}^n))$ (resp. $\hat{\Delta}_{\mathbb{C}} : \mathrm{Spin}^c(n) \to \mathrm{Aut}_{\mathbb{C}}(S_{\mathbb{C}}(\mathbb{R}^n))$) be the spin (resp. spin^c) representation.

Definition 3.31

$$S_{\mathbb{C}}(\tilde{P}) := \tilde{P} \times_{\Delta_{\mathbb{C}}} S_{\mathbb{C}}(\mathbb{R}^n) \text{ (resp. } \tilde{P} \times_{\hat{\Delta}_{\mathbb{C}}} S_{\mathbb{C}}(\mathbb{R}^n))$$

is called the spinor bundle of $\tilde{P}$.

The spinor bundle admits a canonical splitting

$$S_{\mathbb{C}}(\tilde{P}) \cong S^+_{\mathbb{C}}(\tilde{P}) \oplus S^-_{\mathbb{C}}(\tilde{P}),$$

when n is even.

Recall that $\rho : \mathrm{SO}(n) \to \mathrm{Aut}(\mathrm{Cl}(\mathbb{R}^n))$ be the canonical representation.

Definition 3.32 Let $\pi : P \to X$ be a principal $\mathrm{SO}(n)$ bundle. The Clifford algebra bundle $\mathrm{Cl}(P)$ over X is defined by

$$\mathrm{Cl}(P) := P \times_\rho \mathrm{Cl}(\mathbb{R}^n).$$

This is an algebra bundle.

Remark 3.33 Let $E = P \times_{\rho_0} \mathbb{R}^n$ be the associated bundle. By Lemma 3.13, $\mathrm{Cl}(P)$ is isomorphic to $\Lambda^* E$. The algebra structure is generated by (3.6).

Lemma 3.34 *The algebra bundle* $\mathrm{Cl}(P)$ *acts on the spinor bundle* $S_{\mathbb{C}}(\tilde{P})$ *of a spin or spin*^{c} *structure* $\tilde{P}$.

Proof First, we consider the spin case. It follows from Remark 3.4 that the equality $\tilde{\rho} = \rho \circ \xi_0 : \mathrm{Spin}(n) \to \mathrm{Aut}(\mathrm{Cl}(\mathbb{R}^n))$ holds, where $\xi_0 : \mathrm{Spin}(n) \to SO(n)$ is the double covering, and $\tilde{\rho} : \mathrm{Spin}(n) \to \mathrm{Aut}(\mathrm{Cl}(\mathbb{R}^n))$ is given by the conjugation $\tilde{\rho}(q)(w) := q \cdot w \cdot q^{-1}$.

Then, the algebra-bundle isomorphism

$$\mathrm{Cl}(P) = P \times_\rho \mathrm{Cl}(\mathbb{R}^n) \cong \tilde{P} \times_{\tilde{\rho}} \mathrm{Cl}(\mathbb{R}^n) =: \mathrm{Cl}(\tilde{P})$$

holds. Hence, it is enough to check that $\mathrm{Cl}(\tilde{P})$ acts on the spinor bundle $S_{\mathbb{C}}(\tilde{P})$. To see this, take elements $[\tilde{p}, w] \in \mathrm{Cl}(\tilde{P})_p$ and $[\tilde{p}, s] \in S_{\mathbb{C}}(\tilde{P})_p$ for $\tilde{p} \in \tilde{P}$. Then we set the action by $[\tilde{p}, w] \cdot [\tilde{p}, s] := [\tilde{p}, w \cdot s]$, where $w \cdot s$ is the Cllifford action. It is easy to check well-definedness by noting the equality $w \cdot s = (x \cdot w \cdot x^{-1}) \cdot (x \cdot s)$ for $x \in \mathrm{Spin}(n)$. The spin^c case is similar. □

By Proposition 3.11, Corollary 3.12 and Proposition 3.19, we obtain:

Proposition 3.35 *The spinor bundle has a canonical Hermitian inner product* $(\cdot, \cdot)$ *such that the Clifford multiplication by a vector* $v \in P \times_\rho \mathbb{R}^n$ *is skew-symmetric,*

$$(v \cdot \phi, \psi) = -(\phi, v \cdot \psi).$$

3.3 Connections

3.3.1 Connections on Vector Bundles

Let X be a smooth manifold, and $E \to X$ be a real vector bundle. We denote by $\Omega^i(E)$ the set of smooth sections of $\Lambda^i(T^*X) \otimes E$ for $0 \le i \le \dim X$.

Definition 3.36 A connection ∇ on E is a linear functional

$$\nabla : \Omega^0(E) \to \Omega^1(E)$$

such that the Leibniz rule $\nabla(f \cdot \sigma) = f \cdot \nabla(\sigma) + df \otimes \sigma$ holds for any $f \in C^\infty(X)$ and $\sigma \in \Omega^0(E)$.

A connection is also called a covariant derivative.

Example
If $E = X \times \mathbb{R}^n$ is the trivial bundle, then

$$\nabla_0 \begin{pmatrix} \sigma_1 \\ \vdots \\ \sigma_n \end{pmatrix} := \begin{pmatrix} d\sigma_1 \\ \vdots \\ d\sigma_n \end{pmatrix}$$

is a connection for $\sigma_i \in C^\infty(X)$.

Lemma 3.37 *There exists a connection ∇ for any $E \to X$.*

In fact, the set of connections forms an infinite-dimensional affine space.

Proof Take a locally finite open covering $\{U_i\}_i$ of X such that the restriction $E|_{U_i}$ admits a trivialization $\varphi_i : E|_{U_i} \cong U_i \times \mathbb{R}^n$.

By choosing a subdivision of $\{U_i\}_i$, if necessary, one can assume that $\{U_i\}_i$ admits a partition of unity $\rho_i \in C_c^\infty(U_i)$ with $\rho_i : U_i \to [0, 1]$ such that $\sum_i \rho_i \equiv 1$.

Then, for any $\sigma \in \Omega^0(E)$, we set

$$\nabla(\sigma) := \sum_i \varphi_i^{-1} \circ \nabla_0(\varphi_i(\rho_i \cdot \sigma)),$$

where ∇_0 is as above.

Let us check that ∇ satisfies the Leibniz rule. Take $f \in C^\infty(X)$. Then,

$$\begin{aligned} \nabla(f \cdot \sigma) &= \sum_i \varphi_i^{-1} \circ \nabla_0(\varphi_i(\rho_i \cdot f \cdot \sigma)) \\ &= \sum_i \varphi_i^{-1} \circ \nabla_0(f \cdot \varphi_i(\rho_i \cdot \sigma)) \\ &= \sum_i \varphi_i^{-1} \circ f \cdot \nabla_0(f \cdot \varphi_i(\rho_i \cdot \sigma)) + \sum_i \varphi_i^{-1} \circ df \otimes \varphi_i(\rho_i \cdot \sigma) \\ &= f \cdot \sum_i \varphi_i^{-1} \circ \nabla_0(f \cdot \varphi_i(\rho_i \cdot \sigma)) + df \otimes \sum_i \varphi_i^{-1} \circ \varphi_i(\rho_i \cdot \sigma) \\ &= f \cdot \nabla(\sigma) + df \otimes \sigma. \end{aligned}$$

This finishes the proof. □

Let $(E, \langle \ \ \rangle)$ be a vector bundle over X equipped with an inner product. A connection ∇ is *compatible* with the inner product, if the condition

$$d\langle \sigma_1, \sigma_2 \rangle = \langle \nabla(\sigma_1), \sigma_2 \rangle + \langle \sigma_1, \nabla(\sigma_2) \rangle$$

holds. We say that the compatible connection as an $O(n)$-connection on E.

Lemma 3.38 *There exists a compatible connection ∇ for any $E \to X$ with an inner product.*

The proof is parallel to Lemma 3.37, by choosing an isometric trivialization $\varphi_i|_{U_i} \cong U_i \times \mathbb{R}^n$.

Definition 3.39 Let $E \to X$ be a vector bundle with a connection ∇. The covariant exterior derivative

$$\nabla : \Omega^i(E) \to \Omega^{i+1}(E)$$

is given by

$$\omega \otimes \sigma \to d\omega \otimes \sigma + (-1)^{\deg \omega} \omega \wedge \nabla(\sigma).$$

Lemma 3.40 *The covariant exterior derivative is well-defined.*

Proof Exercise Problem 3.91. □

Once a connection is extended to the covariant exterior derivative, one can apply it twice. The result is called the *curvature*.

Lemma 3.41 *The composition*

$$\nabla^2 : \Omega^i(E) \to \Omega^{i+2}(E)$$

is a tensor (i.e., it contains no derivative terms).

Hence, it defines 2-form-valued endomorphism

$$R_\nabla := \nabla^2 \in \Omega^2(\mathrm{End}(E))$$

that is called the *curvature* of (E, ∇).

Proof

$$\begin{aligned}\nabla^2(f \cdot \sigma) &= \nabla(f \cdot \nabla(\sigma) + df \otimes \sigma)\\ &= df \wedge \nabla(\sigma) + f \cdot \nabla^2(\sigma) - df \wedge \nabla^2(\sigma)\\ &= f \cdot \nabla^2(\sigma).\end{aligned}$$

This implies that ∇^2 is a tensor. □

Remark 3.42 $\nabla^3 \equiv 0$ holds, called the Bianchi identity. Hence, the compositions make sense only up to twice (Problem 3.92).

Let $p \in X$ be a point and choose any local trivialization

$$\varphi_p : E|_{U_p} \cong U_p \times \mathbb{R}^n$$

around $p \in U_p$. Consider the local frame $\{e_i\}_{i=1}^n$, where

$$e_i(m) := \varphi_p^{-1}(m, (0, \cdots, 1, 0, \cdots, 0)).$$

Let

$$\nabla(e_i) = \sum_{j=1}^n \omega_i^j \cdot e_j$$

where $\omega_i^j \in \Omega^1(U_p)$. The 1-form-valued matrix $(\omega_i^j)_{0 \le i, j \le n}$ is called the *connection matrix*.

Lemma 3.43 *The formula*

$$R_\nabla(e_i) = \sum_{j=1}^n \Omega_{ij} \cdot e_j$$

holds, where

$$\Omega_{ij} = d\omega_i^j - \sum_{k=1}^n \omega_i^k \wedge \omega_k^j.$$

Proof

$$R_\nabla(e_i) = \nabla\left(\sum_{j=1}^n \omega_i^j \cdot e_j\right) = \sum_{j=1}^n \left(d\omega_i^j \cdot e_j - \omega_i^j \wedge \nabla(e_j)\right)$$
$$= \sum_{j=1}^n \left(d\omega_i^j \cdot e_j - \sum_{k=1}^n \omega_i^j \wedge \omega_j^k \cdot e_k\right) = \sum_{j=1}^n \left(d\omega_i^j - \sum_{k=1}^n \omega_i^k \wedge \omega_k^j\right) \cdot e_j.$$

This completes the proof. □

Let $p \in X$ be a point and choose two local trivializations

$$\varphi_p^i : E|_{U_p} \cong U_p \times \mathbb{R}^n$$

around $p \in U_p$ for $i = 0, 1$. Let

$$\varphi_p^{01} := \varphi_p^0 \circ (\varphi_p^1)^{-1} : U_p \times \mathbb{R}^n \cong U_p \times \mathbb{R}^n$$

be the transition function. It can be expressed as a smooth map

$$\varphi_p^{01} : U_p \to Gl_k(\mathbb{R}).$$

Consider the curvature R_∇ and its local forms $\Omega^i := (\Omega^i_{ij})$ by the matrix-valued 2-forms for $i = 0, 1$.

Corollary 3.44 *The formula*

$$\Omega^0 = \varphi^{01} \circ \Omega^1 \circ (\varphi^{01})^{-1}$$

holds.

Proof This follows from the property $R_\nabla \in \Omega^2(\mathrm{End}(E))$ in Lemma 3.41. □

Remark 3.45 Let $(E, \langle\ ,\ \rangle)$ be a vector bundle with an inner product, and take a local orthonormal frame $\{e_i\}_i$. Denote the local curvature by (ω_{ij}). Let us take a compatible connection with the inner product.

1. It follows by definition that (ω_i^j) is a skew-adjoint matrix, $\omega_i^j + \omega_j^i = 0$.
2. The local curvature is also a skew-adjoint matrix, $\Omega_{ij} + \Omega_{ji} = 0$.
3. Let us represent the real vector bundle with an inner product by a complex vector bundle with a Hermitian metric, and consider a compatible connection with the Hermitian metric. A local orthonormal frame $\{e_i\}_i$ is skew-Hermitian, $\omega_i^j + \overline{\omega_j^i} = 0$.
4. The local curvature (Ω_{ij}) is also skew-Hermitian. $\Omega_{ij} + \overline{\Omega_{ji}} = 0$.

3.3.2 Levi-Civita Connection

Let X be an n-dimensional smooth manifold, and TX be the tangent bundle. A *Riemannian metric* g on X is an inner product on TX.

Recall that a connection ∇ on TX is *g-compatible*, if the equality

$$dg(V, W) = g(\nabla V, W) + g(V, \nabla W)$$

holds.

Lemma 3.46 *Let (X, g) be a Riemannian manifold. Then, there exists a unique torsion-free and g-compatible connection ∇ on TX, where the former condition implies the equality $\nabla_V W - \nabla_W V = [V, W]$ for any vector fields V and W.*

It is called the *Levi-Civita connection.*

Proof On each local coordinate on $U \subset X$, there is a local frame $e = \{e_1, \cdots, e_n\}$ such that it is a pointwisely orthonormal basis with respect to g. Then g can be expressed by the dual basis $\theta = \{\theta^1, \cdots, \theta^n\}$ as $g = \sum_{i=1}^{n} \theta^i \theta^i$.

Let $\omega = (\omega_i^j)_{i,j}$ be the connection matrix of ∇ on the local frame. On the local coordinate, the condition that ∇ is g-compatible is equivalent to that the local frame is alternative, i.e., $\omega_i^j + \omega_j^i = 0$ for any i, j. The condition of torsion-freeness is equivalent to the equalities

$$d\theta^i + \sum_j \omega_j^i \wedge \theta^j = 0$$

for any j. Let us set

$$\omega_j^i = \sum_k c_{jk}^i \cdot \theta^k, \qquad c_{jk}^i + c_{ik}^j = 0,$$

$$d\theta^i = \frac{1}{2} \sum_{jk} a_{jk}^i \cdot \theta^j \wedge \theta^k, \qquad a_{jk}^i + a_{kj}^i = 0.$$

It follows from the torsion-freeness that we obtain the equalities

$$a_{jk}^i = c_{jk}^i - c_{kj}^i$$

for any i, j, k. Then, we can induce the equalities

$$a_{jk}^i + a_{ik}^j - a_{ji}^k = 2c_{ij}^k.$$

This implies that the connection matrix uniquely exists.

Let g be a $O(n)$-valued transition matrix on the local coordinate such that the equalities $e' = g \cdot e$, $\theta = g \cdot \theta'$ and $\omega' = g^{-1} \cdot dg + g^{-1} \cdot \omega \cdot g$ hold. Then, ω satisfies the two conditions of g-compatibility and torsion-freeness with respect to e, if and only if ω' does with respect to e'.

It follows uniqueness that the local connection defined by ω gives a globally defined connection on TX. □

3.3.3 Connections on Principal Bundles

Let $\pi : P \to X$ be a principal G-bundle. There is a canonical subbundle $T^{\vee}P$ defined by

$$T_p^{\vee}P := \ker \pi_* : T_pP \to T_{\pi(p)}X.$$

The group G acts on $T_p^{\vee}P$. Let $\mathfrak{g}$ be the Lie algebra of G. Note that there is the canonical isomorphism $V_p \colon \mathfrak{g} \overset{\cong}{\to} T_p^{\vee}P$ given by $a \mapsto (p \cdot \exp(ta))'_{t=0}$.

A *horizontal distribution* $H = \sqcup_{p\in P} H_p$ is a splitting $TP = T^{\vee}P \oplus H$. A horizontal distribution H gives a *connection* on P if H is G-invariant, that is,

$$g_*(H_p) = H_{p\cdot g}$$

holds for any $g \in G$.

A *connection* 1*-form* is a $\mathfrak{g}$-valued 1-form ω on P defined by the composition of the projection $TP = T^{\vee}P \oplus H \to T_p^{\vee}P$ with V_p^{-1}:

$$\omega_p : T_pP \overset{\text{proj.}}{\longrightarrow} T_p^{\vee}P \overset{V_p^{-1}}{\longrightarrow} \mathfrak{g}.$$

Note that $H = \ker \omega$, and the following hold:

$$g^*\omega = g^{-1} \cdot \omega \cdot g \text{ for } g \in G, \tag{3.8}$$

$$\omega_p \circ V_p = \mathrm{id}_{\mathfrak{g}}. \tag{3.9}$$

(See Problem 3.93.)

Lemma 3.47 *There is one-to-one correspondence between* $\mathfrak{g}$*-valued* 1*-forms satisfying* (3.8) *an* (3.9) *and* G*-invariant horizontal distributions, which define connections.*

Proof As mentioned above, for a G-invariant horizontal distribution, the corresponding connection form satisfies (3.8) and (3.9). Conversely, if a 1-form ω satisfying (3.8) and (3.9) is given, then $H = \ker \omega$ defines a G-invariant horizontal distribution whose connection 1-form is ω. □

Let ω be a connection on P.

Definition 3.48 The curvature R_ω of ω is defined by

$$R_\omega := d\omega + \omega \wedge \omega,$$

where $\omega \wedge \omega \in \Omega^2(P \times_{ad} \mathfrak{g})$ is given by

$$\omega \wedge \omega(\tau_1, \tau_2) := \frac{1}{2}[\omega(\tau_1), \omega(\tau_2)].$$

Let G and H be two Lie groups and $\mu : G \to H$ be a covering homomorphism. Let $\pi_P : P \to X$ and $\pi_Q : Q \to X$ be principal G and H bundles respectively, and let $F : P \to Q$ be a bundle map that factors through μ. Hence, the following diagram

$$\begin{array}{ccc} P & \xrightarrow{F} & Q \\ \cdot g \downarrow & & \downarrow \cdot \mu(g) \\ P & \xrightarrow{F} & Q \end{array}$$

commutes for any $g \in G$.

Lemma 3.49 *Let ω be a connection on P. Then, it lifts canonically to a connection on Q.*

Proof This is clear, because $\mu : G \to H$ is a covering map. □

Examples
For a $U(1)$-bundle P, $P \times_{Ad} \mathfrak{g} \cong X \times \sqrt{-1}\mathbb{R}$, and hence, the curvature R is an element in $\sqrt{-1}\Omega^2(X)$.

3.3.3.1 Connections on Frame Bundles

Let $\pi : P \to X$ be the frame bundle of a vector bundle $E \to X$. Hence, the structure group $G = \mathrm{GL}(n, \mathbb{R})$ with $n = \operatorname{rank} E$.

Let ∇ be a connection on E. Let $p \in X$ be a point and choose a local trivialization $\varphi_p : E|_{U_p} \cong U_p \times \mathbb{R}^n$ around $p \in U_p$. Let us consider the local frame $\{e_i\}_{i=1}^n$ on $E|U_p$ as in Sect. 3.3.1. One can associate an element $g \in G$ with any other frame $\{e_i'\}$ on U_p, where $e_i' = \sum_j g_{ij} \cdot e_j$. Hence, the local trivialization φ_p also gives another one

$$P|_{U_p} \cong U_p \times G$$

for the frame bundle. We also express it by the same φ_p.

Let ∇ be a connection on E, and $\nabla(e_i) = \sum_{j=1}^{n} \omega_i^j \cdot e_j$. Let $\omega_p := (\omega_i^j)_{i,j}$ be the connection matrix. Let us pull back ω_p on $U_p \times G$, and introduce another 1-form-valued matrix

$$\tilde{\omega}_p := g^{-1} \cdot \omega_p \cdot g + g^{-1} \cdot dg$$

on $U_p \times G$. Through the local trivialization, we can regard $\tilde{\omega}_p$ as a 1-form-valued matrix on $P|U_p$. The 1-form-valued matrix is regarded as a $\mathfrak{g}$-valued 1-form, where $\mathfrak{g}$ is the Lie algebra of G.

Proposition 3.50 *The $\mathfrak{g}$-valued 1-form $\tilde{\omega}_p$ defines a connection ω on P.*

Moreover, this assignment $\nabla \leftrightarrow \omega$ gives one-to-one correspondence between the set of connections on E and the one on P.

Proof It is enough to check Lemma 3.51 below. □

Lemma 3.51 *The $\mathfrak{g}$-valued 1-form $\tilde{\omega}_p$ is independent of the choice of the local trivialization φ_p. In particular, it defines a globally defined $\mathfrak{g}$-valued 1-form ω on P.*

Proof Exercise (Problem 3.94). □

Corollary 3.52 *The curvature R_ω corresponds compatibly by*

$$R_\omega(X, Y) = R_\omega(X^H, Y^H) = R_\nabla(\pi_*(X), \pi_*(Y)),$$

where both X and Y are vector fields on P, X^H is the horizontal projection and $\pi : P \to X$.

Proof Recall that the local connections ω on P and ω_p on E satisfy the equation $\omega = g^{-1} \cdot \omega_p \cdot g^{-1} + g^{-1} \cdot dg$. It is straightforward to check the equality

$$R_\omega = g^{-1} \cdot R_{\omega_p} \cdot g^{-1}$$

between their local curvatures. Hence, $R_\omega(V, W) = 0$ holds, when one of V or W is vertical. Because R_{ω_p} is the local expression of ∇^2, we have verified the equality $R_\omega(X, Y) = R_\omega(X^H, Y^H) = R_\nabla(\pi_*(X), \pi_*(Y))$. □

3.3.4 Connections on Induced Bundles

Let $\rho : G \to \mathrm{Aut}(V)$ be a representation to a vector space V. There is an associated vector bundle $E := P \times_\rho V$.

Let $TP \cong T^\vee P \oplus H$. be a horizontal distribution that defines a *connection* on P. We define a connection ∇ on E in the following way:

Take $\sigma \in \Gamma(E)$, $\tau \in T_xX$ and $p \in P$ with $\pi(p) = x$. Choose a smooth curve $\gamma : (-\epsilon, \epsilon) \to X$ such that $\gamma(0) = x$ and $\left.\frac{d\gamma}{dt}\right|_{t=0} = \tau$. There are a unique horizontal lift $p(t)$ such that $p(0) = p$, $\pi(p(t)) = \gamma(t)$ and $\frac{dp}{dt}(t) \in H_{\gamma(t)}$, and a unique curve $v(t) \in V$ such that $\sigma(\gamma(t)) = [p(t), v(t)]$.

Then, we can set

$$\nabla_\tau(\sigma) = \left[p, \left.\frac{dv}{dt}\right|_{t=0} \right].$$

This assigns the connection (E, ∇) from (P, ω). The connection (E, ∇) is called the *induced connection*. Combining the previous processes, we have the following diagram:

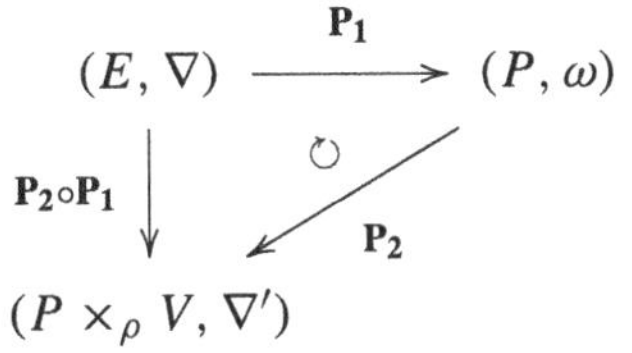

(3.10)

Lemma 3.53 *Let P be the frame bundle of E, and $\rho : G \to Gl(n, \mathbb{R})$ be the identity. Then, $(P \times_G \mathbb{R}^n, \nabla')$ coincides with (E, ∇), under the canonical isomorphism $E \cong P \times_G \mathbb{R}^n$.*

Proof Let $\varphi_p : P|_{U_p} \cong U_p \times G$ be a local trivialization. Let $V \in T_pX$ and $\gamma : [0, 1] \to (p_t, g_t)$ be a horizontal curve with $p_0 = p$ and $g_0 = 1$ $\gamma'(0) = \pi_*(\gamma'(0)) =: V$. Take a local section $\xi = (x, \xi(x))$ on $x \in U_p$. Then by definition, $\nabla'_V(\xi) = (p, (g_t^{-1} \cdot \xi)(0))$.

The horizontal component is given by $\ker \omega$. It follows from the equation $\omega = g^{-1} \cdot \omega_p \cdot g + g^{-1} \cdot dg$ that the restriction on the horizontal component satisfies the equalities

$$g \cdot dg^{-1} = -g^{-1} \cdot dg = \omega_p.$$

Note that the left-hand side is the coordinate of the Lie algebra. Hence, we have the equalities

$$(p, (g_t^{-1} \cdot \xi)(0)) = (p, g \cdot dg^{-1} \otimes \xi(0)) = (p, \omega_p \otimes \xi(0)),$$

when ξ is a linear combination of the local frame $\{e_i\}_i$. This completes the proof.

□

3.3.5 Gauge Transformations

Let $\pi : P \to X$ be a principal G-bundle.

Definition 3.54 A self-diffeomorphism $\varphi : P \xrightarrow{\cong} P$ is called a *gauge transformation*, if the following two properties

$$(1)\ \varphi(p \cdot h) = \varphi(p) \cdot h, \quad (2)\ \pi(\varphi(p)) = \pi(p)$$

hold for any $p \in P$ and $h \in G$.

The set of all gauge transformations on P forms a group that is called the *gauge transformation group* and is denoted by $\mathrm{Aut}(P)$.

Lemma 3.55 *The group* $\mathrm{Aut}(P)$ *is identified with the space of smooth sections of the bundle* $P \times_{Ad} G$*, where the Adjoint action of* G *is given by* $g \to Ad(g)(h) := ghg^{-1}$.

Proof For $\varphi \in \mathrm{Aut}(P)$, define the map $g_\varphi \colon P \to G$ by the relation $\varphi(u) = u \cdot g_\varphi(u)$. Since $\varphi(u \cdot h) = \varphi(u) \cdot h$, we have $g_\varphi(u \cdot h) = h^{-1} g_\varphi(u) h$, and therefore $(u, g_{\varphi(u)})$ defines a section of $P \times_{Ad} G$. It is easy to see that the correspondence gives the required identification. □

Let $\mathfrak{s}$ be a spin^c structure on an $\mathrm{SO}(n)$ bundle P over X with lift $\xi : \tilde{P} \to P$. We define the gauge transformation group $\mathcal{G}$ of $\mathfrak{s}$ as

$$\mathcal{G} = \{\varphi \in \mathrm{Aut}(\tilde{P}) \mid \xi \circ \varphi = \xi\}.$$

Lemma 3.56 *The group* $\mathcal{G}$ *is identified with* $C^\infty(X, \mathrm{U}(1))$.

Proof Exercise (Problem 3.95). □

A gauge transformation $\varphi \in \mathrm{Aut}(P)$ naturally induces an automorphism of the vector bundle $E = P \times_\rho V$ by

$$[u, v] \mapsto [\varphi(u), v] = [u \cdot g_\varphi(u), v] = [u, \rho(g_\varphi(u))v],$$

where $g_\varphi(u)$ is the map in the proof of Lemma 3.55.

For a spin^c structure $\mathfrak{s}$, the $\mathcal{G}$-action on the spinor bundles $S^\pm$ is given by fiberwise multiplication of $g \in \mathcal{G} = C^\infty(X; \mathrm{U}(1))$, and the $\mathcal{G}$-action on the determinant line bundle $L = \det(\mathfrak{s})$ is given by fiberwise multiplication of g^2.

The following is straightforward to check:

Lemma 3.57 *Let* ∇ *be a connection on* E*. Then,* $\varphi_*(\nabla) := \varphi \circ \nabla \circ \varphi^{-1}$ *gives another connection on* E.

Proof Exercise (Problem 3.96). □

Lemma 3.58 *Let ω be a connection 1-form on P with horizontal distribution H and $\varphi \in \mathrm{Aut}(P)$.*

(1) The horizontal distribution $\varphi_*(H)$ gives a connection whose connection 1-form is $(\varphi^{-1})^*\omega$.
(2) Let ∇ be the connection on $E = P \times_\rho V$ induced from ω. Then the connection on E induced from $\varphi_*(H)$ is $\varphi_*(\nabla) = \varphi \circ \nabla \circ \varphi^{-1}$.

Proof

(1) Since φ commutes with the G-action, $\varphi_*(H)$ is also G-invariant, and therefore defines a connection on P. The corresponding connection 1-form is $(\varphi^{-1})^*\omega$. (It is easy to check directly that $(\varphi^{-1})^*\omega$ satisfies (3.8) and (3.9).)
(2) For $\sigma \in \Gamma(E)$, $\tau \in T_xX$, choose $p \in P$ with $\pi(p) = x$ and a smooth curve $\gamma : (-\epsilon, \epsilon) \to X$ such that $\gamma(0) = x$ and $\left.\frac{d\gamma}{dt}\right|_0 = \tau$. There are a unique horizontal lift $p(t)$ such that $p(0) = p$, $\pi(p(t)) = \gamma(t)$ and $\frac{dp}{dt}(t) \in H_{\gamma(t)}$, and a unique curve $v(t) \in V$ such that $\sigma(\gamma(t)) = [p(t), v(t)]$. The covariant derivative $\nabla_\tau(\sigma)$ is given by

$$\nabla_\tau(\sigma) = \left[p, \left.\frac{dv}{dt}\right|_{t=0}\right].$$

The curve $\varphi(p(t))$ is a horizontal lift of γ for $\varphi_*(H)$, and represented by

$$\varphi(p(t)) = p(t) \cdot g(t), \quad g(t) \in G.$$

Then $\sigma(\gamma(t))$ is written as $\sigma(\gamma(t)) = [p(t) \cdot g(t), g(t)^{-1} \cdot v(t)]$. Let ∇^φ be the covariant derivative associated with $\varphi_*(H)$. Then

$$\nabla^\varphi_\tau(\sigma) = \left[p \cdot g(0), \left.\frac{d(g^{-1}v)}{dt}\right|_{t=0}\right] = \left[p, g(0)\left.\frac{d(g^{-1}v)}{dt}\right|_{t=0}\right],$$

and hence $\nabla^\varphi = \varphi \circ \nabla \circ \varphi^{-1}$. □

3.4 Chern–Weil Theory

3.4.1 Classifying Space of Vector Bundles

We refer to [128] on the contents of this subsection. Let $G_k(\mathbb{C}^n)$ be the complex Grassmannian defined by the space of k-dimensional $\mathbb{C}$-linear subspaces. It admits a complex manifold of $\mathbb{C}$-dimension $k(n-k)$. Consider the canonical $\mathbb{C}$-linear embedding $\mathbb{C}^n \hookrightarrow \mathbb{C}^{n+1}$ by $(z_1, \cdots, z_n) \to (z_1, \cdots, z_n, 0)$. This induces the embedding $G_k(\mathbb{C}^n) \hookrightarrow G_k(\mathbb{C}^{n+1})$. By taking the direct limit of the embeddings, we obtain an infinite-dimensional Grassmannian

$$G_k(\mathbb{C}^\infty) := \lim_{n\to\infty} G_k(\mathbb{C}^n).$$

There is a canonical complex vector bundle $\gamma_k \to G_k(\mathbb{C}^\infty)$ of rank k given by

$$\gamma_k(\mathbb{C}) := \{\ ([V], v) : v \in V, [V] \in G_k(\mathbb{C}^\infty)\ \}$$

that is called the universal bundle. The topological space $G_k(\mathbb{C}^\infty)$ is known as the classifying space of complex vector bundles in the sense below.

Lemma 3.59 *Let $E \to M$ be a complex vector bundle over a compact manifold. Then there is a continuous map called the classifying map*

$$f : M \to G_k(\mathbb{C}^\infty)$$

unique up to homotopy such that E is isomorphic to the pull-back $E \cong f^*(\gamma_k(\mathbb{C}))$.

There are two real versions, oriented or unoriented cases. In this book, we always consider the oriented case, otherwise stated. Let $G_k(\mathbb{R}^\infty)$ be the oriented Grassmannian and $\gamma_k(\mathbb{R}) \to G_k(\mathbb{R}^\infty)$ be the oriented universal bundle over $G_k(\mathbb{R}^\infty)$ in the obvious sense. The parallel statement to Lemma 3.59 also holds in the oriented real case.

Proposition 3.60 *The cohomology ring of $G_k(\mathbb{C}^\infty)$ is the polynomial ring*

$$H^*(G_k(\mathbb{C}^\infty); \mathbb{Z}) \cong \mathbb{Z}[c_1, c_2, \cdots, c_m],$$

where $c_i \in H^{2i}(G_k(\mathbb{C}^\infty); \mathbb{Z})$.

Each c_i is called the *i-Chern class* of the universal bundle $\gamma_k(\mathbb{C})$.

Definition 3.61 The i-th Chern class $c_i(E)$ of $E \to M$ is defined by

$$f^*(c_i) \in H^{2i}(M; \mathbb{Z}).$$

Note that the cohomology class $c_i(E)$ is well-defined and determined by the isomorphism class of E by Lemma 3.59.

Let us consider the oriented Grassmannian. Let $E \to M$ be an oriented vector bundle of real rank $2m$.

Proposition 3.62 *The rational cohomology ring of $G_{2m}(\mathbb{R}^\infty)$ is the polynomial ring*

$$H^*(G_{2m}(\mathbb{R}^\infty); \mathbb{Q}) \cong \mathbb{Q}[p_1, p_2, \cdots, p_{m-1}, e_{2m}],$$

where $p_i \in H^{4i}(G_{2m}(\mathbb{R}^\infty); \mathbb{Q})$ and $e_{2m} \in H^{2m}(G_k(\mathbb{R}^\infty); \mathbb{Q})$.

Each p_i is called the *i-Pontryagin class* and e_{2m} is the Euler class of the oriented universal bundle $\gamma_k(\mathbb{R})$. The square $p_m := e_{2m}^2 \in H^{4m}(G_k(\mathbb{R}^\infty); \mathbb{Q})$ is also called the m-th Pontryagin class.

Definition 3.63 Let $E \to M$ be an oriented vector bundle of real rank $2m$, and $f : M \to G_k(\mathbb{R}^\infty)$ be the classifying map. The i-th rational Pontryagin class $p_i(E)$ of $E \to M$ is defined by

$$f^*(p_i) \in H^{4i}(M; \mathbb{Q}).$$

The treatment of the torsion subgroup in $H^*(G_k(\mathbb{R}^\infty); \mathbb{Z})$ is subtle. We will not consider the torsion classes in this book.

Let us state the basic relation between the Chern classes and the Pontryagin classes.

Lemma 3.64 *Let $E \to M$ be an oriented vector bundle of real rank $2m$, and let $E \otimes_{\mathbb{R}} \mathbb{C}$ be the complexification. Then, the equality $p_i(E) = (-1)^i c_{2i}(E \otimes_{\mathbb{R}} \mathbb{C})$ holds in $H^{4i}(M; \mathbb{Q})$ for $0 \leq i \leq m$.*

Moreover, the odd degree Chern classes $c_{2i+1}(E \otimes_{\mathbb{R}} \mathbb{C}) = 0$ all vanish.

3.4.1.1 Euler Class

Let V be an oriented vector space of dimension k, and set $V_0 \equiv V \backslash \{0\}$. Then, there is a generator $u \in H^k(V, V_0; \mathbb{Z}) \cong \mathbb{Z}$. For example, take a compactly supported n-form $\alpha \in \Omega_c^k(V)$ with $\int_V \alpha = 1$. Then, its cohomology class $u = [\alpha]$ suffices.

Let $E \to M$ be an oriented real vector bundle of rank k, and set $E_0 = E \backslash M$, where we embed $I : M \hookrightarrow E$ by the zero section.

Let $i_m : (E_x, (E_x)_0) \hookrightarrow (E, E_0)$ be the embedding of the fiber for each $m \in M$.

Lemma 3.65 *There is a fiberwise compactly supported smooth k-form $u \in H^k(E, E_0; \mathbb{Z})$ such that each restriction $u_m := i_m^*(u) \in H^k(E_x, (E_x)_0; \mathbb{Z})$ represents the generator.*

Moreover, the multiplication by u induces an isomorphism

$$u\wedge : H^*(M; \mathbb{Z}) \cong H^k(E, E_0; \mathbb{Z}).$$

Hence, u has compact support, if M is compact. The cohomology class u is called the *Thom class*.

Definition 3.66 The restriction $e(E) := I^*(u) \in H^k(M; \mathbb{Z})$ is called the Euler class of E.

3.4.2 Chern–Weil Theory

3.4.2.1 Chern Class

Let $E \to M$ be a complex vector bundle and ∇ be a connection on E. Recall Lemma 3.41, and the curvature $R_\nabla = \nabla^2 \in \Omega^2(\mathrm{End}(E))$. Let $p \in X$ be a point and choose two local trivializations

$$\varphi_p^i : E|_{U_p} \cong U_p \times \mathbb{R}^n$$

around $p \in U_p$ for $i = 0, 1$. Let us express the curvature R_∇ by the local form $\Omega = (\Omega_{ij})$ as in Lemma 3.43. Let λ be a real indeterminate, and consider

$$\det\left(\lambda I - \frac{(\Omega_{ij})}{2\pi i}\right).$$

It follows from Corollary 3.44 that it is well-defined polynomial of differential forms on U_p independent of the choice of local trivializations, because:

- the determinant is invariant under taking conjugation, and
- even degree forms are mutually commute.

In particular, it extends a polynomial of differential forms on M.

Hence, we can have the following polynomial

$$c(E, \nabla) := \det\left(\lambda I - \frac{R_\nabla}{2\pi i}\right) = \sum_{i=0}^{\mathrm{rank}\, E} c_i(E, \nabla) \cdot \lambda^i,$$

where $c_i(E, \nabla) \in \Omega^{2i}(X)$.

We call $c_i(E, \nabla)$ as the i-th Chern form.

Proposition 3.67 *Each Chern form $c_i(E, \nabla)$ is a closed form, and its cohomology class is independent of the choice of a connection ∇.*

Hence we can express the cohomology class of i-th Chern form by $c_i(E) \in H^{2i}(M; \mathbb{C})$. Let us equip it with a Hermitian metric on E, and consider a compatible connection ∇. It follows from Remark 3.45 (4) that the local curvature (Ω_{ij}) is skew-Hermitian. Then, we have the equalities

$$\begin{aligned} c(E, \nabla) &= \det\left(\lambda I - \frac{R_\nabla}{2\pi i}\right) = \det\left(\lambda I - \frac{R_\nabla}{2\pi i}\right)^t \\ &= \det\left(\lambda I + \frac{\overline{R_\nabla}}{2\pi i}\right) = \overline{\det\left(\lambda I - \frac{R_\nabla}{2\pi i}\right)} = \overline{c(E, \nabla)}. \end{aligned}$$

Because the cohomology classes $c_i(E)$ are all independent of the choice of connections, it follows that the i-th Chern class is in the real coefficient

$$c_i(E) \in H^{2i}(M; \mathbb{R}).$$

Recall that the class $f^*(c_i)$ in Definition 3.61. We have the following fact:

Lemma 3.68 *Two cohomology classes $f^*(c_i)$ and $c_i(E)$ coincide. In particular, $c_i(E) \in H^{2i}(M; \mathbb{Z})$ is in the integral cohomology class.*

We will not verify this. The proof uses an axiomatic characterization of the characteristic classes [96].

Definition 3.69 The cohomology class

$$c(E) := 1 + c_1(E) + \cdots + c_k(E) \in H^{ev}(M; \mathbb{Z})$$

is called the total Chern class of E. We call $c_i(E) \in H^{2i}(M; \mathbb{Z})$ as the i-th Chern class of E.

3.4.2.2 Chern Character

So far we have seen that the invariance of the determinant under conjugation played an important role in defining the Chern classes. It is immediate to notice that the trace also satisfies such a property.

Let us consider the following local polynomial of differential forms:

$$\operatorname{tr}\left(\exp \frac{(-\Omega_{ij})}{2\pi i}\right) = \operatorname{tr}\left(I - \frac{(\Omega_{ij})}{2\pi i} + \frac{(\Omega_{ij})^2}{2!(2\pi i)^2} - \cdots\right).$$

It is immediate to see that this is well-defined and independent of the choice of local trivializations. hence, we have a polynomial of even degree differential forms on M:

$$ch(E, \nabla) := \operatorname{tr}\left(\exp \frac{-R_\nabla}{2\pi i}\right) = \sum_{i=0}^{\infty} \mathrm{ch}_i(E, \nabla),$$

where $ch_i(E, \nabla) = \frac{1}{i!} \operatorname{tr}\left(\left(-\frac{R_\nabla}{2\pi i}\right)^i\right)$. Note that the sum is finite, depending on $\dim M$. We say that $ch(E, \nabla)$ is the total Chern character form of (E, ∇), and $ch_i(E, \nabla)$ is the i-th Chern character form.

Similarly to the case of Chern classes, we can verify the following:

1. The differential forms $ch_i(E, \nabla)$ are all closed, and hence they define cohomology classes.

2. Each cohomology class of $ch_i(E, \nabla)$ is independent of the choice of a connection ∇.
3. Each cohomology class $ch_i(E, \nabla)$ is in $H^{ev}(M; \mathbb{Q})$.

Definition 3.70 The cohomology class $ch(E, \nabla) \in H^{ev}(M; \mathbb{Q})$ is called the total Chern character of E. We call $ch_i(E) \in H^{2i}(M; \mathbb{Q})$ the i-th Chern character of E.

In fact, we can express each i-th Chern character by a rational polynomial of Chern classes in some universal way. On the other hand, the total Chern character behaves well under algebraic operations.

Lemma 3.71 *The following properties hold:*

- $ch(E \oplus E') = ch(E) + ch(E') \in H^{ev}(M; \mathbb{Q})$.
- $ch(E \otimes E') = ch(E) \cdot ch(E') \in H^{ev}(M; \mathbb{Q})$.

The proofs of these are easy. In fact, the formulas hold for the Chern character forms.

3.4.2.3 Pontryagin Class

Let $E \to M$ be a real vector bundle, and ∇ be a connection. We have the real version of the Chern class and the Chern character.

Consider the following polynomial:

$$\det\left(\lambda I - \frac{R_\nabla}{2\pi}\right).$$

By arguments similar to the case of Chern class, we can see the following:

1. The polynomial of the differential forms $\det\left(\lambda I - \frac{R_\nabla}{2\pi}\right)$ is closed, and the cohomology class of each coefficient of λ^i is independent of the choice of connections.
2. If we choose an inner product and a compatible connection with it, then we obtain the equality

$$\det\left(\lambda I - \frac{R_\nabla}{2\pi}\right) = \det\left(\lambda I + \frac{R_\nabla}{2\pi}\right).$$

In particular, all the odd degree coefficients vanish, and we can express this as

$$p(E, \nabla) := \det\left(I - \frac{R_\nabla}{2\pi}\right) = \sum_j (-1)^i p_i(E, \nabla)\lambda^{2i}.$$

The polynomial $p(E, \nabla)$ is called the total Pontryagin form, and each $p_i(E, \nabla)$ is the i-th Pontryagin form.

3. The cohomology class $p_i(E) := [p_i(E, \nabla)] \in H^{4i}(M; \mathbb{R})$ coincides with $f^*(p_i) \in H^{4i}(M; \mathbb{Q})$ in Definition 3.63. It is called the i-th Pontryagin class.

3.4.2.4 Euler Class

Let $(V, \langle\ ,\ \rangle)$ be a real vector space of even rank $k = 2m$ with an inner product, and choose an oriented orthonormal basis $\{e_i\}_{i=1}^k$. Define

$$f : \wedge^2 V \to \mathbb{R}$$

by $w^m = f(w) \cdot e_1 \wedge \cdots \wedge e_k$ for $w \in \wedge^2 V$. We can check the formula

$$f(w)^2 \cdot e_1 \wedge \cdots \wedge e_k = (m!)^2 \det(w).$$

The function $p(w) := \frac{1}{m!} f(w)$ is called the *Pfaffian*.

Let $E \to M$ be an oriented real vector bundle of even rank $k = 2m$ with an inner product. Let ∇ be a compatible connection with the inner product. We can consider the polynomial

$$e(E, \nabla) := \frac{1}{2^m \pi^m m!} f(R_\nabla).$$

We have the following properties:

1. The differential form $e(E, \nabla)$ is closed, and the cohomology class is independent of the choice of connections.
2. The cohomology class $[e(E, \nabla)]$ coincides with $e(E)$ in Definition 3.66. In particular, its cohomology class is in $H^k(M; \mathbb{Z})$. It is called the Euler class.

3.5 Spin and Spin^c Connections and Dirac Operators

3.5.1 *Spin and* Spin^c *Connections*

Let (X, g) be an oriented Riemannian manifold.

Lemma 3.72 *Let ω be the Levi-Civita connection on the oriented frame bundle P.*

1. *Suppose $\xi : \tilde{P} \to P$ is a spin structure. Then, ω induces a* Spin(n) *connection on $\tilde{P}$.*
2. *Suppose $\xi^c : \tilde{P} \to P$ is a spin^c structure. Let ω' be a* U(1) *connection on $P_{\mathrm{U}(1)} = \tilde{P}/\mathrm{Spin}(n)$. Then ω and ω' induce a* Spin$^c(n)$ *connection on $\tilde{P}$.*

Proof We assume ω and ω' are connection 1-forms.

1. Note that the Lie algebras of $\mathrm{SO}(n)$ and $\mathrm{Spin}(n)$ are isomorphic. The pull-back $\xi^*\omega$ gives a $\mathrm{Spin}(n)$ connection on $\tilde{P}$.

2. For the spin^c case, ω and ω' induce a $\mathrm{SO}(n) \times \mathrm{U}(1)$ connection $\omega \oplus \omega'$ on the fiber sum $P \times_X P_{\mathrm{U}(1)}$. The pull-back $(\xi^c)^*(\omega \oplus \omega')$ gives a $\mathrm{Spin}^c(n)$ connection on $\tilde{P}$. □

Note the following properties:

- Let ω be a connection on the principal $\mathrm{SO}(n)$-bundle $P \to X$. Then, it induces a connection ∇ on $\mathrm{Cl}(P) = P \times_\rho \mathrm{Cl}(\mathbb{R}^n)$ (see Definition 3.32).
- Let $\tilde{P}$ be a spin (resp. spin^c) structure on P and $\tilde{\omega}$ be the connection on $\tilde{P}$ induced from ω (resp. $\omega \oplus \omega'$). Then the connection ∇ on $\mathrm{Cl}(P) = P \times_\rho \mathrm{Cl}(\mathbb{R}^n)$ is equal to the connection on $\mathrm{Cl}(P) = \tilde{P} \times_{\tilde{\rho}} \mathrm{Cl}(\mathbb{R}^n)$ (resp. $\tilde{P} \times_{\tilde{\rho}^c} \mathrm{Cl}(\mathbb{R}^n)$) induced from $\tilde{\omega}$. The connection $\tilde{\omega}$ induces a connection $\tilde{\nabla}$ on $S_{\mathbb{C}}(\tilde{P}) := \tilde{P} \times_{\Delta_{\mathbb{C}}} S_{\mathbb{C}}(\mathbb{R}^n)$ (resp. $\tilde{P} \times_{\hat{\Delta}_{\mathbb{C}}} S_{\mathbb{C}}(\mathbb{R}^n)$).

Lemma 3.73 *The equalities*

$$\nabla(\lambda_1 \cdot \lambda_2) = \nabla(\lambda_1) \cdot \lambda_2 + \lambda_1 \cdot \nabla(\lambda_2), \tag{3.11}$$

$$\tilde{\nabla}(\lambda \cdot \sigma) = \nabla(\lambda) \cdot \sigma + \lambda \cdot \tilde{\nabla}(\sigma) \tag{3.12}$$

hold for any $\lambda_1, \lambda_2 \in \Gamma(\mathrm{Cl}(P))$ *and* $\sigma \in \Gamma(S_{\mathbb{C}}(\tilde{P}))$.

Proof Recall that $\mathrm{Cl}(P) = \Lambda^* E$, where $E = P \times_{\rho_0} \mathbb{R}^n$ (Remark 3.33). Note that the Clifford action of $v \in \Gamma(E)$ on $\Gamma(\Lambda^* E)$ is given by

$$v \cdot (v_1 \wedge \cdots \wedge v_k) = v \wedge (v_1 \wedge \cdots \wedge v_k) - v \lrcorner (v_1 \wedge \cdots \wedge v_k). \tag{3.13}$$

(See (3.6).) The equality (3.11) is easily checked by differentiating (3.13). The equality (3.12) is obvious since $\mathrm{Cl}(V)$ is assumed as a subbundle of $\mathrm{End}(S_{\mathbb{C}}(\tilde{P}))$. □

3.5.2 Dirac Operators

Let (X, g) be an n-dimensional Riemannian manifold and $P \to X$ be the frame bundle.

Let $\tilde{P} \to P$ be a spin, or spin^c structure. We equip $\tilde{P}$ with the $\mathrm{Spin}(n)$ connection induced from the Levi-Civita connection in the spin case. In the spin^c case, we choose a $U(1)$ connection A on the determinant line bundle of the spinor bundle and equip $\tilde{P}$ with the $\mathrm{Spin}^c(n)$ connection induced from the Levi-Civita connection and the $U(1)$ connection A . The spin case can be considered as a spin^c

case with a trivial determinant line bundle and a trivialization $A = d$. In both cases we denote the associated connection by $\tilde{\nabla}$ on $\tilde{P}$. Note that for two choices of $U(1)$-connections A, A', their difference $A - A' =: \alpha$ is an element in $\sqrt{-1}\Omega^1(X)$.

Definition 3.74 The Dirac operator

$$D_A : \Gamma(S_{\mathbb{C}}(\tilde{P})) \to \Gamma(S_{\mathbb{C}}(\tilde{P}))$$

is defined by

$$D_A(\sigma)(x) := \sum_{i=1}^{n} e_i \cdot \tilde{\nabla}_{e_i}(\sigma)(x)$$

for any $x \in X$, where $\{e_1, \cdots, e_n\}$ is an orthonormal basis of $T_x X$.

Lemma 3.75 *D_A is independent of the choice of any orthonormal basis.*

Proof This is an easy exercise in linear algebra (Problem 3.97). □

The L^2-inner product of sections $\phi, \psi \in \Gamma(S_{\mathbb{C}}(\tilde{P}))$ is given by

$$\langle \phi, \psi \rangle_{L^2} = \int_X (\phi, \psi) dvol_X,$$

where $(\cdot, \cdot)$ is the fiberwise inner product on the spinor bundle $S_{\mathbb{C}}(\tilde{P})$.

Lemma 3.76

- *Let $A - A' = \alpha$. Then, the equality*

$$D_A(\sigma) = D_{A'}(\sigma) + \frac{1}{2}\alpha \cdot \sigma$$

holds for any $\sigma \in \Gamma(S_{\mathbb{C}}(\tilde{P}))$, where $\alpha\cdot$ is the Clifford multiplication.
- *If X has no boundary, then $D_A : \Gamma(S_{\mathbb{C}}(\tilde{P})) \to \Gamma(S_{\mathbb{C}}(\tilde{P}))$ is formally self-adjoint, i.e.,*

$$\langle D_A \phi, \psi \rangle_{L^2} = \langle \phi, D_A \psi \rangle_{L^2},$$

for all compactly supported sections ϕ and ψ.

Proof The former property is obvious. Let us verify the latter property. For a point $x \in X$, choose a local orthonormal frame $(e_1, \ldots, e_n)$ such that $\nabla_{e_i} e_j = 0$ at x.

$$
\begin{aligned}
(D_A\phi, \psi)_x &= \left(\sum_i e_i \cdot \tilde{\nabla}_{e_i}\phi, \psi\right)_x = -\sum_i (\tilde{\nabla}_{e_i}\phi, e_i\psi)_x \\
&= \sum_i \left\{(\phi, \tilde{\nabla}_{e_i}(e_i\psi))_x - e_i(\phi, e_i\psi)_x\right\} \\
&= \sum_i \left\{(\phi, e_i\tilde{\nabla}_{e_i}\psi)_x - e_i(\phi, e_i\psi)_x\right\} \\
&= (\phi, D_A\psi)_x - \sum_i e_i(\phi, e_i\psi)_x.
\end{aligned}
$$

Define the vector field V by the relation

$$(V(x), W(x))_x = (\phi(x), W(x) \cdot \psi(x))_x$$

for arbitrary vector filed W. Then

$$
\begin{aligned}
(\operatorname{div} V)_x &= \sum_i (\nabla_{e_i} V, e_i)_x = \sum_i \left\{e_i(V, e_i)_x - (V, \nabla_{e_i} e_i)_x\right\} \\
&= \sum_i e_i(V, e_i)_x = \sum_i e_i(\phi, e_i\psi)_x.
\end{aligned}
$$

Therefore $(D_A\phi, \psi) = (\phi, D_A\psi) - \operatorname{div} V$. Since V is compactly supported, $\int_X \operatorname{div} X \operatorname{vol}_X = 0$. We obtain $\langle D_A\phi, \psi\rangle_{L^2} = \langle\phi, D_A\psi\rangle_{L^2}$. □

Remark 3.77 If $\dim X$ is even, then the Dirac operator has the splitting

$$D_A = \begin{pmatrix} 0 & D_A^- \\ D_A^+ & 0 \end{pmatrix} : \Gamma(S_{\mathbb{C}}^+(\tilde{P}) \oplus S_{\mathbb{C}}^-(\tilde{P})) \to \Gamma(S_{\mathbb{C}}^+(\tilde{P}) \oplus S_{\mathbb{C}}^-(\tilde{P})).$$

By Lemma 3.76, the operators D_A^+ and D_A^- are mutually adjoint if X is closed.

Example
In the case of $X = \mathbb{R}^4$, the spin and spin^c structure are unique up to isomorphism and their spinor bundles $S_{\mathbb{C}}^{\pm}$ are trivial. We trivialize $S_{\mathbb{C}}^{\pm}$ by using the bases in (3.3). The spin Dirac operator $D^+ \colon \Gamma(S_{\mathbb{C}}^+) = \Gamma(\mathbb{R}^4, \mathbb{H}^+) \to \Gamma(S_{\mathbb{C}}^-) = \Gamma(\mathbb{R}^4, \mathbb{H}^-)$ is written as

$$D^+ = e_1\frac{\partial}{\partial x^1} + e_2\frac{\partial}{\partial x^2} + e_3\frac{\partial}{\partial x^3} + e_4\frac{\partial}{\partial x^4} = \mathbf{1}'\frac{\partial}{\partial x^1} - \boldsymbol{i}'\frac{\partial}{\partial x^2} - \boldsymbol{j}'\frac{\partial}{\partial x^3} - \boldsymbol{k}'\frac{\partial}{\partial x^4}.$$

Similarly, The spin Dirac operator $D^- \colon \Gamma(S_{\mathbb{C}}^-) = \Gamma(\mathbb{R}^4, \mathbb{H}^-) \to \Gamma(S_{\mathbb{C}}^+) = \Gamma(\mathbb{R}^4, \mathbb{H}^+)$ is written as

$$D^- = -\mathbf{1}\frac{\partial}{\partial x^1} - \boldsymbol{i}\frac{\partial}{\partial x^2} - \boldsymbol{j}\frac{\partial}{\partial x^3} - \boldsymbol{k}\frac{\partial}{\partial x^4}.$$

Let $I = -\boldsymbol{i}'$ and $J = -\boldsymbol{j}'$. Then $K = \boldsymbol{k}'$. Let $z = x^1 + Ix^2$ and $w = x^3 + Ix^4$. Then $D^+ \colon \Gamma(S_{\mathbb{C}}^+) \to \Gamma(S_{\mathbb{C}}^-)$ is rewritten as

$$D^+ = \frac{\partial}{\partial \bar{z}} + J\frac{\partial}{\partial \bar{w}}.$$

Then we obtain

$$\left(\frac{\partial}{\partial \bar{z}} + J\frac{\partial}{\partial \bar{w}}\right)(f + Jg) = \frac{\partial f}{\partial \bar{z}} - \frac{\partial g}{\partial w} + J\frac{\partial f}{\partial \bar{w}} + J\frac{\partial g}{\partial z}. \tag{3.14}$$

We consider the Dirac operator $D_A : \Gamma(S_{\mathbb{C}}^+(\tilde{P})) \to \Gamma(S_{\mathbb{C}}^-(\tilde{P}))$. (We will often write D_A for its restrictions $D_A^{\pm}$.) The Atiyah–Singer index theorem is given in this case:

Theorem 3.78 *Suppose* $\dim X = 4$ *and X is closed.*

- *If $\tilde{P}$ is spin, then the index of the Dirac operator is given by*

$$\operatorname{ind} D := \dim\ker D - \dim\operatorname{coker} D = -\frac{1}{24}\int_X p_1(X) = -\frac{1}{8}\sigma(X)$$

  where $\sigma(X)$ is the signature of X.
- *If $\tilde{P}$ is spin^c, then the index of the Dirac operator is*

$$\operatorname{ind} D_A = \frac{1}{8}\{\, c_1(\det(S_{\mathbb{C}}^{\pm}(\tilde{P}))) - \sigma(X)\,\}.$$

By the integrality of the index, we can prove Rokhlin's theorem.

Theorem 3.79 (Rokhlin's Theorem [148]) *For a closed spin 4-manifold X, the signature $\sigma(X)$ is divisible by* 16.

Proof Since the spin representation in dimension 4 is the representation of $\mathrm{Spin}(4) = \mathrm{Sp}(1) \times \mathrm{Sp}(1)$ on $\mathbb{H} \oplus \mathbb{H}$ by multiplication, the spinor bundles have quaternionic structures and the Dirac operator is quaternionic linear. Therefore $\operatorname{ind} D/2 = -\sigma(X)/16$ is an integer. □

Let us state the Bochner–Weitzenböck formula.

Lemma 3.80 *Suppose $\tilde{P}$ is spin. Then, the formula*

$$D^2 = \nabla^* \circ \nabla + \frac{1}{4}\kappa$$

holds, where κ is the scalar curvature, and ∇^* is the L^2-adjoint of ∇.

Suppose $\tilde{P}$ is spin^c. Then, the formula

$$D_A^2 = \nabla^* \circ \nabla + \frac{1}{4}\kappa + \frac{1}{2}F_A \tag{3.15}$$

holds.

Proof Let us choose $x \in X$, and take an orthonormal frame $\{e_1, \cdots, e_n\}$ near x. One can assume $\nabla e_i(x) = 0$ holds (at one point). Let $\nabla_i := \nabla_{e_i}$. Then,

$$\begin{aligned} D^2(\sigma)(x) &= \sum_{i,j} e_j \cdot \nabla_j(e_i \cdot \nabla_i \sigma))(x) = \sum_{i,j} e_j \cdot e_i \cdot \nabla_j \nabla_i(\sigma)(x) \\ &= -\sum_i \nabla_i^2(\sigma) + \sum_{j<i} e_j \cdot e_i \cdot (\nabla_j \nabla_i - \nabla_i \nabla_j)(\sigma)(x) \\ &=: -\sum_i \nabla_i^2(\sigma) + \sum_{j<i} K(e_j, e_i)(\sigma)(x) \end{aligned}$$

where $K(e_j, e_i)$ is a tensor. It turns out that the total tensor $-\sum_i \nabla_i^2(\sigma) + \sum_{j<i} K(e_j, e_i) = \frac{1}{4}\kappa$ holds at $x \in X$.

Hence, at $x \in$ with the frame, we have the formula $D^2(\sigma)(x) = \nabla^* \circ \nabla(\sigma) + \frac{1}{4}\kappa \cdot \sigma(x)$. Because the operators D^2, $\nabla^* \circ \nabla$ and κ are all globally defined over X, the conclusion must hold.

The argument for the spin^c case is parallel. □

Corollary 3.81 $\ker D = \operatorname{coker} D = 0$ *must hold whenever the scalar curvature $\kappa > 0$ takes positive values everywhere.*

In particular, $\operatorname{ind} D = 0$ *vanishes in this case, and hence, both $\int_X p_1(X) = \sigma(X) = 0$ must hold.*

This is well-known as the Bochner technique.

Proof Suppose $\kappa > 0$ holds. Choose σ with $D(\sigma) = 0$. Then,

$$\begin{aligned} 0 = \langle D^2(\sigma), \sigma\rangle_{L^2} &= \langle \nabla^* \circ \nabla(\sigma), \sigma\rangle_{L^2} + \left\langle \frac{1}{4}\kappa \cdot \sigma, \sigma \right\rangle_{L^2} \\ &= \langle \nabla(\sigma), \nabla(\sigma)\rangle_{L^2} + \left\langle \frac{1}{4}\kappa \cdot \sigma, \sigma \right\rangle_{L^2}. \end{aligned}$$

The right-hand side is nonnegative, and can be equal to zero only when $\sigma = 0$. □

3.6 Spin^c Structures on Almost Complex Manifolds

Let V be a finite-dimensional vector space with a complex structure $J: V \to V$, $J^2 = -1$. Suppose V has a real inner product $(\cdot, \cdot)_{\mathbb{R}}$ such that J is an orthogonal transformation. The real inner product $(\cdot, \cdot)_{\mathbb{R}}$ induces a unique Hermitian inner product $(\cdot, \cdot)_{\mathbb{C}}$ on $V \otimes_{\mathbb{R}} \mathbb{C}$ extending $(\cdot, \cdot)_{\mathbb{R}}$. We assume the Hermitian inner product is complex linear in the first factor. Decompose $V \otimes_{\mathbb{R}} \mathbb{C}$ into the i and $-i$ eigenspaces $V^{1,0}$ and $V^{0,1}$ of $J \otimes \mathrm{id}_{\mathbb{C}}$. Let $\pi^{1,0}$ and $\pi^{0,1}$ be the projection to $V^{1,0}$ and $V^{0,1}$. The decomposition induces the bigrading of the complex exterior algebra $\Lambda^*_{\mathbb{C}}(V \otimes \mathbb{C})$. We define the action of V on $\Lambda^*_{\mathbb{C}} V^{1,0}$ by

$$v \cdot (v_1 \wedge \cdots \wedge v_k) = \sqrt{2}\left(\pi^{1,0}(v) \wedge v_1 \wedge \cdots \wedge v_k - \pi^{1,0}(v) \lrcorner (v_1 \wedge \cdots \wedge v_k)\right),$$

where

$$\pi^{1,0}(v) \lrcorner (v_1 \wedge \cdots \wedge v_k) = \sum_{i=1}^{k} (-1)^{i-1} (v_i, \pi^{1,0}(v)) \cdot v_1 \wedge \cdots \wedge \hat{v}_i \wedge \cdots \wedge v_k.$$

Remark 3.82 Note that

$$(\lambda v)\lrcorner = \bar{\lambda}(v \lrcorner) \quad \text{for } \lambda \in \mathbb{C}, v \in V^{1,0},$$

because our Hermitian inner product is complex anti-linear in the second factor.

The action of v is a complex linear endomorphism of $\Lambda^*_{\mathbb{C}} V^{1,0}$.

Lemma 3.83 *The following relation holds:*

$$v \cdot (v \cdot (v_1 \wedge \cdots \wedge v_k)) = -|v|^2 v_1 \wedge \cdots \wedge v_k.$$

Then the action of V on $\Lambda^*_{\mathbb{C}} V^{1,0}$ extends to a complex linear action of $\mathrm{Cl}(V) \otimes \mathbb{C}$ *which gives the complex spin representation.*

Let $V^* = \mathrm{Hom}(V, \mathbb{R})$ be the dual space of V. The correspondence

$$V \ni v \mapsto v^* = (v, \cdot) \in V^*$$

gives a natural isomorphism $\iota\colon V \to V^*$. Note that

$$J^* v^* = -(Jv)^*.$$

Hence the isomorphism ι is complex anti-linear. We extend ι to its complexifications $\iota \otimes \mathbb{C}\colon V \otimes_{\mathbb{R}} \mathbb{C} \to V^* \otimes_{\mathbb{R}} \mathbb{C}$. We also decompose $V^* \otimes_{\mathbb{R}} \mathbb{C}$ into the i and $-i$ eigenspaces $(V^*)^{1,0}$ and $(V^*)^{0,1}$ of $J^* \otimes \mathrm{id}_{\mathbb{C}}$. Then

$$\left(V^{1,0}\right)^* = (V^*)^{0,1}, \quad \left(V^{0,1}\right)^* = (V^*)^{1,0}.$$

The action of V on $\Lambda^*_{\mathbb{C}} V^{1,0}$ induces an action of V^* on $\Lambda^*_{\mathbb{C}} \left(V^{1,0}\right)^* = \Lambda^*_{\mathbb{C}} (V^*)^{0,1}$ defined by

$$v^* \cdot (v_1^* \wedge \cdots \wedge v_k^*) = \sqrt{2} \left(\pi^{0,1}(v^*) \wedge v_1^* \wedge \cdots \wedge v_k^* - \pi^{0,1}(v^*) \lrcorner (v_1^* \wedge \cdots \wedge v_k^*)\right). \tag{3.16}$$

Corollary 3.84 *The action of V^* on $\Lambda^*_{\mathbb{C}} (V^*)^{0,1}$ extends to a complex linear action of $\mathrm{Cl}(V^*) \otimes \mathbb{C}$ which gives the complex spin representation.*

From now on, we consider the case of dimension 4. Note that the embedding $\mathrm{U}(2) \hookrightarrow \mathrm{SO}(4)$ factors through the embedding

$$\mathrm{U}(2) = \frac{\mathrm{U}(1) \times \mathrm{Sp}(1)}{\{\pm 1\}} \hookrightarrow \frac{\mathrm{Sp}(1) \times \mathrm{Sp}(1) \times \mathrm{U}(1)}{\{\pm 1\}} = \mathrm{Spin}^c(4),$$

$$(z, q) \mapsto (z^{-1}, q, z^{-1}).$$

Let X be a compact Riemannian 4-manifold with an almost complex structure J. Then the $\mathrm{SO}(4)$-frame bundle of TX has a $\mathrm{U}(2)$-reduction Q. Decompose $TX \otimes_{\mathbb{R}} \mathbb{C}$ and $T^*X \otimes_{\mathbb{R}} \mathbb{C}$ into the direct sums of bundles of the i and $-i$ eigenspaces of J.

$$TX \otimes_{\mathbb{R}} \mathbb{C} = T^{1,0} \oplus T^{0,1}, \quad T^*X \otimes_{\mathbb{R}} \mathbb{C} = \left(T^*\right)^{1,0} \oplus \left(T^*\right)^{0,1}.$$

Let $T^{p,q} = \Lambda^p T^{1,0} \otimes \Lambda^q T^{0,1}$ and $(T^*)^{p,q} = \Lambda^p (T^*)^{1,0} \otimes \Lambda^q (T^*)^{0,1}$.

Lemma 3.85 *The $\mathrm{Spin}^c(4)$ bundle $P = Q \times_{\mathrm{U}(2)} \mathrm{Spin}^c(4)$ defines a spin^c structure on X. Its spinor bundles S^+ and S^- are given by*

$$S^+ = T^{0,0} \oplus T^{2,0} = \left(T^*\right)^{0,0} \oplus \left(T^*\right)^{0,2}, \quad S^- = T^{1,0} = \left(T^*\right)^{0,1}.$$

Its determinant line bundle is $K_X^{-1} = T^{2,0} = (T^*)^{0,2}$.

Proof Note that $Q \times_{\mathrm{U}(2)} \mathbb{C}^2 = T^{1,0}$. Via the identification (3.7), $\mathrm{U}(2) \subset \mathrm{Spin}^c(4)$ acts on $\mathbb{H}^+$ by

$$z\begin{pmatrix} z^{-1} & 0 \\ 0 & z \end{pmatrix} = \begin{pmatrix} 1 & 0 \\ 0 & z^2 \end{pmatrix},$$

and on $\mathbb{H}^-$ by $z\begin{pmatrix} \alpha & \beta \\ -\bar{\beta} & \bar{\alpha} \end{pmatrix}$. Therefore

$$S^+ = Q \times_{\mathrm{U}(2)} \mathbb{H}^+ = T^{0,0} \oplus T^{2,0}, \quad S^- = Q \times_{\mathrm{U}(2)} \mathbb{H}^+ = T^{1,0}.$$

The proof is completed. □

The spin^c structure in Lemma 3.85 is called the *canonical spin^c structure* determined by the almost complex structure.

Remark 3.86 The canonical spin^c structure is defined on almost complex manifolds of every even dimension. (See e.g. [131, §3.4].) Recall that the set $\mathcal{S}$ of isomorphism classes of spin^c structures is identified with $H^2(X; \mathbb{Z})$ (Lemma 3.26). However, this identification is not canonical in general. On an almost complex manifold, every spin^c structure is obtained by twisting the canonical spin^c structure by a complex line bundle. This gives a canonical identification between $\mathcal{S}$ and $H^2(X; \mathbb{Z})$.

3.7 Holomorphic Vector Bundles

A reference for this section is [40, §2.1.5]. Let Z be a complex manifold. A complex vector bundle $\mathcal{E}$ over Z is a *holomorphic vector bundle* if the total space is a complex manifold and the projection $\pi : \mathcal{E} \to Z$ is holomorphic. A holomorphic vector bundle can be given by a system of holomorphic transition functions

$$g_{\alpha\beta} : U_\alpha \cap U_\beta \to \mathrm{GL}(n, \mathbb{C}).$$

Let $\Omega^{p,q}(Z) = \Gamma\left((T^*)^{p,q}\right)$. Recall that the complexified de Rham complex $(\Omega(Z) \otimes \mathbb{C}, d)$ is decomposed into a double complex $(\Omega^{p,q}(Z), \partial, \bar{\partial})$ such that $d = \partial + \bar{\partial}$ and

$$\partial : \Omega^{p,q}(Z) \to \Omega^{p+1,q}(Z), \quad \bar{\partial} : \Omega^{p,q}(Z) \to \Omega^{p,q+1}(Z).$$

For a complex vector bundle E, let $\Omega^{p,q}(E)$ be the space of E-valued (p, q)-forms. Suppose a holomorphic structure $\mathcal{E}$ on E is given. Then the holomorphic structure naturally induces a linear operator

$$\bar{\partial}_{\mathcal{E}} : \Omega^{0,q}(E) \to \Omega^{0,q+1}(E)$$

as follows. Choosing a holomorphic trivialization τ, we represent each section s of $\Omega^{0,q}(E)$ as a vector-valued function s_τ. We define $\bar\partial_{\mathcal{E}} s$ locally by $\bar\partial s_\tau$. This is independent of the local holomorphic trivializations. If two different holomorphic trivializations are related by a holomorphic transition function g, then

$$\bar\partial(gs) = (\bar\partial g)s + g(\bar\partial s) = g(\bar\partial s).$$

Hence the locally defined $\bar\partial s$'s form a global section.

The operator $\bar\partial_{\mathcal{E}}$ satisfies and is characterized by the following properties.

1. $\bar\partial_{\mathcal{E}}(f \cdot s) = (\bar\partial f)s + f(\bar\partial_{\mathcal{E}} s)$ for $f \in \Omega^{0,0}(Z)$ and $s \in \Omega^{0,*}(E)$.
2. $\bar\partial_{\mathcal{E}} s = 0$ on an open set $U \subset Z$ if and only if s is holomorphic over U.

The above property 1 is analogous to the Leibniz rule of the covariant derivative.

We consider operators

$$\bar\partial_\alpha : \Omega^0(E) \to \Omega^{0,1}(E),$$

on the C^∞ bundle E which satisfy property 1. In a local C^∞ trivialization τ of E, $\bar\partial_\alpha$ is expressed as

$$\bar\partial_\alpha = \bar\partial + \alpha^\tau,$$

where α^τ is a matrix of (0, 1)-forms. The operator $\bar\partial_\alpha$ naturally extends to operators on $\Omega^{0,q}(E)$. Then we can consider $\bar\partial_\alpha^2$ as an element of $\Omega^{0,2}(\mathrm{End}(E))$. In a local trivialization, we have

$$\bar\partial_\alpha^2 = \bar\partial\alpha^\tau + \alpha^\tau \wedge \alpha^\tau.$$

We say $\bar\partial_\alpha$ is *integrable* if $\bar\partial_\alpha$ comes from a holomorphic structure on E.

Lemma 3.87 *The operator $\bar\partial_\alpha$ is integrable if and only if, for each point $z \in Z$, there is a C^∞ trivialization of E on a neighborhood of z such that $\alpha = 0$.*

Proof Suppose there are two trivializations given by $\mathrm{GL}(n, \mathbb{C})$-valued C^∞ functions τ_1 and τ_2 such that

$$\tau_i(\bar\partial + \alpha)\tau_i^{-1} = \bar\partial \quad (i = 1, 2).$$

These are equivalent to

$$\bar\partial\tau_i = \tau_i\alpha. \tag{3.17}$$

These relations imply that the transition function $g = \tau_2\tau_1^{-1}$ between two trivializations is holomorphic. A collection of such transition functions gives a holomorphic structure on E. □

Theorem 3.88 *The operator $\bar\partial_\alpha$ is integrable if and only if $\bar\partial_\alpha^2 = 0$.*

Proof In this book, we only use the rank $E = 1$ case. If rank $E = 1$, then $\bar\partial_\alpha^2 = \bar\partial\alpha + \alpha \wedge \alpha = \bar\partial\alpha$. If $\bar\partial\alpha = 0$, then we can find η with $\bar\partial\eta = \alpha$ by the $\bar\partial$-Poincaré lemma (see [65]). Then $\tau = \exp(\eta)$ solves the Eq. (3.17). For higher rank cases, see [40, §2.2]. □

Before concluding this section, we state the relation between the Dirac operators and $\bar\partial$-operators in the simple case when $Z = \mathbb{C}^2$. By using (3.16) and (3.14), we can prove the following.

Proposition 3.89 *For $Z = \mathbb{C}^2$ with standard metric, the spin Dirac operator $D\colon \Gamma(S_{\mathbb{C}}^+) \to \Gamma(S_{\mathbb{C}}^-)$ is identified with*

$$\sqrt{2}(\bar\partial + \bar\partial^*)\colon \Omega^{0,1} \to \Omega^{0,2} \oplus \Omega^{0,0}.$$

The proof is an exercise (Problem 3.99). (It is known that $D = \sqrt{2}(\bar\partial + \bar\partial^*)$ holds for the canonical spin^c structures on Kähler manifolds. See [131, §3.4].)

Problems

3.90 Verify Lemma 3.13.

3.91 Verify Lemma 3.40.

3.92 Verify the Bianchi identity in Remark 3.42.

3.93 Verify (3.8) and (3.9) hold for a connection 1-form ω.

3.94 Verify Lemma 3.51.

3.95 Verify Lemma 3.56.

3.96 Verify Lemma 3.57.

3.97 Verify Lemma 3.75.

3.98 (Bochner Technique) Consider the formula (3.15) on a compact spin^c manifold. Suppose that it admits a Riemannian metric of positive scalar curvature. Then, verify that there is positive constant $\epsilon > 0$ such that if $||F_A||_{L^\infty} < \epsilon$, then $\mathrm{ind}(D_A) = 0$ vanishes.

3.99 Verify Proposition 3.89.

Chapter 4
Seiberg–Witten Theory

The purpose of this chapter is to introduce the basics of the Seiberg–Witten theory. Especially the analysis of the moduli spaces and the construction of the Seiberg–Witten invariants will be discussed in detail.

4.1 Spin^c 4-Manifolds and the Seiberg–Witten Equations

4.1.1 Review of Spin^c *Representations*

We recall the spin^c representation $\hat{\Delta}^{\pm}_{\mathbb{C}} : \mathrm{Spin}(V) \to \mathrm{Aut}(S^{\pm}_{\mathbb{C}}(V))$ in the case when $V = \mathbb{R}^4$ summarized in the Example in Sect. 3.1.3.2. As real vector spaces, the spaces $V = \mathbb{H}_T$ and $S^{\pm}_{\mathbb{C}}(V) = \mathbb{H}^{\pm}$ are identified with the copies of $\mathbb{H}$. The $\mathrm{Spin}^c(V) = \mathrm{Spin}^c(4)$ module structures on V and $\mathbb{H}^{\pm}$ are given as follows. The actions of $[q_+, q_-, u] \in \mathrm{Spin}^c(4) = (\mathrm{Sp}(1) \times \mathrm{Sp}(1)) \times_{\{\pm 1\}} \mathrm{U}(1)$ on $v \in \mathbb{H}_T$, $\phi \in \mathbb{H}^+$ and $\psi \in \mathbb{H}^-$ are defined by $q_+ v q_-^{-1}$, $q_+ \phi u^{-1}$ and $q_- \psi u^{-1}$ respectively. The complex structures on $\mathbb{H}^{\pm}$ is defined by multiplication of complex numbers by right. The Clifford multiplication is given by the $\mathrm{Spin}^c(4)$-equivariant maps $c_- : V \otimes \mathbb{H}^- \to \mathbb{H}^+$ and $c_- : V \otimes \mathbb{H}^+ \to \mathbb{H}^-$ defined by $c_-(v \otimes \psi) = v\psi$ and $c_+(v \otimes \phi) = -\bar{v}\phi$.

It will be convenient to represent these $\mathrm{Spin}^c(4)$-module structures of $\mathbb{H}^{\pm}$ by complex matrices. Define the "Pauli matrices" by

$$e_0 = \begin{pmatrix} 1 & 0 \\ 0 & 1 \end{pmatrix}, \quad e_1 = \begin{pmatrix} i & 0 \\ 0 & -i \end{pmatrix}, \quad e_2 = \begin{pmatrix} 0 & 1 \\ -1 & 0 \end{pmatrix}, \quad e_3 = \begin{pmatrix} 0 & i \\ i & 0 \end{pmatrix}.$$

The respective correspondence $1, i, j, k \mapsto e_0, e_1, e_2, e_3$ defines an $\mathbb{R}$-linear homomorphism I from the quaternion algebra $\mathbb{H}$ to the complex matrix ring $M_2(\mathbb{C})$.

T. Kato, N. Nakamura, *Monopole Maps and L^2-cohomology*, Springer Monographs in Mathematics, https://doi.org/10.1007/978-981-96-8472-4_4

Remark 4.1 For $x, y \in \mathbb{H}$,

$$I(x)^* \circ I(y) + I(y)^* \circ I(x) = 2(x, y)\mathrm{id}_{\mathbb{C}^2},$$

where $(\cdot, \cdot)$ is the standard Euclidean inner product on $\mathbb{H} \cong \mathbb{R}^4$.

Identifying $\mathbb{H}^\pm$ with $\mathbb{C}^2$ by the correspondence $\phi = \phi_1 - j\phi_2 \mapsto \begin{pmatrix}\phi_1\\ \phi_2\end{pmatrix}$, we see that, for $q = \alpha + \beta j$, $\phi = \phi_1 - j\phi_2 \in \mathbb{H}$, and $z \in \mathbb{C}$, $q\phi z^{-1}$ is written as

$$z^{-1}\begin{pmatrix}\alpha & \beta\\ -\bar{\beta} & \bar{\alpha}\end{pmatrix}\begin{pmatrix}\phi_1\\ \phi_2\end{pmatrix}.$$

There is a natural Hermitian structure on $\mathbb{C}^2$. Endow Hermitian structures on $\mathbb{H}^\pm$ via the identifications $\mathbb{H}^\pm$ with $\mathbb{C}^2$. Then the Clifford multiplication

$$\rho(v) = \begin{pmatrix} 0 & c_-(v \otimes \cdot)\\ c_+(v \otimes \cdot) & 0 \end{pmatrix} = \begin{pmatrix} 0 & v\\ -\bar{v} & 0\end{pmatrix}$$

for $v \in V$ is a trace-free skew-Hermitian endomorphism of $\mathbb{H}^+ \oplus \mathbb{H}^-$. We extend $c_\pm$ to $\tilde{c}_\pm \colon (V \otimes \mathbb{C}) \otimes \mathbb{H}^\pm \to \mathbb{H}^\mp$ by $\tilde{c}_-((v \otimes u) \otimes \psi) = v\psi u$ and $\tilde{c}_+((v \otimes u) \otimes \phi) = -\bar{v}\phi u$. Then ρ is also extended to $V \otimes \mathbb{C}$, and $\rho(\sqrt{-1}v)$ is a Hermitian endomorphism.

4.1.2 Spin^c *Structures on* 4-*Manifolds*

A spin^c structure $\mathfrak{s}$ on a 4-manifold X is given by a $\mathrm{Spin}^c(4)$ bundle P over X and an isomorphism $\xi \colon T = P \times_{\mathrm{Spin}^c(4)} V \to TX$. (See Sect. 3.2.) The identification $T \cong TX$ induces an orientation and a Riemannian metric of X. We also identify T with the cotangent bundle T^*X via the natural identification $TX \cong T^*X$.

Remark 4.2 Note the following exact sequence:

$$1 \longrightarrow \mathrm{U}(1) \longrightarrow \mathrm{Spin}^c(4) \longrightarrow \mathrm{SO}(4) \longrightarrow 1.$$

For a spin^c structure $\mathfrak{s}$, the quotient bundle $P/\mathrm{U}(1)$ is naturally identified with the $\mathrm{SO}(4)$-frame bundle of TX.

The associated bundles $S^+ = P \times_{\mathrm{Spin}^c(4)} \mathbb{H}_+$ and $S^- = P \times_{\mathrm{Spin}^c(4)} \mathbb{H}_-$ are the positive and negative spinor bundles of $\mathfrak{s}$, and $L = P \times_{\mathrm{Spin}^c(4)} \mathbb{C}_2$ is the determinant line bundle where $\mathbb{C}_2$ is the complex one-dimensional $\mathrm{Spin}^c(4)$ module such that the action of $[q_+, q_-, u] \in \mathrm{Spin}^c(4)$ is given by multiplication of u^2. Let $S = S^+ \oplus S^-$.

The Clifford multiplications $\gamma_{\pm}: T \otimes S^{\pm} \to S^{\mp}$ are induced from $c_{\pm}$. For $\xi \in T_x$, we have a trace-free skew-Hermitian endomorphism $\rho(\xi)$ of $S_x = (S^+ \oplus S^-)_x$ as

$$\rho(\xi) = \begin{pmatrix} 0 & \rho_-(\xi) \\ \rho_+(\xi) & 0 \end{pmatrix},$$

where $\rho_{\pm}$ is defined by $\gamma_{\pm}(\xi \otimes \phi) = \rho_{\pm}(\xi)\phi$. By Remark 4.1.1, we obtain the following.

Proposition 4.3 *For $u, v \in T_x$, we have the Clifford identity*

$$\rho(u) \circ \rho(v) + \rho(v) \circ \rho(u) = -2(u, v)\mathrm{id}_{S_x}.$$

Since $V \oplus \mathbb{C}_2$ is a faithful representation of the Lie algebra of Spin$^c(4)$, the Levi-Civita connection on TX and a U(1)-connection A_0 on L determine a Spin$^c(4)$-connection on P (Lemma 3.72), which induces the covariant derivative ∇_0 on $\Gamma(S^+)$. The Dirac operator $D_{A_0}: \Gamma(S^+) \to \Gamma(S^-)$ is defined as the composition

$$\Gamma(S^+) \xrightarrow{\nabla_0} \Gamma(S^+ \otimes T) \xrightarrow{\gamma_+} \Gamma(S^-).$$

For $a \in \Omega^1(X, \sqrt{-1}\mathbb{R})$, $A = A_0 + a$ is another U(1)-connection on L. The Dirac operator for A is

$$D_A = D_{A_0} + \frac{1}{2}\rho(a).$$

4.1.3 Seiberg–Witten Equations

Consider Spin$^c(4)$ modules $\Lambda_T^{\pm}$ such that, as vector spaces, these are copies of the imaginary part $\mathrm{im}\,\mathbb{H}$ of $\mathbb{H}$ and the action of $[q_+, q_-, z] \in \mathrm{Spin}^c(4)$ is given by $\omega \mapsto q_{\pm}\omega q_{\pm}^{-1}$. The following proposition is an exercise (Problem 4.107).

Proposition 4.4 *Let $V = \mathbb{R}^4$. Decompose the 2nd exterior power $\Lambda^2 V$ into*

$$\Lambda^2 V = \Lambda^+ V \oplus \Lambda^- V,$$

where $\Lambda^{\pm} V$ are the (± 1)-eigenspaces of the Hodge star operator. Then $\Lambda^{\pm} V$ are naturally identified with $\Lambda_T^{\pm}$ as Spin$^c(4)$*-modules.*

Then it follows from Lemma 3.14, that

$$\mathbb{C}\left(\frac{1 \pm \omega_{\mathbb{C}}}{2}\right) \oplus (\Lambda_T^{\pm} \otimes \mathbb{C}) \cong (\mathrm{Cl}_0(V) \otimes \mathbb{C})^{\pm} \cong \mathrm{End}_{\mathbb{C}}(\mathbb{H}^{\pm}).$$

In particular, $\sqrt{-1}\Lambda_T^{\pm}$ is identified with the space of trace-free Hermitian endomorphisms of $\mathbb{H}^{\pm}$.

Remark 4.5 Concretely, the action of $\Lambda_T^{+} \otimes \mathbb{C}$ on $\mathbb{H}^{+}$ is given by

$$(\Lambda_T^{+} \otimes \mathbb{C}) \times \mathbb{H}^{+} \ni (\omega \otimes z, \phi) \mapsto \omega\phi z \in \mathbb{H}^{+}.$$

Corollary 4.6 *We have the following identifications:*

$$P \times_{\mathrm{Spin}^c(4)} \Lambda_T^{\pm} = \Lambda^{\pm} T^* X.$$

Define a quadratic map $q_0 \colon \mathbb{H}_{+} \to \Lambda_T^{+}$ by $q_0(\phi) = \phi i \bar{\phi}$ for $\phi \in \mathbb{H}_{+}$. For $\phi = \phi_1 - j\phi_2$, the matrix representation of $q_0(\phi)$ is

$$\sqrt{-1}\begin{pmatrix} |\phi_1|^2 - |\phi_2|^2 & 2\phi_1\bar{\phi}_2 \\ 2\bar{\phi}_1\phi_2 & -|\phi_1|^2 + |\phi_2|^2 \end{pmatrix}. \tag{4.1}$$

(See Problem 4.108.) We define $q \colon \mathbb{H}^{+} \to \sqrt{-1}\Lambda^{+} V$ by $q = \dfrac{1}{2\sqrt{-1}} q_0$. This is extended to the map

$$q \colon \Gamma(S^{+}) \to \sqrt{-1}\Omega^{+}(X).$$

Now we introduce the Seiberg–Witten equations.

Definition 4.7 Let $\mathcal{A}(L)$ be the space of U(1)-connections on the determinant line bundle L. The *Seiberg–Witten equations* are equations for $(A, \Phi) \in \mathcal{A}(L) \times \Gamma(S^{+})$ as follows:

$$\begin{cases} D_A \Phi = 0, \\ F_A^{+} = q(\Phi), \end{cases} \tag{4.2}$$

where F_A^{+} is the self-dual part of the curvature F_A.

The gauge transformation group $\mathcal{G}$ of a spin^c structure (P, φ) is given as the space of Spin^c(4)-equivariant diffeomorphisms of P covering the identity map of the quotient $P/\,\mathrm{U}(1)$. By Lemma 3.56, $\mathcal{G}$ is identified with

$$\Gamma(P \times_{\mathrm{ad}} \mathrm{U}(1)) = C^{\infty}(X, \mathrm{U}(1)).$$

The $\mathcal{G}$-action on $\mathcal{A}(L) \times \Gamma(S^+)$ is given by $g(A, \Phi) = (A - 2g^{-1}dg, g\Phi)$ for $g \in \mathcal{G}$ and $(A, \Phi) \in \mathcal{A}(L) \times \Gamma(S^+)$. (See Lemma 3.58. Note that $g \in \mathcal{G}$ acts on L by multiplication of g^2.)

Lemma 4.8 *The Seiberg–Witten equations* (4.2) *are* $\mathcal{G}$*-equivariant. Hence,* $\mathcal{G}$ *acts on the space of solutions to* (4.2).

Proof The proof is an exercise (Problem 4.109). □

The *moduli space* of solutions to the Seiberg–Witten equations is defined as the quotient space:

$$\mathcal{M} = \mathcal{M}(\mathfrak{s}) = \{\text{solutions to (4.2)}\}/\mathcal{G}. \tag{4.3}$$

4.2 Analysis

The purpose of this section is to give an analytical set up for analysing the moduli spaces.

From now on in this chapter, we assume X is a closed connected oriented Riemannian 4-manifold with spin^c structure $\mathfrak{s}$. Fix a reference C^∞ U(1)-connection A_0 on the determinant line bundle L of $\mathfrak{s}$. Then A_0 and the Levi-Civita connection determine the connection on the Spin^c(4) bundle of $\mathfrak{s}$ (Lemma 3.72) and it induces a unique C^∞ connection ∇_{A_0} on the spinor bundles. By using the Riemannian metric and the pointwise inner product of the spinor bundles, we define the L^2-inner product and the L^p-norms as follows:

$$\langle a, b\rangle_{L^2} = \int_X a \wedge *b, \quad \langle \phi, \psi\rangle_{L^2} = \int_X (\phi, \psi)\,\mathrm{vol},$$

$$\|a\|_{L^p} = \left(\int_X |a|^p\,\mathrm{vol}\right)^{\frac{1}{p}}, \quad \|\Phi\|_{L^p} = \left(\int_X |\Phi|^p\,\mathrm{vol}\right)^{\frac{1}{p}}.$$

The L^∞-norm is also defined.

$$\|f\|_{L^\infty} = \inf\{C;\ |f(x)| \le C \text{ a.e. on } X\}.$$

We introduce the Sobolev k-inner product for a nonnegative integer k by

$$\langle a, b\rangle_{L^2_k} = \sum_{l=0}^{k} \langle \nabla^l a, \nabla^l b\rangle_{L^2}, \quad \langle \phi, \psi\rangle_{L^2_k} = \sum_{l=0}^{k} \langle \nabla^l_{A_0}\phi, \nabla^l_{A_0}\psi\rangle_{L^2}.$$

The Sobolev norm is given by

$$\|a\|^p_{L^p_k} = \sum_{l=0}^{k} \|\nabla^l a\|^p_{L^p}, \quad \|\Phi\|^p_{L^p_k} = \sum_{l=0}^{k} \|\nabla^l_{A_0}\Phi\|^p_{L^p}. \tag{4.4}$$

In the Appendix, we extend the definition of L^2_k inner product to the case when k is a nonnegative real number by using a Fourier transform. The operators $d^* + d^+$ and D_{A_0} are operators of first order. They are extended to the bounded maps from L^2_{k+1} to L^2_k for arbitrary nonnegative real number k. One of our goals of this section is to verify that the operators $d^* + d^+$ and D_{A_0} are Fredholm.

> Sobolev Analysis

Definition 4.9 A linear operator $L\colon E \to F$ between Banach spaces is a *Fredholm operator* if it satisfies:

- Ker L is finite dimensional;
- Im L is closed and has finite codimension.

Lemma 4.10 *For sections of a bundle over a closed* 4*-manifold* X*, if* $k > 2 + l$*, then* $L^2_k \to C^l$ *is defined and continuous.*

Lemma 4.11 (Rellich) *If* $k' > k$*, then the natural injection* $L^2_{k'} \to L^2_k$ *is compact, i.e., every sequence bounded in* $L^2_{k'}$ *has a subsequence convergent in* L^2_k*.*

Lemma 4.12 (Cf. [19, Theorem 7.3]) *Suppose* s *and* s_i $(i = 1, 2)$ *are real numbers satisfying*

1. $s_i \geq s \geq 0$,
2. $s_1 + s_2 - s > \frac{n}{2}$.

Then $L^2_{s_1} \otimes L^2_{s_2} \to L^2_s$ *is defined and continuous.*

Lemmas 4.10, 4.11 and 4.12 are proved in the Appendix.

Weitzenböck formula for p-forms:

Proposition 4.13 *Let* $D = d + d^*$*. For a* p*-form* ω*,*

$$D^2\omega = \nabla^*\nabla\omega + \frac{1}{2}\sum_{i,j} \theta^i \cdot \theta^j \cdot R_\nabla(E_i, E_j)\omega,$$

where E_i is a frame, θ^i is its dual coframe, R_∇ is the curvature of the Levi-Civita connection, and $\cdot$ means the Clifford multiplication given by (3.6).

Proof (See [145, Chapter 7, Theorem 50].) □

Definition 4.14 Let P be a differential operator. A differential operator P^* is a L^2-*formal adjoint of* P if, for every compactly supported sections u, v, the equality holds:

$$\langle Pu, v\rangle_{L^2} = \langle u, P^*v\rangle_{L^2}.$$

Examples

- The formal adjoint of $d^* + d^+$ on $\sqrt{-1}\Omega^1(X)$ is $d \oplus d^*$ on $\sqrt{-1}(\Omega^0 \oplus \Omega^+)(X)$.
- In dimension 4, the Dirac operator D_{A_0} splits into $D^\pm_{A_0} : \Gamma(S^\pm) \to \Gamma(S^\mp)$. Then $D^\pm_{A_0}$ are formal adjoints of each other (Lemma 3.76 and Remark 3.77). (We will often omit $\pm$ from the notation $D^\pm_{A_0}$.)

Proposition 4.15 (A Priori Estimate) *For each nonnegative integer k, there exist constants C_i, C'_i $(i = 1, 2)$ such that*

$$\|a\|_{L^2_{k+1}} \leq C_1\left(\|a\|_{L^2_k} + \|(d^* + d^+)a\|_{L^2_k}\right),$$

$$\|b\|_{L^2_{k+1}} \leq C'_1\left(\|b\|_{L^2_k} + \|(d^* + d^+)^*b\|_{L^2_k}\right),$$

$$\|\Phi\|_{L^2_{k+1}} \leq C_2\left(\|\Phi\|_{L^2_k} + \|D^+_{A_0}\Phi\|_{L^2_k}\right),$$

$$\|\Psi\|_{L^2_{k+1}} \leq C'_2\left(\|\Psi\|_{L^2_k} + \|D^-_{A_0}\Psi\|_{L^2_k}\right),$$

where $a \in L^2_k\left(X; \sqrt{-1}\Lambda^1\right)$, $b \in L^2_k\left(X; \sqrt{-1}(\Lambda^0 \oplus \Lambda^+)\right)$, $\Phi \in L^2_k(X; S^+)$ and $\Psi \in L^2_k(X; S^-)$.

Proof The proof uses the Bochner–Weitzenböck formulae (Lemma 3.80 and Proposition 4.13). We prove the inequality for Φ. The proofs of the other cases are similar. Taking the inner product of (3.15) with Φ, we have

$$\|D^+_{A_0}\Phi\|^2_{L^2} = \|\nabla\Phi\|^2_{L^2} + \frac{\kappa}{4}\|\Phi\|_{L^2} + \left\langle\frac{\rho(F^+_{A_0})}{2}\Phi, \Phi\right\rangle.$$

From this, we obtain

$$\|\nabla\Phi\|^2_{L^2} \leq \text{const.}(\|\Phi\|^2_{L^2} + \|D^+_{A_0}\Phi\|^2_{L^2}).$$

Then $D^+_{A_0}$ extends to

$$D^+_{A_0} = D^+_{A_0} \otimes 1 \colon \Gamma(S^+ \otimes T^*X^{\otimes k}) \to \Gamma(S^- \otimes T^*X^{\otimes k}).$$

Differentiating $D^+_{A_0}\Phi$ by ∇^j, we obtain

$$\nabla^j(D^+_{A_0}\Phi) = D^+_{A_0}\nabla^j\Phi + \sum_{l\le j-1} T_l\nabla^l\Phi, \tag{4.5}$$

where T_l are tensors independent of Φ. By using (4.5), we obtain

$$\begin{aligned}
\|\nabla^j\Phi\|_{L^2_1} &\le \text{const.}(\|\nabla^j\Phi\|_{L^2} + \|D^+_{A_0}(\nabla^j\Phi)\|_{L_2}) \\
&\le \text{const.}\left(\|\nabla^j\Phi\|_{L^2} + \|\nabla^j(D^+_{A_0}\Phi)\|_{L_2} + \text{const.}\sum_{l\le j-1}\|\nabla^l\Phi\|_{L_2}\right) \\
&\le \text{const.}\left(\|\Phi\|_{L^2_j} + \|D_{A_0}\Phi\|_{L^2_j}\right).
\end{aligned}$$

Then we can find a constant C_2 such that

$$\|\Phi\|_{L^2_{k+1}} \le C_2(\|\Phi\|_{L^2_k} + \|D^+_{A_0}\Phi\|_{L^2_k}).$$

Thus the estimate for D_{A_0} is proved. □

Remark 4.16 The inequalities in Proposition 4.15 are special cases of Gårding's inequalities for elliptic operators. (See e.g. [168, §6.9].)

Theorem 4.17 *Proposition 4.15 holds even when k is a nonnegative real number.*

Theorem 4.17 will be proved in the Appendix.

Lemma 4.18 *Let L be the L^2-orthogonal complement of* $\mathrm{Ker}(d^* + d^+)$ *in $L^2\left(X; \sqrt{-1}\Lambda^1\right)$. Then there exists a constant C such that, for any $a \in L \cap L^2_{k+1}$,*

$$\|a\|_{L^2_k} \le C\|(d^* + d^+)a\|_{L^2_k}.$$

Similar statements hold for $(d^* + d^+)^*$ and $D^{\pm}_{A_0}$.

Proof Suppose the lemma is not true. Then there exists a sequence $a_i \in L^2_{k+1}$ such that

$$\lim_{i\to\infty} \frac{\|(d^* + d^+)a_i\|_{L^2_k}}{\|a_i\|_{L^2_k}} = 0.$$

We may assume $\|a_i\|_{L^2_k} = 1$ by normalizing. Then

$$\|a_i\|_{L^2_{k+1}} \leq C_1 \left(\|a_i\|_{L^2_k} + \|(d^* + d^+)a_i\|_{L^2_k} \right) \leq 2C_1.$$

By Rellich's lemma (Lemma 4.11), there is a subsequence converging in L^2_k. Then

$$\|a_i - a_j\|_{L^2_{k+1}} \leq C_1 \left(\|a_i - a_j\|_{L^2_k} + \|(d^* + d^+)a_i\|_{L^2_k} + \|(d^* + d^+)a_j\|_{L^2_k} \right) \to 0.$$

Therefore a_i converges in L^2_{k+1}. Let $a = \lim_{i\to\infty} a_i$. Then $(d^* + d^+)a = 0$. However, a_i is orthogonal to $\ker(d^* + d^+)$ and $a_i \neq 0$. This is a contradiction. □

Theorem 4.19 *The extensions of $d^* + d^+$, $(d^* + d^+)^*$ and $D^{\pm}_{A_0}$ to the maps from L^2_{k+1} to L^2_k are Fredholm for $k \geq 1$. Any element in their kernels is smooth.*

Proof The proofs are same for every operator. We give a proof for $P = d^* + d^+$. First, we prove that $\ker P$ is finite dimensional. Suppose this is not true. Then there exists a bounded sequence $\{a_i\}$ in $\ker P$ which has no subsequence converging in L^2_{k+1}. By Theorem 4.17,

$$\|a_i\|_{L^2_{k+2}} \leq C_1 \left(\|a_i\|_{L^2_{k+1}} + \|Pa_i\|_{L^2_{k+1}} \right) = C_1 \|a_i\|_{L^2_{k+1}}.$$

By Rellich's lemma (Lemma 4.11), there is a subsequence converging in L^2_{k+1}. This is a contradiction. Note that a similar proof proves that $\ker P^*$ is also finite dimensional.

Let us prove that the elements in $\ker P$ are in C^∞. If $Pa = 0$, then Theorem 4.17 implies $\|a\|_{L^2_{k+2}} \leq C_1\|a\|_{L^2_{k+1}}$ and therefore $a \in L^2_{k+2}$. Then, by Theorem 4.17 again, we have $\|a\|_{L^2_{k+3}} \leq C_1\|a\|_{L^2_{k+2}}$ and therefore $a \in L^2_{k+3}$. Repeating this process, we obtain $a \in L^2_m$ for all m. By Theorem 4.10, $a \in C^\ell$ for all ℓ.

We prove that $\operatorname{Im} P$ is a closed subspace of L^2_k. Suppose there are a sequence $\{a_i\}$ in $L^2_{k+1}(X; \sqrt{-1}\Lambda^1)$ and $v \in L^2_k(X; \sqrt{-1}(\Lambda^0 \oplus \Lambda^+))$ such that $\lim_{i\to\infty} \|Pa_i - v\|_{L^2_k} = 0$. We have a unique decomposition $a_i = b_i + k_i$ such that $b_i \in L$ and $k_i \in \operatorname{Ker} P$. Since k_i is smooth, k_i is in L^2_{k+1}. By Lemma 4.18,

$$\|b_i\|_{L^2_k} \leq C\|Pb_i\|_{L^2_k} = C\|Pa_i\|_{L^2_k}.$$

By Theorem 4.17,

$$\|b_i\|_{L^2_{k+1}} \le C_1\left(\|b_i\|_{L^2_k} + \|Pb_i\|_{L^2_k}\right) \le C_1(1+C)\|Pa_i\|_{L^2_k}.$$

By Lemma 4.11, a subsequence of $\{b_i\}$ converges in L^2_k. By Theorem 4.17,

$$\begin{aligned}\|b_i - b_j\|_{L^2_{k+1}} &\le C_1\left(\|b_i - b_j\|_{L^2_k} + \|Pb_i - Pb_j\|_{L^2_k}\right)\\ &= C_1\left(\|b_i - b_j\|_{L^2_k} + \|Pa_i - Pa_j\|_{L^2_k}\right) \to 0.\end{aligned}$$

Then b_i converges in L^2_{k+1}. Let $b = \lim_{i\to\infty} b_i$. Then $Pb = v$. Therefore $v \in \operatorname{Im} P$.

We prove that $\operatorname{Im} P$ has finite codimension. Suppose $u \in L^2_k$ is in the L^2-orthogonal complement of $\operatorname{Im} P$. Then for any $a \in L^2_{k+1}$,

$$0 = \langle Pa, u\rangle_{L^2} = \langle a, P^*u\rangle_{L^2},$$

where we consider P^* is an operator form L^2_k to L^2_{k-1}. Hence u is in $\operatorname{Ker} P^*$. Since $\operatorname{Ker} P^*$ is also finite dimensional, $\operatorname{Im} P$ has finite codimension. □

Proposition 4.20 (Hodge Decomposition) *There is an orthogonal decomposition,*

$$L^2_k(X;\Lambda^1) = R(d) \oplus R(d^*) \oplus H^1, \tag{4.6}$$

where

$$\begin{aligned}R(d) &= \operatorname{im}(d\colon L^2_{k+1}(X;\Lambda^0) \to L^2_k(X;\Lambda^1)),\\ R(d^*) &= \operatorname{im}(d^*\colon L^2_{k+1}(X;\Lambda^2) \to L^2_k(X;\Lambda^1)),\end{aligned}$$

and $H^1 = \ker(d + d^*)$ *is the space of harmonic* 1*-forms.*

The proof is left to the readers as an exercise (Problem 4.110).

4.3 Configuration Space

We introduce the following spaces:

$$\begin{aligned}\mathcal{A}_k &= \left\{A_0 + a \,\middle|\, a \in L^2_k\left(X;\sqrt{-1}\Lambda^1\right)\right\},\\ C_k &= \mathcal{A}_k \times L^2_k(X;S^+),\end{aligned}$$

(continued)

$$\mathcal{G}_k = \{g \in L_k^2(X, \mathbb{C}) \mid |g| = 1 \text{ a.e. }\}.$$

The space C is called the *configuration space*. Note that C_k is a Hilbert affine space.

4.3.1 Quotient Space

Proposition 4.21 *If $k > 2$, then $\mathcal{G}_k$ is an abelian Hilbert Lie group, whose Lie algebra is $L_k^2\left(X; \sqrt{-1}\mathbb{R}\right)$.*

Proof First, we prove that $\mathcal{G}_k$ has a structure of a Hilbert manifold. Note that, if $k > 2$, then an L_k^2 section f has a unique continuous representative by Lemma 4.10, and the value $f(x)$ of each point $x \in X$ is meaningful. Note also that, if $k > 2$, then L_k^2 is an algebra by Lemma 4.12. Then, for $f \in L_k^2\left(X; \sqrt{-1}\mathbb{R}\right)$,

$$u = \exp(2\pi f) = \sum_{n=0}^{\infty} \frac{(2\pi f)^n}{n!}$$

is in L_k^2, and satisfies $|u(x)| = 1$ for every $x \in X$. Therefore $u \in \mathcal{G}_k$. For $g \in \mathcal{G}_k$,

$$\psi_g = g \exp(2\pi \cdot) \colon L_k^2\left(X; \sqrt{-1}\mathbb{R}\right) \to \mathcal{G}_k \subset L_k^2(X; \mathbb{C})$$

is a smooth map. The restriction of ψ_g to a neighborhood of the origin of $L_k^2\left(X; \sqrt{-1}\mathbb{R}\right)$ is a bijective map to a neighborhood U_g of g in $\mathcal{G}_k$. Let $\varphi_g = (\psi_g|_{U_g})^{-1}$. Then the collection $\{(U_g, \varphi_g)\}$ gives a smooth atlas of $\mathcal{G}_k$. (It is easy to check that $\varphi_g \varphi_{g'}^{-1}|_{U_g \cap U_{g'}}$ is a smooth isomorphism.) Next, we prove that the product and inverse in $\mathcal{G}_k$ are smooth. Since L_k^2 is an algebra, the product $f(x)g(x)$ defines a bilinear map

$$L_k^2(X; \mathbb{C}) \times L_k^2(X; \mathbb{C}) \to L_k^2(X; \mathbb{C}).$$

Hence the product structure of $\mathcal{G}_k$ is the restriction of the bilinear map to $\mathcal{G}_k$. Similarly, the inverse $g \mapsto g^{-1}$ of $\mathcal{G}_k$ is induced from the linear map of the complex conjugation $g \mapsto \overline{g}$. Since (bi)linear maps are smooth, the product and inverse are also smooth. □

Proposition 4.22 *If $k > 1$, then $\mathcal{G}_{k+1}$ acts smoothly on $\mathcal{C}_k$.*

Proof The map $\mathcal{G}_{k+1} \to L^2_k(X; \Lambda^1)$ given by $g \mapsto -2g^{-1}dg$ is bounded for $k > 1$. The $\mathcal{G}_{k+1}$-action on $\mathcal{A}_k$ is a translation by $-2g^{-1}dg$, and therefore smooth. The multiplication

$$L^2_{k+1}(X; \mathbb{C}) \times L^2_k(X; S^+) \to L^2_k(X; S^+), \quad (f, \phi) \mapsto f \cdot \phi$$

is a bounded bilinear map by Lemma 4.12. Hence, the $\mathcal{G}_{k+1}$-action on $L^2_k(X; S^+)$ is also smooth. □

For $k > 1$, let $\mathcal{B}_k = \mathcal{C}_k / \mathcal{G}_{k+1}$ with quotient topology.

Proposition 4.23 *The quotient space $\mathcal{B}_k$ ($k > 2$) is a Hausdorff space.*

We need a lemma to prove Proposition 4.23.

Lemma 4.24 *Let $k \geq 2$. Suppose $\{(A_n, \Phi_n)\}$ and $\{(B_n, \Psi_n)\}$ are sequences in $\mathcal{C}_k$ converging to (A, Φ) and (B, Ψ) respectively. Suppose there is a $g_n \in \mathcal{G}_{k+1}$ for each n such that*

$$g_n(A_n, \Phi_n) = (B_n, \Psi_n).$$

Then there is a subsequence of $\{g_n\}$ converging to g in $\mathcal{G}_{k+1}$ and

$$g(A, \Phi) = (B, \Psi).$$

Proof We prove the case when $k = 2$. Since g_n is in L^2_3, g_n is in C^0. Since

$$dg_n = \frac{1}{2} g_n (B_n - A_n), \tag{4.7}$$

We have

$$\|dg_n\|_{L^2} \leq \text{const.} \|g_n\|_{C^0} \|B_n - A_n\|_{L^2}.$$

By Proposition 4.15, g_n is L^2_1-bounded. By Lemma 4.12,

$$\|dg_n\|_{L^2_{2/3}} \leq \text{const.}\|g_n\|_{L^2_1}\|B_n - A_n\|_{L^2_2}$$

Hence, g_n is $L^2_{1+2/3}$-bounded. Proceeding similarly, we see that g_n is $L^2_{2+1/3}$-bounded. By Rellich's lemma (Lemma 4.11), there is a subsequence of g_n converging to some g in $L^2_{2+1/5}$, and we have

$$dg = \frac{1}{2}g(B - A).$$

By Lemma 4.12 again, we see that dg is in L^2_2, and hence g is in L^2_3, and that dg_n converges to dg in L^2_2. Finally, we see that g_n converges to g in L^2_3. □

Proof of Proposition 4.23 Suppose $\mathcal{B}_k$ is not Hausdorff. Then there exist a sequence $\{(A_n, \Phi_n)\}$ in C_k and a sequence $\{g_n\}$ in $\mathcal{G}_{k+1}$ such that $\{(A_n, \Phi_n)\}$ converges to (A, Φ) and $\{g_n(A_n, B_n)\}$ converges to (B, Ψ) and (A, Φ) and (B, Ψ) are not in the same orbit. By Lemma 4.24, we find a subsequence of $\{g_n\}$ which converges to g such that $(B, \Psi) = g(A, \Phi)$. This is a contradiction. □

The following proposition is obvious from the definition of the $\mathcal{G}$-action.

Proposition 4.25 *For $(A, \Phi) \in C_k$, let $\Gamma_{(A,\Phi)}$ be the stabilizer of (A, Φ). Then $\Gamma_{(A,\Phi)}$ is trivial if and only if $\Phi \not\equiv 0$. When $\Phi \equiv 0$, $\Gamma_{(A,\Phi)}$ is the group of constant maps of X to* $\mathrm{U}(1)$, *which is identified with* $\mathrm{U}(1)$.

Definition 4.26 A configuration $(A, \Phi) \in C_k$ is *irreducible* if $\Phi \not\equiv 0$. If otherwise, it is *reducible*.

4.3.2 Slice Theorems

Let k be an integer greater than 2. The set of irreducible configurations is denoted by C^*_k. Let $\mathcal{B}^*_k = C^*_k/\mathcal{G}_{k+1}$. Since $\mathcal{G}$ acts on C^*_k freely, $\mathcal{B}^*_k$ has a natural structure of a Hilbert manifold. We identify the tangent space of C_k at a configuration (A, Φ) with $L^2_k\left(X; \sqrt{-1}\Lambda^1 \oplus S^+\right)$.

Then, the linearization of the $\mathcal{G}$-action at (A, Φ) is given by

$$L_{(A,\Phi)} : L^2_{k+1}\left(X; \sqrt{-1}\mathbb{R}\right) \to L^2_k\left(X; \sqrt{-1}\Lambda^1 \oplus S^+\right),$$

(continued)

$$f \mapsto (-2df, f \cdot \Phi).$$

Let $L^*_{(A,\Phi)}$ be the formal adjoint of $L_{(A,\Phi)}$. Then the map $L^*_{(A,\Phi)}$ is given by

$$L^*_{(A,\Phi)}(a, \psi) = -2d^*a + \sqrt{-1}\,\mathrm{Im}(\langle \psi, \Phi \rangle_{L^2}), \tag{4.8}$$

for $(a, \psi) \in L^2_k\left(X; \sqrt{-1}\Lambda^1 \oplus S^+\right)$. (Problem 4.111.)

Lemma 4.27 *The subspace* $\mathrm{Im}\, L_{(A,\Phi)}$ *is closed in* $L^2_k\left(X; \sqrt{-1}\Lambda^1 \oplus S^+\right)$.

Proof Let L be the L^2-orthogonal complement of $\ker L_{(A,\Phi)}$ in $L^2\left(X; \sqrt{-1}\mathbb{R}\right)$. (Note that $\ker L_{(A,\Phi)} = 0$ if (A, Φ) is an irreducible, and $\ker L_{(A,\Phi)} \cong \sqrt{-1}\mathbb{R}$ if (A, Φ) is a reducible.) Then we can easily check the following estimate holds:

$$\|f\|_{L^2_k} \leq \text{const.} \|L_{(A,\Phi)}(f)\|_{L^2_k}$$

for $f \in L^2_{k+1}\left(X; \sqrt{-1}\mathbb{R}\right) \cap L$. Then, as in the proof of Theorem 4.19, we can prove that $\mathrm{Im}\, L_{(A,\Phi)}$ is a closed subspace. □

Lemma 4.28 *We have an identification*

$$T_w \mathcal{B}^*_k \cong \ker L^*_{(A,\Phi)}. \tag{4.9}$$

Proof By Lemma 4.27, there is an L^2-orthogonal decomposition

$$L^2_k\left(X; \sqrt{-1}\Lambda^1 \oplus S^+\right) = \mathrm{Im}\, L_{(A,\Phi)} \oplus \ker L^*_{(A,\Phi)}.$$

The tangent space $T_w \mathcal{B}^*_k$ of $\mathcal{B}^*_k$ at the gauge equivalence class $w = [(A, \Phi)]$ of an irreducible (A, Φ) is identified with $L^2_k\left(X; \sqrt{-1}\Lambda^1 \oplus S^+\right) / \mathrm{Im}\, L_{(A,\Phi)} \cong \ker L^*_{(A,\Phi)}$. □

For $\varepsilon > 0$, we set

$$T_{(A,\Phi),\varepsilon} = \{(a, \psi) \in \ker L^*_{(A,\Phi)} \mid \|(a, \psi)\|_{L^2_k} < \varepsilon\}.$$

For a small ε, $T_{(A,\Phi),\varepsilon}$ is diffeomorphic to a neighborhood of $[(A, \Phi)]$ in $\mathcal{B}^*_k$, and $T_{(A,\Phi),\varepsilon} \times \mathcal{G}_{k+1}$ is diffeomorphic to a neighborhood of the orbit of (A, Φ) in $\mathcal{C}^*_k$. These facts are generalized to the following slice theorem on the whole $\mathcal{C}_k$.

Definition 4.29 Let G be a smooth transformation group of a smooth manifold Y. A subspace $S \subset Y$ is a *slice* at a point $x \in Y$ if the following conditions hold:

1. $G_x(S) = S$ where G_x is the stabilizer of x.
2. $G(S)$ is an open neighborhood of the orbit $G(x)$.
3. The map $G \times_{G_x} S \to G(S)$, taking $[g, s] \mapsto gs$, is a G-equivariant diffeomorphism.

Theorem 4.30 (Local Slice Theorem) *For each $(A, \Phi) \in C_k$, there is a slice for the $\mathcal{G}_{k+1}$-action on C_k.*

Corollary 4.31 *The projection map from C_k to $\mathcal{B}_k$ induces a homeomorphism from the quotient $U/\Gamma_{(A,\Phi)}$ to a neighborhood of $[(A, \Phi)]$ in $\mathcal{B}_k$.*

Proof of Theorem 4.30 Note that $\ker L^*_{(A,\Phi)}$ is invariant under the $\Gamma_{(A,\Phi)}$-action for any (A, Φ). Choose a small $\Gamma_{(A,\Phi)}$-invariant neighborhood U of the origin in $\ker L^*_{(A,\Phi)}$. Consider the following $\mathcal{G}_{k+1}$-equivariant map

$$\Xi \colon \mathcal{G}_{k+1} \times_{\Gamma_{(A,\Phi)}} U \to C_k, \quad [g, u] \mapsto g \cdot ((A, \Phi) + u).$$

The differential at $[(0, 1)]$ of Ξ is identified with the map

$$L_{(A,\Phi)} \oplus \mathrm{Id} \colon \left(L^2_{k+1}\left(X; \sqrt{-1}\Lambda^0\right) / \operatorname{Lie}\Gamma_{(A,\Phi)}\right) \oplus \operatorname{Ker} L^*_{(A,\Phi)} \to \operatorname{Im} L_{(A,\Phi)} \oplus \operatorname{Ker} L^*_{(A,\Phi)},$$

where $\operatorname{Lie}\Gamma_{(A,\Phi)}$ is the tangent space of $\Gamma_{(A,\Phi)}$ at 1. (Note that $\operatorname{Lie}\Gamma_{(A,\Phi)} = \{0\}$ if (A, Φ) is irreducible, and $\operatorname{Lie}\Gamma_{(A,\Phi)} = \sqrt{-1}\mathbb{R}$ if (A, Φ) is reducible.) Since $\operatorname{Lie}\Gamma_{(A,\Phi)} = \ker L_{(A,\Phi)}$, the above map is a linear isomorphism. By the inverse function theorem, we can choose a small U and a small $\Gamma_{(A,\Phi)}$-invariant neighborhood V of 1 in $\mathcal{G}_{k+1}$ such that the map

$$V \times_{\Gamma_{(A,\Phi)}} U \to C_k, \quad [g, u] \mapsto g \cdot ((A, \Phi) + u)$$

is a diffeomorphism onto an open neighborhood of (A, Φ). Since Ξ is $\mathcal{G}_{k+1}$-equivariant, this implies Ξ is a local diffeomorphism.

We shall prove that Ξ is a diffeomorphism onto an open neighborhood of the orbit of (A, Φ) by choosing a smaller U if necessary. Since Ξ is a local diffeomorphism, it suffices to prove that Ξ is bijective for a sufficiently small U. Suppose there is no such U where Ξ is bijective. Then we can find sequences $\{x_n\}, \{y_n\} \subset U$ converging to 0, and $\{g_n\} \subset \mathcal{G}_{k+1}$ such that $[g_n, a_n] \neq [1, b_n]$ in $\mathcal{G}_{k+1} \times_{\Gamma_{(A,\Phi)}} U$ and

$$g_n \cdot ((A, \Phi) + x_n) = (A, \Phi) + y_n.$$

By Lemma 4.24, we can choose a subsequence of $\{g_n\}$ converging to some $g \in \mathcal{G}_{k+1}$ such that $g \cdot (A, \Phi) = (A, \Phi)$, i.e., $g \in \Gamma_{(A,\Phi)}$. This implies that, for sufficiently large n, $[g_n, a_n]$, $[1, b_n]$ are contained in $V \times_{\Gamma_{(A,\Phi)}} U$. This is a contradiction because the restriction of Ξ to $V \times_{\Gamma_{(A,\Phi)}} U$ is a diffeomorphism onto its image. Thus the theorem is proved. □

By the local slice theorem (Theorem 4.30), a neighborhood of 0 in $\operatorname{Ker} d^* \times \Gamma(S^+)$ gives a local slice around a reducible $(A_0, 0)$. On the other hand, we will prove that, for every U(1)-connection A, there is an element $g \in \mathcal{G}_{k+1}$ such that $gA = A_0 + a$ with $d^*a = 0$.

Proposition 4.32 *For $k > 1$, there are constants $K, C > 0$ depending only on X, A_0, k such that the following hold. For any* U(1) *connection A in L^2_k on L, there exists an L^2_{k+1} gauge transformation g such that $g(A) = A_0 + a$ where $a \in L^2_k(X; \sqrt{-1}\Lambda^1)$ satisfies*

$$d^*a = 0, \quad \|a\|_{L^2_k} \leq C\|F_A^+\|_{L^2_{k-1}} + K. \tag{4.10}$$

Proof Let $a_0 = A - A_0 \in L^2_k(X; \sqrt{-1}\Lambda^1)$. By the Hodge decomposition (Proposition 4.20), we find f, b and h such that

$$a_0 = df + d^*b + h,$$

where $df \in R(d)$, $d^*b \in R(d^*)$ and $h \in H^1$. Note that $\operatorname{Ker} d^* = R(d^*) \oplus H^1$. Let $g_1 = \exp(-2\pi\sqrt{-1}f/2) \in \mathcal{G}_{k+1}$. Then $g_1(A) = A_0 + d^*b + h \in \operatorname{Ker} d^*$. Take $h_0 \in H^1$ with integral periods such that $h - h_0$ is in a fixed fundamental domain of the $H^1(X; \mathbb{Z})$ action on H^1. Then $\|h - h_0\|_{L^2_k}$ is bounded by a constant K_1. Let $g = g_{h_0}g_0$ where g_{h_0} is a gauge transformation as in Remark 4.33. Then $g(A) = A_0 + d^*b + (h - h_0)$. Let $a = g(A) - A_0 = d^*b + (h - h_0)$. Note that $(d^* + d^+)a = (d^* + d^+)d^*b = F_A^+ - F_{A_0}^+$. By Lemma 4.18,

$$\|d^*b\|_{L^2_k} \leq C\|(d^* + d^+)d^*b\|_{L^2_{k-1}} \leq C\|F_A^+\|_{L^2_{k-1}} + C\|F_{A_0}^+\|_{L^2_{k-1}}.$$

Setting $K_2 = \|F_{A_0}^+\|_{L^2_{k-1}}$, we obtain

$$\|a\|_{L^2_k} \leq C\|F_A^+\|_{L^2_{k-1}} + CK_2 + K_1.$$

This finishes the proof. □

Proposition 4.32 means that we can project the whole of C onto $\operatorname{Ker} d^* \times \Gamma(S^+)$ by the $\mathcal{G}$-action. We will see that $\operatorname{Ker} d^* \times \Gamma(S^+)$ can be assumed as a global slice for the $\mathcal{G}$-action as follows.

First, we notice that $\operatorname{Ker} d^* \times \Gamma(S^+)$ has a symmetry larger than $\Gamma_{(A_0,0)} = \mathrm{U}(1)$. Note that

$$\operatorname{Im}(d^* \colon \Omega^1(X) \to \Omega^0(X)) \cong \left\{ f \in \Omega^0(X) \;\middle|\; \int_X f \,\mathrm{vol} = 0 \right\},$$

$$\Omega^0(X) \cong \mathbb{R} \oplus \operatorname{Im} d^*,$$

$$\exp(2\pi\sqrt{-1}\Omega^0(X))/\exp(2\pi\sqrt{-1}\operatorname{Im} d^*) \cong \mathrm{U}(1).$$

We have an exact sequence

$$1 \longrightarrow \operatorname{Im} d^* \xrightarrow{\exp(2\pi\sqrt{-1}\cdot)} \mathcal{G} \longrightarrow \mathrm{U}(1) \times H^1(X;\mathbb{Z}) \longrightarrow 1. \tag{4.11}$$

Define $\delta \colon \mathcal{G} \to \Omega^1(X)$ by $\delta(g) = \frac{-2}{\sqrt{-1}} g^{-1} dg$. Then the following diagram is commutative:

$$\begin{array}{ccc} \operatorname{Im} d^* & \xrightarrow{\exp} & \mathcal{G} \\ & \nwarrow^{d^*} & \downarrow \delta \\ & & \Omega^1(X) \end{array} \tag{4.12}$$

Then $d^* \circ \delta \colon \mathcal{G} \to \operatorname{Im} d^*$ is surjective and gives a splitting of the exact sequence (4.11) and

$$\operatorname{Ker}(d^* \circ \delta) \cong \mathrm{U}(1) \times H^1(X;\mathbb{Z}). \tag{4.13}$$

Since the $\mathcal{G}$-action on $\Omega^1(X)$ is given via δ, $\operatorname{Ker}(d^* \circ \delta)$ naturally acts on $\operatorname{Ker} d^*$.

Remark 4.33

(1) A way of realizing $H^1(X;\mathbb{Z})$ as a subgroup of $\mathcal{G}$ is as follows. For a cohomology class $h \in H^1(X;\mathbb{Z})$, assume h as a harmonic 1-form with integral periods. Let $\pi \colon \tilde{X} \to X$ be the universal covering of X. Choose a base point $x_0 \in \tilde{X}$. The integrals $\int_\gamma \pi^* h$ over paths γ from x_0 to $x \in X$ define a well-defined map $\varphi \colon \tilde{X} \to \mathbb{R}$. Since the periods are integral, $\exp(2\pi\sqrt{-1}\varphi) \colon \tilde{X} \to \mathrm{U}(1)$ descends to a map $g_h \colon X \to \mathrm{U}(1)$. It is easy to see that $\delta g_h = h$.

(2) Note that $\operatorname{Im} \delta \cong \operatorname{Im} d \oplus H^1(X;\mathbb{Z})$.

Theorem 4.34 (Global Slice Theorem) *Let* $K = \operatorname{Ker}(d^* \circ \delta)$. *For a fixed* $\mathrm{U}(1)$*-connection* A_0 *on* L, *the following map is a diffeomorphism:*

$$\mathcal{G}_{k+1} \times_K \left(\operatorname{Ker} d^* \times L^2_k(X; S^+)\right) \to C_k, \quad [g, (a, \psi)] \mapsto g \cdot ((A_0, 0) + (a, \psi)),$$

where $\operatorname{Ker} d^*$ *is the kernel of* $d^* : L^2_k\left(X; \sqrt{-1}\Lambda^1\right) \to L^2_{k-1}\left(X; \sqrt{-1}\Lambda^0\right)$.

Proof By Proposition 4.32, for a $\mathrm{U}(1)$-connection A, there is a unique element $g = \exp(2\sqrt{-1}f) \in \mathcal{G}_{k+1}$ such that $gA = A_0 + a$ with $d^*a = 0$, where $f \in L^2_{k+1}\left(X; \sqrt{-1}\Lambda^0\right)$ with $\int_X f \operatorname{vol}_X = 0$. The correspondence $(A, \psi) \mapsto [g^{-1}, (a, g\psi)]$ gives the inverse map. □

4.3.3 *Topology of* $\mathcal{B}^*_k$

Since $C^*_k = \mathcal{A}_k \times \left(L^2_k(\Gamma(S^+)) \setminus \{0\}\right)$ is homotopic to an infinite-dimensional sphere which is contractible, and $\mathcal{G}_{k+1}$ acts on C^*_k freely, the space $\mathcal{B}^*_k = C^*_k/\mathcal{G}_{k+1}$ is a classifying space of the group $\mathcal{G}_{k+1}$.

Theorem 4.35 *The space* $\mathcal{B}^*_k$ *is an Eilenberg–MacLane space* $K(\mathrm{U}(1) \times H^1(X; \mathbb{Z}))$. *In particular,*

$$\mathcal{B}^*_k \simeq \mathbb{C}P^\infty \times T^{b_1},$$

where b_1 *is the 1st Betti number of* X *and* T^{b_1} *is a* b_1*-dimensional torus.*

Proof By the global slice theorem (Theorem 4.34), there is a natural identification

$$\mathcal{B}^* \cong \operatorname{Ker} d^* \times \left(L^2_k(X; S^+) \setminus \{0\}\right) / (\mathrm{U}(1) \times H^1(X; \mathbb{Z})).$$

Since $\operatorname{Ker} d^* \times \left(L^2_k(X; S^+) \setminus \{0\}\right)$ is contractible and the action of $\mathrm{U}(1) \times H^1(X; \mathbb{Z})$ on it is free, the theorem follows. □

We introduce the universal line bundle $\mathbb{L} \to X \times \mathcal{B}^*$ by

$$\mathbb{L} = (X \times C^* \times \mathbb{C})/\mathcal{G},$$

where the action of $g \in \mathcal{G}$ on $(x, (A, \Phi), z) \in X \times C^* \times \mathbb{C}$ is given by

$$g(x, (A, \Phi), z) = (x, g(A, \Phi), g^{-1}(x)z).$$

Definition 4.36 The *μ-map*

$$\mu\colon H_j(X; \mathbb{Z}) \to H^{2-j}(\mathcal{B}^*\mathbb{Z}), \quad j = 0, 1$$

is given by

$$\mu(\alpha) = c_1(\mathbb{L})/\alpha, \quad \alpha \in H_j(X; \mathbb{Z}),$$

where / denotes the slant product $H^j(\mathcal{B}^* \times X) \times H_i(X) \to H^{j-i}(\mathcal{B}^*)$.

Remark 4.37 Note that the slant product $c_1(\mathbb{L})/\alpha$ is characterized by the property that

$$\langle c_1(\mathbb{L})/\alpha, \beta\rangle = \langle c_1(\mathbb{L}), \alpha \times \beta\rangle,$$

for any homology class $\beta \in H_*(\mathcal{B}^*; \mathbb{Z})$, where $\alpha \times \beta$ is the cross product of α and β.

Choose a point $x_0 \in X$. Let $[x_0] \in H_0(X; \mathbb{Z})$ be its homology class. Then $\mu([x_0])$ is given as follows. Restricting $\mathbb{L}$ to $\{x_0\} \times \mathcal{B}^*$, we obtain a line bundle $\mathbb{L}_{x_0}$ over $\mathcal{B}^*$. Then we have

$$u := c_1(\mathbb{L}_{x_0}) = c_1(\mathbb{L})/[x_0] = \mu([x_0]).$$

Remark 4.38 Let γ be the tautological line bundle over $\mathbb{C}P^\infty$. Then the generator u corresponds to $-c_1(\gamma)$ via the natural identification $H^*(\mathcal{B}^*) \cong H^*(\mathbb{C}P^\infty \times T^{b_1})$.

For $\gamma \in H_1(X; \mathbb{Z})$, $\mu(\gamma)$ is given as follows. Let $\tilde{\gamma}\colon S^1 \to X$ be a loop in X representing γ. For a U(1) connection A, the holonomy $\mathrm{Hol}_{\tilde{\gamma}}(A)$ of A around $\tilde{\gamma}$ is invariant under gauge transformation. Then the correspondence $A \mapsto \mathrm{Hol}_{\tilde{\gamma}}(A)$ induces a map

$$\mathrm{Hol}_{\tilde{\gamma}}\colon \mathcal{B}^* \to \mathrm{U}(1).$$

Proposition 4.39 *Let η be the generator of $H^1(\mathrm{U}(1); \mathbb{Z})$. Then*

$$\mu(\gamma) = (\mathrm{Hol}_{\tilde{\gamma}})^*(\eta).$$

Proof It suffices to show that, for every $\beta \in H_1(\mathcal{B}^*; \mathbb{Z})$,

$$\langle c_1(\mathbb{L}), \gamma \times \beta\rangle = \deg(\mathrm{Hol}_{\tilde{\gamma}} \circ \tilde{\beta}),$$

where $\tilde{\beta}$ is a loop in $\mathcal{B}^*$ representing β. Let π^*L be the pull-back of the determinant line bundle L by the projection $X \times \mathcal{B}^* \to X$. Then $\mathbb{L} \otimes \pi^*L$ has a tautological connection $\mathbb{A}$ along X. The pull-back $(\tilde{\gamma} \times \tilde{\beta})^*(\mathbb{L} \otimes \pi^*L)$ over $S^1 \times S^1$ has a connection $\mathbb{A}'$ induced from the connection $\mathbb{A}$. Then

$$\mathrm{Hol}_{S^1 \times \{t\}}(\mathbb{A}') = \mathrm{Hol}_{\tilde{\gamma}}(\tilde{\beta}(t)). \tag{4.14}$$

Note that $(\tilde{\gamma} \times \tilde{\beta})^*(\mathbb{L} \otimes \pi^*L) \cong (\tilde{\gamma} \times \tilde{\beta})^*\mathbb{L}$ because $\tilde{\gamma}^*L$ is trivial. Fix such an isomorphism. Then the line bundle $(\tilde{\gamma} \times \tilde{\beta})^*\mathbb{L}$ also inherits a connection, also denoted by $\mathbb{A}'$, from $\mathbb{A}$ and the relation (4.14) also holds. The Chern number $\langle c_1((\tilde{\gamma} \times \tilde{\beta})^*\mathbb{L}), [S^1 \times S^1]\rangle$ is given by the degree of the map

$$S^1 \ni t \mapsto \mathrm{Hol}_{S^1 \times \{t\}}(\mathbb{A}') \in \mathrm{U}(1).$$

(See Problem 4.112.) By (4.14), this degree is equal to $\deg(\mathrm{Hol}_{\tilde{\gamma}} \circ \tilde{\beta})$. Thus we obtain

$$\langle c_1(\mathbb{L}), \gamma \times \beta\rangle = \langle c_1((\tilde{\gamma} \times \tilde{\beta})^*\mathbb{L}), [S^1 \times S^1]\rangle = \deg(\mathrm{Hol}_{\tilde{\gamma}} \circ \tilde{\beta}).$$

This completes the proof. □

Corollary 4.40 *The map*

$$\mathbb{Z}[u] \otimes \Lambda^*(H_1(X;\mathbb{Z})/\mathrm{Tor}) \to H^*(\mathcal{B}^*;\mathbb{Z})$$
$$p(u) \otimes (\gamma_1 \wedge \cdots \wedge \gamma_k) \mapsto p(u) \otimes (\mu(\gamma_1) \wedge \cdots \wedge \mu(\gamma_k))$$

is an isomorphism.

The proof of Corollary 4.40 is left to the readers (Problem 4.114).

4.4 Compactness of the Moduli Spaces

One of the most salient features of the Seiberg–Witten theory is the compactness of the moduli spaces $\mathcal{M}$. The purpose of this section is to prove this property of $\mathcal{M}$.

Proposition 4.41 *Let X be a closed Riemannian 4-manifold and $\mathfrak{s}$ a spin^c structure on X. There exists a constant M which depends only on X and the Riemannian metric such that, for every solution (A, Φ) to the Seiberg–Witten equations (4.2), the following estimate holds:*

$$\|\Phi\|_{C^0} \leq M.$$

Proof We basically follow the proof by Jost et al. [82]. Choose a smooth function $\rho_1 \colon \mathbb{R} \to \mathbb{R}$ such that for every t, the uniform bounds $0 \leq \rho_1'(t) \leq 2$ hold on the differential of ρ_1. Moreover, we suppose

$$\rho_1(t) \begin{cases} = t, & (t \geq 3) \\ > 0, & (t > 1) \\ = 0. & (t \leq 1) \end{cases}$$

For $M > 0$, let $\rho_M(t) = M\rho_1\left(\dfrac{t}{M}\right)$. For $\phi \in \Gamma(S^+)$, let

$$\rho_M[\phi] = \begin{cases} \rho_M(|\phi|)e, & (\phi \neq 0) \\ 0. & (\phi = 0) \end{cases}$$

where $e = \dfrac{\phi}{|\phi|}$. For a Seiberg–Witten solution (A, Φ), the Bochner–Weitzenböck formula (Lemma 3.80) implies

$$0 = D_A \circ D_A \Phi = \nabla_A^* \circ \nabla_A \Phi + \frac{\kappa}{4}\Phi + \frac{q(\Phi)}{2}\Phi. \tag{4.15}$$

Taking the inner product with $\rho_M[\Phi]$, we obtain

$$0 = \int_X \langle \nabla_A \rho_M[\Phi], \nabla_A \Phi \rangle \,\mathrm{vol} + \int_X \frac{\rho_M(|\Phi|)}{|\Phi|}\left(\left\langle \Phi, \frac{q(\Phi)}{2}\Phi \right\rangle + \frac{\kappa}{4}|\Phi|^2\right) \mathrm{vol}. \tag{4.16}$$

Then, we have the equalities

$$\nabla_A \rho_M[\Phi] = \nabla_A (\rho_M(|\Phi|)e) = \rho_M'(|\Phi|)e \otimes d|\Phi| + \rho_M(|\Phi|)\nabla_A e,$$
$$\nabla_A \Phi = \nabla(|\Phi| e) = e \otimes d|\Phi| + |\Phi|\nabla_A e.$$

Note that, since $|e| = 1$,

$$0 = d|e|^2 = 2\langle \nabla_A e, e \rangle.$$

From these, we obtain the nonnegativity of the inner product

$$\langle \nabla_A \rho_M[\Phi], \nabla_A \Phi \rangle = \rho_M'(|\Phi|)\, |d|\Phi||^2 + \rho_M(|\Phi|)|\nabla_A e|^2 \geq 0. \tag{4.17}$$

The Eq. (4.16) and the inequality (4.17) imply the estimate

$$\int_X \frac{\rho_M(|\Phi|)}{|\Phi|} \left(\left\langle \Phi, \frac{q(\Phi)}{2} \Phi \right\rangle + \frac{\kappa}{4} |\Phi|^2 \right) \mathrm{vol} \leq 0. \tag{4.18}$$

Note that

$$\left\langle \Phi, \frac{q(\Phi)}{2} \Phi \right\rangle + \frac{\kappa}{4} |\Phi|^2 = \frac{1}{4} (|\Phi|^4 + \kappa |\Phi|^2).$$

Let $m = \min_{x \in X} \kappa(x)$. Choose $C_0, C_1 > 0$ such that, for every $t \in \mathbb{R}$,

$$t^4 + mt^2 \geq C_0 t^2 - C_1.$$

For ε such that $0 < \varepsilon < C_0$, let $M = \sqrt{\dfrac{C_1}{C_0 - \varepsilon}}$. Then $C_0 t^2 - C_1 \geq \varepsilon t^2$ if $t \geq M$. Then we have

$$\frac{\rho_M(|\Phi|)}{|\Phi|} \left(\left\langle \Phi, \frac{q(\Phi)}{2} \Phi \right\rangle + \frac{\kappa}{4} |\Phi|^2 \right) \begin{cases} = 0 & \text{if } |\Phi| \leq M, \\ \geq \dfrac{\rho_M(|\Phi|)}{|\Phi|} \cdot \varepsilon |\Phi|^2 & \text{if } |\Phi| > M. \end{cases}$$

Combining with (4.18), we obtain $|\Phi| \leq M$. □

Remark 4.42 We can also obtain the C^0 bound on Φ by using the maximum principle for the Laplace–Beltrami operator [99, 131, 165].

Proposition 4.43 *Let X be a closed Riemannian 4-manifold and $\mathfrak{s}$ a spin^c structure on X. There exist constants M, M' and M'' which depend on $(X, \mathfrak{s})$ such that, for every L^1_{loc} solution (A, Φ) to the Seiberg–Witten equations (4.2), the following estimates hold:*

$$\|\Phi\|_{L^\infty} \leq M, \quad \|F_A^+\|_{L^\infty} \leq \frac{M^2}{2}, \quad \|\nabla_A \Phi\|_{L^2} \leq M', \quad \|F_A^+\|_{L^2_1} \leq M''.$$

Proof The L^∞ bound on Φ is an immediate consequence of Proposition 4.41. By the equality $F_A^+ = q(\Phi)$, we have

$$|F_A^+(x)| \leq |\Phi(x)|^2/2,$$

and hence the L^∞ bound on F_A^+ is obtained. The L^2 bound on $\nabla_A\Phi$ is obtained from (4.21) and the L^∞-bound on Φ. The L^2 bound on F_A^+ is obtained from the L^∞ bound on F_A^+. Taking the covariant derivative of the equation $F_A^+ = q(\Phi)$, we obtain

$$\nabla_{L.C.}F_A^+ = \nabla_A\Phi \otimes \Phi^* + \Phi \otimes \nabla_A\Phi^* - \mathrm{Re}\langle\nabla_A\Phi, \Phi\rangle\mathrm{id},$$

where $\nabla_{L.C.}$ is the Levi-Civita connection. Then the L^2 bound on $\nabla_{L.C.}F_A^+$ is obtained from the L^∞ bound on Φ and the L^2 bound on $\nabla_A\Phi$. Therefore we obtain the L_1^2 bound on F_A^+. □

Theorem 4.44 *Let X be a closed Riemannian 4-manifold with a spin^c structure $\mathfrak{s}$. For $k \geq 2$, every L_k^2 solution (A, Φ) to the Seiberg–Witten equations* (4.2) *is gauge equivalent to a solution which is in L_ℓ^2 for arbitrary $\ell \geq k$ by an L_{k+1}^2-gauge transformation. In particular, (A, Φ) is gauge equivalent to a C^∞ solution.*

Proof It is enough to prove the case when $k = 2$. For an L_2^2 solution (A, Φ), we may assume by Proposition 4.32 that $A = A_0 + a$ with a satisfying (4.10) for $k = 2$ after gauge transformation. We prove the following claim.

Claim There are constants C_k and D_k for each k such that, for every solution $(A_0 + a, \Phi)$ as above, we have the estimates

$$\|a\|_{L_k^2} \leq C_k, \quad \|\Phi\|_{L_k^2} \leq D_k. \tag{4.19}$$

Since $\|F_A^+\|_{L_1^2} \leq M''$ by Proposition 4.43, the estimate (4.10) implies that there is a constant C_1 such that $\|a\|_{L_1^2} \leq C_1$. Furthermore, the elliptic estimate (Proposition 4.15) implies that

$$\begin{aligned}\|a\|_{L_2^2} &\leq \mathrm{const.}(\|a\|_{L_1^2} + \|(d^* + d^+)a\|_{L_1^2})\\ &\leq \mathrm{const.}(\|a\|_{L_1^2} + \|F_A^+\|_{L_1^2} + 1) \leq \mathrm{const.}(C_1 + M'' + 1).\end{aligned}$$

Hence there is a constant C_2 such that $\|a\|_{L_2^2} \leq C_2$. Note that

$$\|\Phi\|_{L^2}^2 \leq \|\Phi\|_{L^\infty}^2 \mathrm{Vol}(X) \leq M^2 \mathrm{Vol}(X),$$
$$\|\nabla_{A_0}\Phi\|_{L^2}^2 \leq \|\nabla_{A-a}\Phi\|_{L^2}^2 \leq \|\nabla_A\Phi\|_{L^2}^2 + \|a\Phi\|_{L^2}^2$$
$$\leq \|\nabla_A\Phi\|_{L^2}^2 + \|a\|_{L^2}^2\|\Phi\|_{L^\infty}^2 \leq (M')^2 + C_1^2 M^2.$$

Thus there is a constant D_1 such that $\|\Phi\|_{L_1^2} \leq D_1$. The Seiberg–Witten equations are rewritten as

$$D_{A_0}\Phi = -\frac{\rho(a)}{2}\Phi,$$
$$d^+a = q(\Phi) - F_{A_0}^+,$$
$$d^*a = 0.$$

By Lemmas 4.10 and 4.12,

$$\|D_{A_0}\Phi\|_{L_{2/3}^2} = \frac{1}{2}\|\rho(a)\Phi\|_{L_{2/3}^2} \leq \text{const.}\|a\|_{L_2^2}\|\Phi\|_{L_1^2} \leq \text{const.}C_2 D_1.$$

By the elliptic estimate (Proposition 4.15), we find a constant $D_{5/3}$ such that $\|\Phi\|_{L_{5/3}^2} \leq D_{5/3}$. Similarly, we obtain

$$\|D_{A_0}\Phi\|_{L_{4/3}^2} = \frac{1}{2}\|\rho(a)\Phi\|_{L_{4/3}^2} \leq \text{const.}\|a\|_{L_2^2}\|\Phi\|_{L_{5/3}^2} \leq \text{const.}C_2 D_{5/3}.$$

We also find a constant $D_{7/3}$ such that $\|\Phi\|_{L_{7/3}^2} \leq D_{7/3}$. Then,

$$\|(d^*+d^+)a\|_{L_2^2} \leq \text{const.}(\|q(\Phi)\|_{L_2^2}+1) \leq \text{const.}(\|\Phi\|_{L_{7/3}^2}^2+1) \leq \text{const.}(D_{7/3}+1).$$

Therefore there is a constant C_3 such that $\|a\|_{L_3^2} \leq C_3$. Then,

$$\|D_{A_0}\Phi\|_{L_2^2} = \frac{1}{2}\|\rho(a)\Phi\|_{L_2^2} \leq \text{const.}\|a\|_{L_3^2}\|\Phi\|_{L_{7/3}^2} \leq \text{const.}C_3 D_{7/3}.$$

Therefore there is a constant D_3 such that $\|\Phi\|_{L_3^2} \leq D_3$. Since L_k^2 is an algebra if $k - \frac{4}{2} > 0$, then the multiplication of two sections in L_k^2 is also in L_k^2. If a and Φ are in L_k^2 with $k > 2$, then the following estimates hold:

$$\|(d^* + d^+)a\|_{L_k^2} \leq \text{const. } (\|q(\Phi)\|_{L_k^2} + 1) \leq \text{const. } (\|\Phi\|_{L_k^2}^2 + 1)$$
$$\leq \text{const. } (D_k + 1),$$
$$\|D_{A_0}\Phi\|_{L_k^2} = \frac{1}{2}\|\rho(a)\Phi\|_{L_k^2} \leq \text{const. } \|a\|_{L_k^2}\|\Phi\|_{L_k^2} \leq \text{const. } C_k D_k.$$

Therefore we find constants C_{k+1} and D_{k+1} such that $\|a\|_{L^2_{k+1}} \le C_{k+1}$ and $\|\Phi\|_{L^2_{k+1}} \le D_{k+1}$. Thus the claim is proved.

Now a and Φ are in L^2_k for arbitrary large k, and therefore in C^m for all m, that is, in C^∞. □

Let $\mathcal{M}(k) = \{L^2_k \text{ solutions to (4.2) }\}/\mathcal{G}_{k+1}$ for $k \ge 2$.

Theorem 4.45 *The moduli space $\mathcal{M}(k)$ ($k \ge 2$) is compact.*

Proof Let $\{(A_i, \Phi_i)\}$ be a sequence of L^2_k solutions to the Seiberg–Witten equations. For each i, we may assume $A_i = A_0 + a_i$ by a L^2_{k+1} gauge transformation with a_i satisfying (4.10). By the claim in the proof of Theorem 4.44, we have the uniform bounds (4.19) for each $\ell \ge k+1$. By Lemma 4.10, we find a subsequence of $\{(A_i, \Phi_i)\}$ convergent in L^2_k. □

Proposition 4.46 *For $l \ge k \ge 2$, the natural inclusion of the set of L^2_l-solutions into the set of L^2_k-solutions induces a homeomorphism $\mathcal{M}(l) \cong \mathcal{M}(k)$.*

Proof The bijectivity of the map $\mathcal{M}(l) \to \mathcal{M}(k)$ is proved by using Theorem 4.44. The fact that this is a homeomorphism is left as an exercise (Problem 4.115). □

By this proposition, we can safely omit the Sobolev notation from $\mathcal{M}$.

Remark 4.47 The moduli space $\mathcal{M}$ can be considered as a quotient space by $\mathcal{G}$-action of the zero set of a certain map between infinite-dimensional spaces. Recall $\mathcal{B}_k = \mathcal{C}_k/\mathcal{G}_{k+1}$ in Proposition 4.23. Consider the $\mathcal{G}_{k+1}$-equivariant non-linear elliptic map:

$$\mathcal{F}\colon \mathcal{C}_k \to L^2_{k-1}(X; S^- \oplus \sqrt{-1}\Lambda^+) \tag{4.20}$$

given by

$$(A, \Phi) \to (D_A(\Phi), F_A^+ - q(\Phi)).$$

See Definition 4.2. Then the moduli space $\mathcal{M}$ is

$$\mathcal{M} = \mathcal{F}^{-1}(0)/\,\mathcal{G}_{k+1}.$$

In Chap. 5, we will introduce the monopole map which is a more sophisticated version of the map $\mathcal{F}$.

4.4.1 Scalar Curvature

For 4-manifolds with positive or nonnegative scalar curvature metrics, several facts are obtained by the Bochner technique.

Proposition 4.48 *Let X be a closed Riemannian 4-manifold with positive scalar curvature metric. The Seiberg–Witten equations* (4.2) *have no irreducible solutions.*

Proof Suppose (A, Φ) is a solution to the Seiberg–Witten equations (4.2). By using the Bochner–Weitzenböck formula (Lemma 3.80), we obtain an estimate

$$0 \geq \int_X \left(\left\langle \Phi, \frac{q(\Phi)}{2}\Phi \right\rangle + \frac{\kappa}{4}|\Phi|^2 \right) \mathrm{vol} \geq \int_X \frac{1}{4}(|\Phi|^4 + \kappa|\Phi|^2)\,\mathrm{vol},$$

and therefore $\Phi \equiv 0$. So the solution is reducible by Definition 4.26. □

Let $c_1(\mathfrak{s})_{\mathbb{R}}$ be the image of $c_1(\mathfrak{s})$ via the homomorphism $H^2(X;\mathbb{Z}) \to H^2(X;\mathbb{R})$. Identifying $H^2(X;\mathbb{R})$ with the space of harmonic 2-forms, we decompose $H^2(X;\mathbb{R})$ into the (± 1)-eigenspaces $H^{\pm}(X;\mathbb{R})$ of the Hodge star operator. Then $c_1(\mathfrak{s})_{\mathbb{R}}$ is also decomposed into the sum of positive and negative parts:

$$c_1(\mathfrak{s})_{\mathbb{R}} = c_1(\mathfrak{s})_{\mathbb{R}}^{+} \oplus c_1(\mathfrak{s})_{\mathbb{R}}^{-} \in H^+(X;\mathbb{R}) \oplus H^-(X;\mathbb{R}).$$

Proposition 4.49 *Let X be a closed Riemannian 4-manifold with nonnegative scalar curvature κ. For a spin^c structure $\mathfrak{s}$ on X, the Seiberg–Witten equations (4.2) admit a solution if and only if $c_1(\mathfrak{s})_{\mathbb{R}}^{+} = 0$. If a solution to (4.2) exists, then $\Phi \equiv 0$ for every solution (A, Φ) and the moduli space is diffeomorphic to the $b_1(X)$-dimensional torus.*

Proof Suppose there is a solution (A, Φ) to (4.2). Taking the inner product of Φ and (4.15), we obtain

$$0 = \int_X |\nabla_A \Phi|^2\,\mathrm{vol} + \int_X \left(\left\langle \Phi, \frac{q(\Phi)}{2}\Phi \right\rangle + \frac{\kappa}{4}|\Phi|^2 \right) \mathrm{vol}. \tag{4.21}$$

Since the first term of the right-hand side is nonnegative, we have

$$0 \geq \int_X \left(\left\langle \Phi, \frac{q(\Phi)}{2}\Phi \right\rangle + \frac{\kappa}{4}|\Phi|^2 \right) \mathrm{vol} = \int_X \frac{1}{4}(|\Phi|^4 + \kappa|\Phi|^2)\, \mathrm{vol},$$

and therefore $\Phi \equiv 0$. Then $F_A^+ = 0$ by (4.2). Hence $c_1(\mathfrak{s})_{\mathbb{R}}^+ = 0$. In this case, the moduli space is identified with

$$\{a \in \sqrt{-1}\Omega^1(X) \mid d^+a = 0\}/\{-2g^{-1}dg \mid g \in \mathcal{G}\}.$$

For $a \in \Omega^1(X)$,

$$\int_X (|d^+a|^2 - |d^-a|^2)\, \mathrm{vol} = \int_X da \wedge da = \int_X d(a \wedge da) = 0$$

by Stokes' theorem. Therefore $\operatorname{Ker} d^+ = \operatorname{Ker} d = \operatorname{Im} d \oplus H^1(X; \mathbb{R})$. On the other hand, $\{-2g^{-1}dg \mid g \in \mathcal{G}\} \cong \operatorname{Im} \delta \cong \operatorname{Im} d \oplus H^1(X; \mathbb{Z})$ by Remark 4.33(2). Thus the moduli space is isomorphic to $H^1(X; \mathbb{R})/H^1(X; \mathbb{Z})$.

Suppose $c_1(\mathfrak{s})_{\mathbb{R}}^+ = 0$. Choose a U(1)-connection A_0 on the determinant line bundle. Since the harmonic part of $F_{A_0}^+$ is zero, we can find a 1-form a such that $F_{A_0}^+ + d^+a = 0$. Thus $(A_0 + a, 0)$ is a solution to (4.2). □

4.5 Transversality

As we will see later, the Seiberg–Witten invariants are defined by using the fundamental classes of the moduli spaces. In order to ensure that the moduli spaces have fundamental classes, we need in general to perturb the Seiberg–Witten equations.

A simple method of perturbation is introduced by E. Witten [169] as follows:

$$\begin{cases} D_A\Phi = 0, \\ F_A^+ = q(\Phi) + \eta, \end{cases} \tag{4.22}$$

where $\eta \in \mathcal{P}_{k-1} = L_{k-1}^2\left(X; \sqrt{-1}\Lambda^+\right)$.

Let

$$\mathcal{M}_\eta = \mathcal{M}(\mathfrak{s}, \eta) = \{\text{solutions to (4.22)}\}/\mathcal{G},$$
$$\mathcal{M}^*_\eta = \mathcal{M}^*(\mathfrak{s}, \eta) = \{\text{irreducible solutions to (4.22)}\}/\mathcal{G}.$$

Proposition 4.50 *For any Riemannian metric g and any g-self-dual 2-form η, $\mathcal{M}_\eta$ is compact.*

We can easily prove Proposition 4.50 by modifying the proof of Theorem 4.45. In particular, the following holds.

Proposition 4.51 *There exists a constant M which depends only on X, the Riemannian metric and $\|\eta\|_{C^0}$ such that, for every solution (A, Φ) to the perturbed Seiberg–Witten equations (4.22), the following estimate holds:*

$$\|\Phi\|_{C^0} \leq M.$$

We will prove the following properties:

- $\mathcal{M}^*_\eta$ is a smooth manifold for a generic η.
- There is a codimension-b^+ subspace $\mathcal{W} \subset \mathcal{P}_{k-1}$ such that $\mathcal{M}_\eta$ has no reducible solution if $\eta \notin \mathcal{W}$.

Remark 4.52 The word "generic" means that η is in the set of regular values of a certain smooth Fredholm map which is of the second category of Baire. See Theorem 4.55 (a Fredholm version of Sard's theorem due to Smale).

As a consequence, $\mathcal{M}_\eta$ is a compact manifold for a generic choice of η outside $\mathcal{W}$ if $b_2^+ \geq 1$.

We define a map

$$\tilde{m}\colon \mathcal{C}_k \times \mathcal{P}_{k-1} \to L^2_{k-1}\left(X; \sqrt{-1}\Lambda^+ \oplus S^-\right) \tag{4.23}$$

by

$$\tilde{m}(A, \phi, \eta) = (F_A^+ - q(\phi) - \eta, D_A\phi),$$

where $\mathcal{P}_{k-1} = L^2_{k-1}\left(X; \sqrt{-1}\Lambda^+\right)$.

Lemma 4.53 *At every point* (A, Φ, η) *such that* $\tilde{m}(A, \Phi, \eta) = 0$ *and* $\Phi \neq 0$*, the differential*

$$D\tilde{m}\colon L_k^2\left(X; \sqrt{-1}\Lambda^1 \oplus S^+\right) \oplus L_{k-1}^2\left(X; \sqrt{-1}\Lambda^+\right) \to L_{k-1}^2\left(X; \sqrt{-1}\Lambda^+ \oplus S^-\right)$$

is surjective.

Proof The differential $D\tilde{m}$ of $\tilde{m}$ at (A, Φ, η) is given by

$$D\tilde{m}(a, \phi, h) = \begin{pmatrix} d^+a - (Dq)(\phi) - h \\ D_A\phi + \frac{1}{2}a \cdot \Phi \end{pmatrix},$$

where $(a, \phi, h) \in L_k^2\left(X; \sqrt{-1}\Lambda^1 \oplus S^+\right) \oplus L_{k-1}^2\left(X; \sqrt{-1}\Lambda^+\right)$ and Dq is the differential of $q(\Phi)$. Since the restriction of $D\tilde{m}$ to $L_{k-1}^2\left(X; \sqrt{-1}\Lambda^+\right)$ is minus the inclusion, $D\tilde{m}$ is onto the first factor. Hence it suffices to prove that $D\tilde{m}$ is onto to the second factor in the range. For this purpose, we will show the following map is surjective:

$$F\colon (a, \phi) \mapsto D_A\phi + \frac{1}{2}a \cdot \Phi.$$

Suppose that $\psi \in L_{k-1}^2(X; S^-)$ is L^2-orthogonal to the image of F. Then, for any ϕ,

$$0 = \langle F(0, \phi), \psi\rangle_{L^2} = \langle D_A\phi, \psi\rangle_{L^2} = \langle \phi, D_A\psi\rangle_{L^2}$$

since D_A is self-adjoint. Therefore $D_A\psi = 0$. Suppose $\psi \neq 0$. By the unique continuation theorem (Theorem 4.96), ψ does not vanish on any open subset of X. By the assumption $\Phi \neq 0$, Φ also does not vanish on any open subset of X. Choose $x_0 \in X$ such that $\Phi(x_0) \neq 0$ and $\psi(x_0) \neq 0$. Then we can choose a small open neighborhood U of x_0 and a L_k^2-one-form a on U such that $\psi(x), \Phi(x) \neq 0$ and

$$\mathrm{Re}\langle a \cdot \Phi, \psi\rangle > 0 \tag{4.24}$$

for every $x \in U$ as follows. Recall that, for $x \in X$, the Clifford multiplication is identified with the multiplication of quaternion $c_+(iv, \phi) = \bar{v}\phi i$ for $v \in T_xX \cong V$ and $\phi \in S_x^+ \cong \mathbb{H}^+$. Then we can easily find a satisfying (4.24) at x_0. By continuity, the inequality (4.24) holds near x_0.

Choose a cut-off function ρ on X supported in U and extend ρa over X. Then

$$\langle F(2\rho a, 0), \psi\rangle_{L^2} = \int_X \mathrm{Re}\langle \rho a \cdot \Phi, \psi\rangle\, \mathrm{vol} > 0.$$

This is a contradiction. Therefore the orthogonal complement of the image of $D\tilde{m}$ is trivial. □

Corollary 4.54 *The space* $\mathbb{M}^* = (\tilde{m}^{-1}(0)/\mathcal{G}) \cap (\mathcal{B}_k^* \times \mathcal{P}_{k-1})$ *is a smooth submanifold of* $\mathcal{B}_k^* \times \mathcal{P}_{k-1}$.

Proof By Lemma 4.53, $\tilde{m}^{-1}(0) \cap (\mathcal{C}_k^* \times \mathcal{P}_{k-1})$ is a smooth submanifold of $\mathcal{C}_k^* \times \mathcal{P}_{k-1}$. Since the $\mathcal{G}$-action is free and smooth on $\mathcal{C}_k^* \times \mathcal{P}_{k-1}$, $\mathbb{M}^*$ is a smooth submanifold of $\mathcal{B}_k^* \times \mathcal{P}_{k-1}$. □

We invoke the Sard–Smale theorem for smooth Fredholm maps.

Theorem 4.55 (Sard–Smale [156]) *Let* M, V *be Banach manifolds modeled on separable Banach spaces, and* $f : M \to V$ *be a* C^∞ *Fredholm map. Then the set of regular values of* f *is of the second category in the sense of Baire (a countable intersection of open dense sets.)*

The projection $\varpi : \mathcal{B}_k^* \times \mathcal{P}_{k-1} \to \mathcal{P}_{k-1}$ induces a smooth map

$$\pi : \mathbb{M}^* \to \mathcal{P}_{k-1}. \tag{4.25}$$

Proposition 4.56 *For a generic* $\eta \in \mathcal{P}_{k-1}$, *the space* $\mathcal{M}^*(\mathfrak{s}, \eta) = \pi^{-1}(\eta)$ *is a finite-dimensional smooth submanifold of* $\mathcal{B}_k^*$ *whose dimension is given by*

$$\begin{aligned} d(\mathfrak{s}) &= \frac{1}{4}(c_1(\mathfrak{s})^2 - \sigma(X)) - (b_0(X) - b_1(X) + b_2^+(X)) \\ &= \frac{1}{4}(c_1(\mathfrak{s})^2 - (2\chi(X) + 3\sigma(X))), \end{aligned} \tag{4.26}$$

where $c_1(\mathfrak{s})$ *is the first Chern class of the determinant line bundle of* $\mathfrak{s}$, $\chi(X)$ *is the Euler characteristic and* $\sigma(X)$ *is the signature.*

Proof In order to apply the Sard–Smale theorem (Theorem 4.55), we show that π is a Fredholm map. Recall that the tangent space $T_w\mathcal{B}_k^*$ of $w = [(A, \Phi)]$ is identified with $\ker L^*_{(A,\Phi)}$ in (4.9). For a point $(w, \eta) \in \mathbb{M}(\mathfrak{s})$, the differential $(D\pi)_{(w,\eta)}$ of π at (w, η) is identified with

$$\begin{aligned} \ker(Dm \oplus L_w^*) &\to \mathcal{P}_{k-1}, \\ (a, \phi, h) &\mapsto h, \end{aligned}$$

where $(a, \phi, h) \in \ker(D\tilde{m} \oplus L_w^*) \subset L_k^2(X; \sqrt{-1}\Lambda^1 \oplus S^+) \oplus \mathcal{P}_{k-1}$. By Lemma 4.53, $D\tilde{m}$ is surjective if $\Phi \neq 0$. It is easy to see that $L^*_{(A,\Phi)}$ is also surjective if $\Phi \neq 0$. Therefore $Dm \oplus L_w^*$ is surjective if $\Phi \neq 0$. Then the kernel and cokernel of $D\pi$ is identified with those of $D\tilde{m} \oplus L_w^* \oplus D\varpi$ which is given by

$$(D\tilde{m} \oplus L_w^* \oplus D\varpi)(a, \phi, h) = \begin{pmatrix} -2d^*a + \sqrt{-1}\,\mathrm{Im}(\langle \phi, \Phi \rangle) \\ d^+a - (Dq)(\phi) - h \\ D_A\phi + \frac{1}{2}a \cdot \Phi \\ h \end{pmatrix}.$$

This is a compact perturbation of the elliptic operator $-2d^* + d^+ + D_A$, and therefore Fredholm.

By Theorem 4.55, $\mathcal{M}^*(\mathfrak{s}, \eta) = \pi^{-1}(\eta)$ is a smooth manifold for a generic $\eta \in \mathcal{P}_{k-1}$. The dimension is calculated as follows. Since the tangent space $T_w\mathcal{M}^*(\mathfrak{s}, \eta)$ is identified with $\ker(D\tilde{m} \oplus L_w^* \oplus D\varpi)$, we obtain

$$\dim \mathcal{M}^*(\mathfrak{s}, \eta) = \dim \ker(L_{(A,\Phi)}^* \oplus D\tilde{m} \oplus D\varpi) = \mathrm{ind}(L_{(A,\Phi)}^* \oplus D\tilde{m} \oplus D\varpi)$$

$$= \mathrm{ind}(d^* + d^+) + \mathrm{ind}\, D_A = -(b_0(X) - b_1(X) + b^+(X)) + \frac{1}{4}(c_1(\mathfrak{s})^2 - \sigma(X)).$$

Here $\mathrm{ind}\, D_A$ is calculated by the Atiyah–Singer index theorem (Theorem 3.78). □

For a Riemannian metric g on X, the g-self-dual harmonic part of a 2-form η is denoted by $h_g^+(\eta)$. Note that $h_g^+(F_A)$ of the curvature F_A of a connection A is independent of A, because by the Chern–Weil theory, the equality

$$c_1(\mathfrak{s}) = \frac{\sqrt{-1}}{2\pi}[F_A]$$

holds in $H^2(X; \mathbb{R})$ where $[F_A]$ is the de Rham class of F_A. We denote $\frac{\sqrt{-1}}{2\pi}h_g^+(F_A)$ by $c_1(\mathfrak{s})^{+_g}$.

Proposition 4.57 *Define a codimension-b_2^+ subspace $\mathcal{W} \subset \mathcal{P}_{k-1} = L_{k-1}^2\left(X; \sqrt{-1}\Lambda^+\right)$ by*

$$\mathcal{W} = \left\{ \eta \in \mathcal{P}_{k-1} \;\middle|\; c_1(\mathfrak{s})^{+_g} = \frac{\sqrt{-1}}{2\pi}h_g^+(\eta) \right\}.$$

The perturbed Seiberg–Witten equations (4.22) have a reducible solution if and only if $\eta \in \mathcal{W}$. If there is a reducible solution, then the space of gauge equivalence classes of reducible solutions is diffeomorphic to a $b_1(X)$-dimensional torus.

Proof Chose a reference connection A_0. Then $(A_0 + a, 0)$ is a reducible solution of the perturbed Seiberg–Witten equations (4.22) if and only if it satisfies $F_{A_0}^+ + d^+a =$

η. This equation has a solution if and only if $\eta - F^+_{A_0}$ is in the image of d^+, that is, $h^+_g(\eta - F_{A_0}) = 0$.

If $\eta - F^+_{A_0} \in \operatorname{Im} d^+$, then there is a unique L^2_k-one-form a_0 in the L^2-orthogonal complement of $\ker d^+$ such that $d^+ a_0 = \eta - F^+_{A_0}$. The space of reducible solutions is identified with $\{a + a_0 \mid d^+ a = 0\}$. Then the space of gauge equivalence classes of reducible solutions is

$$\begin{aligned}\{a + a_0 \mid d^+ a = 0\}/\mathcal{G} = a_0 + \ker d^+/\mathcal{G} &\cong \ker(d^+ \oplus d^*)/H^1(X;\mathbb{Z})\\ &= H^1(X;\mathbb{R})/H^1(X;\mathbb{Z}),\end{aligned}$$

which is a b_1-dimensional torus. □

Corollary 4.58 *If $b^+_2(X) = 0$, then the perturbed Seiberg–Witten equations (4.22) always have a reducible solution for any self-dual two-form η.*

Corollary 4.59 *If $b^+_2(X) \geq 1$, then we can choose a generic η such that $\mathcal{M}(\mathfrak{s}, \eta)$ contains no reducible and is a smooth $d(\mathfrak{s})$-dimensional compact submanifold of $\mathcal{B}^*_k$.*

Proposition 4.60 *Let X be a closed Riemannian 4-manifold with positive scalar curvature κ. If $\kappa \geq 2\|\eta\|_{C^0}$, the perturbed Seiberg–Witten equations (4.22) have no irreducible solutions.*

Proof The proof is an exercise (Problem 4.116). See the proof of Proposotion 4.48. □

4.6 Orientability of the Moduli Space

The goal of this section is to prove that the moduli spaces are orientable. Choose a generic η so that the irreducible part $\mathcal{M}^*(\mathfrak{s}, \eta)$ of the moduli space is a smooth manifold. For the purpose of discussing the orientability of $\mathcal{M}^*(\mathfrak{s}, \eta)$, we give a description of the tangent bundle of $\mathcal{M}^*(\mathfrak{s}, \eta)$ by using a family of linear operators. For a solution (A, Φ) to (4.22), let $L^{sw}_{(A,\Phi)} = D\tilde{m}|_{h=0}$, that is, for $(a, \phi) \in L^2_k\left(X; \sqrt{-1}\Lambda^1 \oplus S^+\right)$,

$$L^{sw}_{(A,\Phi)}(a, \phi) = D\tilde{m}(a, \phi, 0) = \begin{pmatrix} d^+ a - (Dq)(\phi) \\ D_A\phi + \frac{1}{2} a \cdot \Phi \end{pmatrix}. \tag{4.27}$$

As in the proof of Proposition 4.56, the tangent space $T_w\mathcal{M}^*(\mathfrak{s},\eta)$ at $w = [(A,\Phi)]$ is identified with $\ker(L^*_{(A,\Phi)} \oplus L^{sw}_{(A,\Phi)})$.

Since C^*_k is an affine space modelled on $L^2_k\left(X;\sqrt{-1}\Lambda^1 \oplus S^+\right)$, the tangent bundle TC^*_k of C^*_k is identified with the product $C_k \times L^2_k\left(X;\sqrt{-1}\Lambda^1 \oplus S^+\right)$. The gauge transformation group $\mathcal{G}$ naturally acts on $TC^*_k = C^*_k \times L^2_k\left(X;\sqrt{-1}\Lambda^1 \oplus S^+\right)$. The $\mathcal{G}$-action is given by

$$g \cdot ((A,\Phi),(a,\phi)) = ((A - 2g^{-1}dg, g\Phi),(a, g\phi)).$$

Define the map

$$\Lambda\colon TC^*_k \to L^2_{k-1}\left(X;\sqrt{-1}(\Lambda^0 \oplus \Lambda^+) \oplus S^-\right)$$

by

$$\Lambda((A,\Phi),(a,\phi)) = \left(L^*_{(A,\Phi)} \oplus L^{sw}_{(A,\Phi)}\right)(a,\phi) = \begin{pmatrix} -2d^*a + \sqrt{-1}\,\mathrm{Im}(\langle\phi,\Phi\rangle) \\ d^+a - (Dq)(\phi) \\ D_A\phi + \frac{1}{2}a\cdot\Phi \end{pmatrix}.$$

Then it is easy to check that Λ is $\mathcal{G}$-equivariant. Let $\tilde{\mathcal{M}}^*(\mathfrak{s},\eta)$ be the space of irreducible solutions to (4.22). Due to the choice of η, the restriction of Λ to $\tilde{\mathcal{M}}^*(\mathfrak{s},\eta)$ is fiberwise surjective. Then $\ker\Lambda|_{\tilde{\mathcal{M}}^*(\mathfrak{s},\eta)}$ forms a vector bundle over $\tilde{\mathcal{M}}^*(\mathfrak{s},\eta)$ such that the fiber over (A,Φ) is identified with the tangent space $T_{[(A,\Phi)]}\mathcal{M}^*(\mathfrak{s},\eta)$. Dividing by the $\mathcal{G}$-action, we obtain an identification

$$T\mathcal{M}^*(\mathfrak{s},\eta) = \left(\ker\Lambda|_{\tilde{\mathcal{M}}^*(\mathfrak{s},\eta)}\right)/\mathcal{G}.$$

Hence $\mathcal{M}^*(\mathfrak{s},\eta)$ is orientable if and only if $\bigwedge^{\max} T\mathcal{M}^*(\mathfrak{s},\eta) = \bigwedge^{\max}\left(\ker\Lambda|_{\tilde{\mathcal{M}}(\mathfrak{s},\eta)}\right)/\mathcal{G}$ is a trivial real line bundle.

We can assume Λ as a continuous family of Fredholm maps parametrized by C^*_k. By the construction in Sect. 4.9, the determinant line bundle over C^*_k,

$$\det(\Lambda) = \bigwedge^{\max}\ker\Lambda \otimes \bigwedge^{\max}(\operatorname{coker}\Lambda)^*,$$

is defined. Dividing by $\mathcal{G}$, we obtain a real line bundle $\det(\Lambda)/\mathcal{G}$ over $\mathcal{B}^*$. The restriction of $\det(\Lambda)/\mathcal{G}$ to $\mathcal{M}^*(\mathfrak{s},\eta)$ is $\bigwedge^{\max}\left(\ker\Lambda|_{\tilde{\mathcal{M}}^*(\mathfrak{s},\eta)}\right)/\mathcal{G}$.

Proposition 4.61 *The real line bundle* $\det(\Lambda)/\mathcal{G}$ *over* $\mathcal{B}^*$ *is a trivial real bundle. An orientation of* $\det(\Lambda)/\mathcal{G}$ *is determined by choosing an orientation* O *on*

$$\bigwedge^{\max}\left\{H^1\left(X;\sqrt{-1}\mathbb{R}\right)\oplus\left(H^+\left(X;\sqrt{-1}\mathbb{R}\right)\oplus H^0\left(X;\sqrt{-1}\mathbb{R}\right)\right)^*\right\}, \tag{4.28}$$

where $\left(H^+\left(X;\sqrt{-1}\mathbb{R}\right)\oplus H^0\left(X;\sqrt{-1}\mathbb{R}\right)\right)^*$ *is a dual vector space of* $H^+\left(X;\sqrt{-1}\mathbb{R}\right)\oplus H^0\left(X;\sqrt{-1}\mathbb{R}\right)$.

Proof Deform Λ by a homotopy as

$$\Lambda_t((A,\Phi),(a,\phi)) = \begin{pmatrix} -2d^*a + (1-t)\sqrt{-1}\,\mathrm{Im}(\langle\phi,\Phi\rangle) \\ d^+a - (1-t)(Dq)(\phi) \\ D_A\phi + (1-t)\frac{1}{2}a\cdot\Phi \end{pmatrix}, \qquad (0\le t\le 1).$$

Then $\Lambda_0 = \Lambda$ and $\Lambda_1 = (-2d^*\oplus d^+)\oplus(D_A)_{A\in\mathcal{A}}$. Therefore $\det(\Lambda)$ is isomorphic to

$$\det(\Lambda_1) = \det(-2d^*\oplus d^+)\otimes\det((D_A)_{A\in\mathcal{A}}).$$

Since $(D_A)_{A\in\mathcal{A}}$ is a family of complex operators, $\det((D_A)_{A\in\mathcal{A}})$ is a trivial real line bundle and has an orientation induced from the complex structure. Since $-2d^*\oplus d^+$ is a constant family, $\det(-2d^*\oplus d^+)$ is a trivial line bundle whose constant fiber is (4.28). Thus an orientation of (4.28) determines an orientation of $\det(\Lambda)$. □

Corollary 4.62 *The space* $\mathcal{M}^*(\mathfrak{s},\eta)$ *is orientable, and an orientation is given by choosing an orientation of (4.28).*

Remark 4.63 Since the orientation of X determines the orientation of $H^0(X)$, the orientation of (4.28) is determined by the orientations of $H^1(X)$ and $H^+(X)$.

4.7 Seiberg–Witten Invariants

Definition 4.64 Let $\mathrm{Spin}^c(X)$ be the set of equivalence classes of spin^c structures on X.

Definition 4.65 For a Riemannian metric g on X, we say $\eta \in \mathcal{P}_{k-1}$ is *good* or *g-good* if:

- η is a regular value of π of (4.25), and
- $\eta \in \mathcal{P}_{k-1} \setminus \mathcal{W}$.

If η is good, then the moduli space $\mathcal{M}(\mathfrak{s}, \eta)$ is a smooth compact $d(\mathfrak{s})$-dimensional submanifold of $\mathcal{B}_k^*$.

Definition 4.66 Let X be a closed smooth Riemannian 4-manifold with $b^+(X) \geq 2$. For $\mathfrak{s} \in \mathrm{Spin}^c(X)$, choose a good $\eta \in \mathcal{P}_{k-1}$. Fixing an orientation O of (4.28), we define a homomorphism

$$\mathrm{SW}_{X,\mathfrak{s},O} \colon \mathbb{Z}[u] \otimes \wedge^*(H_1(X;\mathbb{Z})/\mathrm{Tor}) \to \mathbb{Z}$$

by

$$\mathrm{SW}_{X,\mathfrak{s},O}(u^m \otimes (\gamma_1 \wedge \cdots \wedge \gamma_k)) = \langle u^m \otimes (\mu(\gamma_1) \wedge \cdots \wedge \mu(\gamma_k)), [\mathcal{M}(\mathfrak{s}, \eta)] \rangle. \tag{4.29}$$

We call $\mathrm{SW}_{X,\mathfrak{s},O}$ the *Seiberg–Witten invariant* of $(X, \mathfrak{s})$.

Theorem 4.67

1. *$\mathrm{SW}_{X,\mathfrak{s},O}$ depends only on X, $\mathfrak{s}$ and O, and does not depend on a choice of Riemannian metric and η.*
2. *Let $f \colon X_1 \to X_2$ be an orientation-preserving diffeomorphism. Then*

$$\mathrm{SW}_{X_1,f^*\mathfrak{s},f^*O}(u^m \otimes (\gamma_1 \wedge \cdots \wedge \gamma_k)) = \mathrm{SW}_{X_2,\mathfrak{s},O}(u^m \otimes (f_*(\gamma_1) \wedge \cdots \wedge f_*(\gamma_k))),$$

3. $\mathrm{SW}_{X,\mathfrak{s},-O} = -\mathrm{SW}_{X,\mathfrak{s},O}$.

Proof Assertions 2 and 3 are obvious. We will prove the assertion 1 in Sect. 5.3. □

Remark 4.68 The outline of a standard proof of assertion 1 is as follows. Suppose two Riemannian metrics g_0 and g_1 are given, and choose g_i-good η_i for $i = 0, 1$. For a smooth path g_t of Riemannian metrics connecting g_0 and g_1, we can choose a path η_t connecting η_0 and η_1 so that $\eta_t \in (\mathcal{P}_{k-1} \setminus \mathcal{W})_{g_t}$ for $t \in [0, 1]$ if $b^+ \geq 2$. Choosing a generic path as above, we can construct a cobordism between the moduli spaces for (g_0, η_0) and (g_1, η_1). In Sect. 5.3, we will give a proof by using the finite-dimensional approximation.

Corollary 4.69 *If the moduli space* $\mathcal{M}(\mathfrak{s}, \eta)$ *is empty for some Riemannian metric and* η, *then* $\mathrm{SW}_{X,\mathfrak{s},O} = 0$. *In particular, if* $d(\mathfrak{s}) < 0$, *then* $\mathrm{SW}_{X,\mathfrak{s},O} = 0$.

Theorem 4.70 *If a smooth closed 4-manifold with* $b^+(X) \geq 2$ *admits a Riemannian metric of positive scalar curvature, then* $\mathrm{SW}_{X,\mathfrak{s},O} = 0$ *for every* $\mathfrak{s} \in \mathrm{Spin}^c(X)$.

Proof This is a corollary of Proposition 4.60 and Corollary 4.69. □

Proposition 4.71 *If* $b^+(X) \geq 2$, *then the number of equivalence classes* $\mathfrak{s} \in \mathrm{Spin}^c(X)$ *such that* $\mathrm{SW}_{X,\mathfrak{s},O} \neq 0$ *is finite.*

Proof Fix a Riemannian metric on X and a small constant $\varepsilon > 0$. For a spin^c structure $\mathfrak{s}$ on X, choose a self-dual 2-form $\eta \notin \mathcal{W}$ such that $\|\eta\|_{C^0} < \varepsilon$. Suppose $\mathrm{SW}_{X,\mathfrak{s}} \neq 0$. Then there exists an irreducible solution (A, Φ) to the perturbed equations (4.22). By Proposition 4.51, we obtain an estimate

$$|F_A^+(x)| \leq \frac{|\Phi(x)|^2}{2} + |\eta(x)| \leq \frac{M^2}{2} + \varepsilon.$$

Using this estimate, we find a constant C which does not depend on $\mathfrak{s}$ such that $\|F_A^+\|_{L^2}^2 \leq C$. Since $c_1(\mathfrak{s})^2 = \frac{1}{4\pi^2}(\|F_A^+\|_{L^2}^2 - \|F_A^-\|_{L^2}^2)$ and $d(\mathfrak{s}) = \frac{1}{4}(c_1(\mathfrak{s})^2 - (2\chi(X) + 3\sigma(X))) \geq 0$, we have

$$\|F_A^-\|_{L^2}^2 \leq C - 4\pi^2(2\chi(X) + 3\sigma(X)).$$

There are only a finite number of spin^c structures such that $\|F_A^+\|_{L^2}^2$ and $\|F_A^-\|_{L^2}^2$ are bounded by constants. □

4.8 Computations of the Seiberg–Witten Invariants of Four-Torus

In general, it is not easy to compute Seiberg–Witten invariants, because we need to solve the Seiberg–Witten equations (4.2) which are non-linear PDEs. However, a direct computation is possible for a four-torus. This is the topic of this section.

Theorem 4.72 *Let $X = T^4$ be the 4-dimensional torus, and $\mathfrak{s}_0$ be the spin^c structure on X with $c_1(\mathfrak{s}_0) = 0$. Then $\mathrm{SW}_{X,\mathfrak{s},O} \equiv 0$ if $\mathfrak{s} \not\cong \mathfrak{s}_0$ and*

$$\mathrm{SW}_{X,\mathfrak{s}_0,O}(\beta) = \begin{cases} \pm 1 & \beta = 1 \in \mathbb{Z}[u], \\ 0 & \deg \beta \neq 0. \end{cases}$$

Proof First, we prove that $\mathrm{SW}_{X,\mathfrak{s}} \equiv 0$ if $\mathfrak{s} \not\cong \mathfrak{s}_0$. The intersection form of X is isomorphic to $3H$, where H is the hyperbolic form

$$H = \begin{pmatrix} 0 & 1 \\ 1 & 0 \end{pmatrix}.$$

Let $(e_1, f_1, e_2, f_2, e_3, f_3)$ be the basis of $H^2(X; \mathbb{Z})$ with respect to the decomposition into $3H$. Then the set of the isomorphism classes of the spin^c structures on X is identified with

$$\left\{ c_1(\mathfrak{s}) = \sum_{i=1}^{3} (a_i e_i + b_i f_i) \;\middle|\; a_i, b_i \in 2\mathbb{Z} \right\}.$$

Since the Euler characteristic and the signature of X are zero, the virtual dimension $d(\mathfrak{s})$ of the moduli space is given by

$$d(\mathfrak{s}) = \frac{1}{4} c_1(\mathfrak{s})^2 = \frac{1}{2} \sum_{i=1}^{3} a_i b_i.$$

If $\mathrm{SW}_{X,\mathfrak{s}} \neq 0$, then $d(\mathfrak{s}) = \frac{1}{2} \sum_{i=1}^{3} a_i b_i \geq 0$.

Equip X with the standard flat Riemannian metric. If $\mathrm{SW}_{X,\mathfrak{s}} \neq 0$, then the moduli space must be non-empty for every metric and perturbation. Since the scalar curvature vanishes everywhere, Proposition 4.49 implies

$$c_1(\mathfrak{s})^+_{\mathbb{R}} = \frac{1}{2} \sum_{i=1}^{3} (a_i + b_i)(e_i + f_i) = 0,$$

and hence $a_i + b_i = 0$ for every i. In this case, all of the solutions are reducible.

From these, we obtain $a_i = b_i = 0$ for all i if $\mathrm{SW}_{X,\mathfrak{s}} \neq 0$. That is, $\mathrm{SW}_{X,\mathfrak{s}} \equiv 0$ if $\mathfrak{s} \not\cong \mathfrak{s}_0$.

Suppose $\mathfrak{s} = \mathfrak{s}_0$. Since $d(\mathfrak{s}_0) = 0$, $\mathrm{SW}_{X,\mathfrak{s}}(\beta) = 0$ if $\deg \beta \neq 0$. We assume $X = T^4$ as a complex torus $\mathbb{C}^2/\Lambda$, where

$$\Lambda = \left\{ \left(m_1 + n_1\sqrt{-1}, m_2 + n_2\sqrt{-1} \right) \,\middle|\, m_1, m_2, n_1, n_2 \in \mathbb{Z} \right\}.$$

Let (z_1, z_2) be the complex coordinate of $X = \mathbb{C}^2/\Lambda$. Then $\mathfrak{s}_0$ is the canonical spin^c structure. Therefore its spinor bundles are

$$S^+ = \Lambda^0 \oplus \Lambda^{0,2}, \quad S^- = \Lambda^{0,1}.$$

(Lemma 3.85.) Note that $S^\pm$ are trivial bundles. Let A_0 be a trivial flat connection on the determinant line bundle $L = (\Lambda^*)^{0,2}$. By Proposition 3.89, the Dirac operator is expressed as

$$D_{A_0} = \sqrt{2}(\bar{\partial} + \bar{\partial}^*).$$

For a U(1)-connection $A = A_0 + a$ on L where $a \in \sqrt{-1}\Omega^1(X)$, the Dirac operator D_A is expressed as

$$D_A = \sqrt{2}(\bar{\partial}_\alpha + (\bar{\partial}_\alpha)^*),$$

where $\alpha = \pi^{0,1}(a/2)$. For $\phi = (\phi_1, \phi_2) \in \Gamma(S^+) = (\Omega^0 \oplus \Omega^{0,2})(X)$ and $a \in \sqrt{-1}\Omega^1(X)$, the Seiberg–Witten equations (4.2) are equivalent to

$$\begin{cases} \sqrt{2}(\bar{\partial}_\alpha \phi_1 + (\bar{\partial}_\alpha)^* \phi_2) = 0 \\ \pi^{1,1}(d^+ a) = \dfrac{i}{4}(|\phi_1|^2 - |\phi_2|^2)\omega \\ \pi^{0,2}(d^+ a) = \dfrac{1}{2}\bar{\phi}_1 \phi_2, \end{cases} \tag{4.30}$$

where $\alpha = \pi^{0,1}(a/2)$ and $\omega = \frac{i}{2}(dz_1 \wedge d\bar{z}_1 + dz_2 \wedge d\bar{z}_2)$. Since the unperturbed equations have only reducible solutions as mentioned above, we perturb the second equation of (4.30) as follows:

$$\pi^{1,1}(d^+ a) = \frac{i}{4}(|\phi_1|^2 - |\phi_2|^2 - 1)\omega. \tag{4.31}$$

Note that $\pi^{0,2}(d^+ a) = 2\bar{\partial}\alpha = 2\left(\bar{\partial}_\alpha\right)^2$. Applying $\bar{\partial}_\alpha$ to the first equation in (4.30) and using the third equation, we obtain

$$\frac{1}{4}|\phi_1|^2 \phi_2 + \bar{\partial}_\alpha (\bar{\partial}_\alpha)^* \phi_2 = 0.$$

Taking the L^2-inner product of this equation with ϕ_2, we have

$$\int_X \frac{1}{4}|\phi_1|^2|\phi_2|^2 \operatorname{vol} + \|(\bar{\partial}_\alpha)^*\phi_2\|_{L^2}^2 = 0.$$

Then we have $|\phi_1||\phi_2| = 0$ and $(\bar{\partial}_\alpha)^*\phi_2 = 0$ everywhere. This with the third equation implies that $(\bar{\partial}_\alpha)^2 = \frac{1}{4}\bar{\phi}_1\phi_2 = 0$. By Theorem 3.88, $\bar{\partial}_\alpha$ defines a holomorphic structure and ϕ_1 is a holomorphic function. Then $(\bar{\partial}_\alpha)^*\phi_2 = 0$ means that $\bar{\phi}_2$ is also a holomorphic function. Therefore ϕ_1 and ϕ_2 are constant and one of them vanishes identically. It follows from $\bar{\partial}\alpha = (\bar{\partial}_\alpha)^2 = 0$ that α is constant, and therefore $a = 2(\alpha - \bar{\alpha})$ is also constant. By the perturbed equation (4.31), we have

$$0 = \langle c_1(\mathfrak{s}_0) \cup [\omega], [X]\rangle = \int_X c_1(\mathfrak{s}_0) \wedge \omega = \frac{1}{8\pi}\int_X (1 - |\phi_1|^2 + |\phi_2|^2)\operatorname{vol}.$$

Finally, we obtain $\phi_1 \equiv c$ for some $c \in \mathbb{C}$ with $|c| = 1$ and $\phi_2 \equiv 0$. In conclusion, all solutions of the perturbed Seiberg–Witten equations have the form of $(A_0 + a, (c, 0))$ where a and c are constant and $|c| = 1$. Note that these solutions are gauge equivalent to $(A_0, (1, 0))$, and the moduli space consists of a sole equivalence class of it. It can be directly checked that $L^{SW}_{(A,\Phi)}$ in (4.27) is surjective if $(A, \Phi) = (A_0, (1, 0))$. Thus we obtain $\mathrm{SW}_{X,\mathfrak{s}_0}(1) = \pm 1$. □

The computation of the Seiberg–Witten invariants of T^4 in this section can be considered as a prototype of computations of the invariants of Kähler surfaces. In the Kähler case, a kind of Kobayashi–Hitchin type correspondence holds, and the computation of the Seiberg–Witten invariant is reduced to counting effective divisors of the class of the determinant line bundle. We refer the readers to the textbooks, e.g., Morgan [131] and Teleman [165], for the Kähler case.

Taubes' outstanding work [162] on symplectic 4-manifolds can be considered as a spectacular extension of the computations for Kähler surfaces. This work relates the Seiberg–Witten invariants with a kind of Gromov invariants which counts pseudo-holomorphic curves in symplectic 4-manifolds. This is in contrast to the Kähler case where the effective divisors are counted. The following non-vanishing theorem is a part of his results.

Theorem 4.73 (Taubes [162]) *Let (X, ω) be a symplectic 4-manifold with the orientation given by the volume form $\omega \wedge \omega$. Suppose $b^+(X) \geq 2$. Let $\mathfrak{s}_0$ be the canonical spin^c structure associated with the almost complex structure compatible with the symplectic form ω. Then* $\mathrm{SW}_{X,\mathfrak{s}_0}(1) = \pm 1$.

Remark 4.74 There is already a vast amount of research on the applications of the Seiberg–Witten invariants. We refer the readers to the textbooks of gauge theory [130, 131, 139, 151, 165] and 4-manifolds [2, 64] and the references therein.

4.9 The Case When $b_2^+ = 1$

Let us consider the case when X is a closed oriented connected 4-manifold with $b_2^+ = 1$. Since $\mathcal{W}$ in Proposition 4.57 is a codimension-b_2^+ subspace of $\mathcal{P}_{k-1}$, the space $\mathcal{W}$ divides $\mathcal{P}_{k-1} \setminus \mathcal{W}$ into two connected components which are called *chambers*. In general, the fundamental class of the moduli space $\mathcal{M}(\mathfrak{s}, \eta)$ depends on the chamber where η lives. In this case, we can also define the Seiberg–Witten invariants $\mathrm{SW}^{\pm}_{X,\mathfrak{s}}$ corresponding to chambers. In general, $\mathrm{SW}^{+}_{X,\mathfrak{s}}$ and $\mathrm{SW}^{-}_{X,\mathfrak{s}}$ are different. But the difference $\mathrm{SW}^{+}_{X,\mathfrak{s}} - \mathrm{SW}^{-}_{X,\mathfrak{s}}$ is computed from topological data of $(X, \mathfrak{s})$. The formula which gives the relation between $\mathrm{SW}^{\pm}_{X,\mathfrak{s}}$ is called the *wall-crossing formula* (Theorem 4.77). In this section, we follow the description of Teleman's book [165].

Let $H^2_{DR}(X)$ be the 2nd de Rham cohomology group. Consider the hyperboloid of two sheets

$$\mathbf{H} = \{c \in H^2_{DR}(X) \mid c^2 = 1\}.$$

Choose a sheet $\mathfrak{h} \subset \mathbf{H}$. Then $\mathbf{H} = \mathfrak{h} \cup (-\mathfrak{h})$. Choose a Riemannian metric g on X. Let $\omega_g(\mathfrak{h})$ be the unique g-harmonic g-self-dual 2-form whose cohomology class is in $\mathfrak{h}$. It is obvious that $\omega_g(-\mathfrak{h}) = -\omega_g(\mathfrak{h})$. Then $\mathcal{W}$ in Proposition 4.57 is written as

$$\mathcal{W} = \left\{ \eta \;\middle|\; \mathfrak{S}(g, \eta, \mathfrak{s}, \mathfrak{h}) := -\omega_g(\mathfrak{h}) \cup \left(c_1(\mathfrak{s}) - \frac{\sqrt{-1}}{2\pi} h_g^+(\eta) \right) = 0 \right\},$$

and $\mathcal{P}_{k-1} \setminus \mathcal{W}$ is divided into two components according to whether $\mathfrak{S}(g, \eta, \mathfrak{s}, \mathfrak{h})$ is positive or negative. Let $\mathscr{C}^+(\mathfrak{h})$ (resp. $\mathscr{C}^-(\mathfrak{h})$) be the space of parameters (g, η) such that $\mathfrak{S}(g, \eta, \mathfrak{s}, \mathfrak{h})$ is positive (resp. negative). We assume $\mathscr{C}^{\pm}(\mathfrak{h})$ are subspaces of $\mathscr{M}et(X) \times L^2_{k-1}\left(\sqrt{-1}\Omega^2(X)\right)$ where $\mathscr{M}et(X)$ is the space of Riemannian metrics on X with C^∞ topology. Then,

- each of $\mathscr{C}^{\pm}(\mathfrak{h})$ is path-connected, and
- $\mathscr{C}^{\pm}(-\mathfrak{h}) = \mathscr{C}^{\mp}(\mathfrak{h})$.

We can define the Seiberg–Witten invariants which depend on the choice of *chambers* $\mathscr{C}^{\pm}(\mathfrak{h})$. For this purpose, we need to choose an orientation of the line (4.28) to orient the moduli spaces. The orientation of $H^+\left(X; \sqrt{-1}\mathbb{R}\right)$ is determined by $\sqrt{-1}\omega_g(\mathfrak{h})$. Since we have a canonical orientation on $H^0\left(X; \sqrt{-1}\mathbb{R}\right)$ determined by the orientation of X (Remark 4.63), it suffices to choose an orientation O_1 of $H^1\left(X; \sqrt{-1}\mathbb{R}\right)$ to orient the line (4.28).

Definition 4.75 Let X be a closed oriented connected 4-manifold with $b_2^+ = 1$. Choose a sheet $\mathfrak{h}$ of the hyperboloid **H**. For $\mathfrak{s} \in \mathrm{Spin}^c(X)$, choose a good $(g, \eta) \in \mathscr{C}^{\pm}(\mathfrak{h})$. Choose an orientation O_1 of $H^1\left(X; \sqrt{-1}\mathbb{R}\right)$. We define the Seiberg–Witten invariant as follows:

$$\begin{aligned} \mathrm{SW}^{\pm}_{X,\mathfrak{s},\mathfrak{h},O_1} &: \mathbb{Z}[u] \otimes \wedge^*(H_1(X;\mathbb{Z})/\mathrm{Tor}) \to \mathbb{Z}, \\ &\mathrm{SW}^{\pm}_{X,\mathfrak{s},\mathfrak{h},O_1}(u^m \otimes (\gamma_1 \wedge \cdots \wedge \gamma_k)) \\ &= \langle u^m \otimes (\mu(\gamma_1) \wedge \cdots \wedge \mu(\gamma_k)), [\mathcal{M}(\mathfrak{s}, \eta)] \rangle. \end{aligned}$$

Theorem 4.76

1. $\mathrm{SW}^{\pm}_{X,\mathfrak{s},\mathfrak{h},O_1}$ *depends only on* X, $\mathfrak{s}$, $\mathfrak{h}$ *and* O_1 *and does not depend on a choice of* $(g, \eta) \in C^{\pm}(\mathfrak{h})$.
2. *Let* $f : X_1 \to X_2$ *be an orientation-preserving diffeomorphism. Then*

$$\begin{aligned} &\mathrm{SW}^{\pm}_{X_1, f^*\mathfrak{s}, f^*\mathfrak{h}, f^*O_1}(u^m \otimes (\gamma_1 \wedge \cdots \wedge \gamma_k)) \\ &\qquad = \mathrm{SW}^{\pm}_{X_2,\mathfrak{s},\mathfrak{h},O_1}(u^m \otimes (f_*(\gamma_1) \wedge \cdots \wedge f_*(\gamma_k))) \end{aligned}$$

3. $\mathrm{SW}^{\pm}_{X,\mathfrak{s},\mathfrak{h},-O_1} = -\mathrm{SW}^{\pm}_{X,\mathfrak{s},\mathfrak{h},O_1}$, $\mathrm{SW}^{\pm}_{X,\mathfrak{s},-\mathfrak{h},O_1} = -\mathrm{SW}^{\mp}_{X,\mathfrak{s},\mathfrak{h},O_1}$.

Proof Assertions 2 and 3 are obvious. We will prove assertion 1 in Sect. 5.3. □

The wall-crossing formula which computes the difference of $\mathrm{SW}^{+}_{X,\mathfrak{s},\mathfrak{h},O_1}$ and $\mathrm{SW}^{-}_{X,\mathfrak{s},\mathfrak{h},O_1}$ was first proved by Kronheimer and Mrowka [99] in the case when $b_1 = 0$. Later this was generalized to the case when $b_1 > 0$ by Li and Liu [111] and Okonek and Teleman [142]. We state the formula due to Okonek and Teleman [142].

Theorem 4.77 ([142]) *For* $u^m \otimes h \in \mathbb{Z}[u] \otimes \Lambda^k\left(H_1(X;\mathbb{Z})/\mathrm{Tor}\right))$ *such that* $2m + k = d(\mathfrak{s}) \geq 0$ *and* $0 \leq k \leq b_1$, *the following equality holds:*

$$\left(\mathrm{SW}^{+}_{X,\mathfrak{s},\mathfrak{h},O_1} - \mathrm{SW}^{-}_{X,\mathfrak{s},\mathfrak{h},O_1}\right)(u^m \otimes h) = \langle \exp(u_c) \wedge h, l_{O_1} \rangle,$$

where $u_c \in \Lambda^2\left(H_1(X;\mathbb{Z})/\mathrm{Tor}\right))$ *is defined by the skew-symmetric form*

(continued)

Theorem 4.77 (continued)

$$H^1(X;\mathbb{Z}) \times H^1(X;\mathbb{Z}) \ni (u, v) \mapsto \frac{1}{2}\langle c_1(\mathfrak{s}) \cup u \cup v, [X]\rangle,$$

and l_{O_1} *is the generator of* $\Lambda^{b_1}(H_1(X;\mathbb{Z})/\mathrm{Tor})$ *determined by the orientation* O_1.

Corollary 4.78 *Suppose further that* $b_1 = 0$ *and* $d(\mathfrak{s}) \geq 0$. *Then*

$$\left(\mathrm{SW}^{+}_{X,\mathfrak{s},\mathfrak{h},O_1} - \mathrm{SW}^{-}_{X,\mathfrak{s},\mathfrak{h},O_1}\right)(u^{d(\mathfrak{s})/2}) = 1.$$

We will give the proof of this special case (Corollary 4.78) in Sect. 5.3.

Appendix

Sobolev Spaces

The purpose of this subsection is to show several results in Sobolev analysis such as Lemmas 4.10, 4.11, 4.12 and Theorem 4.17. General references are [147, Chapter 5], [163, Chapter 4], [168, Chapter 6]. The argument on the interpolation below is due to Jin Miyazawa. The authors would like to thank him.

First, we discuss the Sobolev spaces on n-dimensional torus $T^n = \mathbb{R}^n/2\pi\mathbb{Z}^n$.

Definition 4.79 For an integrable function $f\colon T^n \to \mathbb{C}$, the *Fourier series* for f is the formal series

$$\sum_{\xi \in \mathbb{Z}^n} a_\xi e^{i\xi\cdot x},$$

where

$$a_\xi = \hat{f}(\xi) = \frac{1}{(2\pi)^n}\int_{T^n} f(x)e^{-i\xi\cdot x}dx.$$

We recall several basic facts in the Fourier analysis (see e.g. [163, Chapter 3]).

- The functions $e_\xi(x) = e^{-i\xi\cdot x}$ for $\xi \in \mathbb{Z}^n$ form an orthonormal basis for the Hilbert space $L^2(T^n)$. For $f \in L^2(T^n)$,

$$\|f\|_{L^2}^2 = \int_{T^n} |f|^2 = (2\pi)^n \sum_{\xi} |\hat{f}(\xi)|^2 \qquad \text{(Parseval's identity)}$$

Hence $\hat{f}$ is a square summable sequence if $f \in L^2(T^n)$. Moreover, if $f \in L^2(T^n)$, then the Fourier series of f converges to f in the L^2-norm.

- If $f \in C^\infty(T^n)$ then $\hat{f}(\xi)$ are rapidly decreasing: for any s, there is a constant C_s such that $|\hat{f}(\xi)| \le C_s(1 + |\xi|)^{-s}$. The Fourier series of $f \in C^\infty(T^n)$ converges to f in the C^∞ Fréchet topology.

Definition 4.80 For a nonnegative real number s, the *Sobolev s-inner product* on $C^\infty(T^n)$ is defined by

$$\langle f, g\rangle_s = (2\pi)^n \sum_{\xi} \hat{f}(\xi)\overline{\hat{g}(\xi)}(1 + |\xi|^2)^s.$$

Note that the sum on the right-hand side converges since $\hat{f}, \hat{g}$ are rapidly decreasing.

The norm induced from the Sobolev s-inner product is called the *Sobolev s-norm.* Let $L_s^2(T^n)$ be the completion of $C^\infty(T^n)$ in the Sobolev s-norm.

- If $t > s$, then $L_t^2(T^n) \subset L_s^2(T^n)$ by definition.
- Let $\partial_j f = \frac{\partial f}{\partial x_j}$. For $\alpha = (\alpha_1, \dots \alpha_n)$, $\alpha_i \in \mathbb{Z}_{\ge 0}$, let $\partial^\alpha = (\partial_1)^{\alpha_1} \cdots (\partial_n)^{\alpha_n}$ and $|\alpha| = \alpha_1 + \cdots + \alpha_n$. Note that $\widehat{\partial^\alpha f} = (i\xi)^\alpha \hat{f}$, where $(i\xi)^\alpha = (i\xi_1)^{\alpha_1} \cdots (i\xi_n)^{\alpha_n}$. If s is a nonnegative integer, then the Sobolev s norm on $C^\infty(T^n)$ is equivalent to the norm $\|\cdot\|'_{L_s^2}$ given by

$$\|f\|'_{L_s^2} = \sum_{|\alpha| \le k} \|\partial^\alpha f\|_{L^2}.$$

  Then $L_s^2(T^n)$ is also the completion of $C^\infty(T^n)$ with the norm $\|\cdot\|'_{L_s^2}$.

By using the characterization of $L_s^2(T^n)$ by the norm $\|\cdot\|'_{L_s^2}$, we can easily verify the following two lemmas.

Lemma 4.81 *Let s be a nonnegative integer. Multiplication by a smooth function defines a bounded operator on L_s^2. Linear differential operator of order l defines a bounded operator from L_{s+l}^2 to L_s^2.*

Lemma 4.82 *Suppose $f \in L^2(T^n)$ has the support in an open subset $U \subset T^n$. Let φ be a diffeomorphism of U into T^n. For a nonnegative integer s, $f \circ \varphi \in L_s^2$ if and only if $f \in L_s^2$.*

We will extend Lemmas 4.81 and 4.82 to the case when s is a nonnegative real number later.

Lemma 4.83 *For a nonnegative integer s, $C^s(T^n) \subset L^2_s(T^n)$, and the inclusion map is continuous.*

Proof Since $\partial^\alpha f$ is continuous for α with $|\alpha| \leq s$, it is in L^2. Then $\sum_\xi \left|\widehat{\partial^\alpha f}(\xi)\right|^2 = \sum_\xi \left|(i\xi)^\alpha \hat{f}(\xi)\right|^2$ converges by Parseval's identity. Therefore $\sum_\xi |\hat{f}(\xi)|^2(1+|\xi|)^s$ converges.

By Parseval's identity,

$$(2\pi)^n \sum_\xi \left|(i\xi)^\alpha \hat{f}(\xi)\right|^2 = \|\partial^\alpha f\|^2_{L^2} \leq \text{const.} \left(\sup_x |\partial^\alpha f(x)|\right)^2.$$

This proves that the inclusion map is continuous. □

Theorem 4.84 (Sobolev Embedding) *For $k \in \mathbb{Z}$ and $s \in \mathbb{R}$ such that $k - \frac{n}{2} > s \geq 0$, $L^2_s(T^n) \subset C^k(T^n)$, and the inclusion map is continuous.*

Proof Let $\delta = k - s$. For $f \in L^2_s(T^n)$,

$$\sum_\xi |\hat{f}(\xi)|^2(1+|\xi|^2)^k(1+|\xi|^2)^\delta < \infty.$$

The Cauchy–Schwartz inequality imlies

$$\sum_\xi |\hat{f}(\xi)|(1+|\xi|)^{k/2}$$
$$\leq \left(\sum_\xi |\hat{f}(\xi)|^2(1+|\xi|^2)^k(1+|\xi|^2)^\delta\right)^{1/2} \left(\sum_\xi (1+|\xi|^2)^{-\delta}\right)^{1/2}.$$

Since $\delta > n/2$, $\sum_\xi (1+|\xi|^2)^{-\delta}$ converges. Therefore $\sum_\xi |\xi|^k|\hat{f}(\xi)|$ is finite. This means the Fourier series $\sum_\xi (i\xi)^\alpha \hat{f}$ converges uniformly to the continuous derivatives $\partial^\alpha f$ for α with $|\alpha| \leq k$. □

Theorem 4.85 (Rellich's Lemma) *If $t > s$, then the inclusion map $L^2_t \to L^2_s$ is a compact operator, that is, every sequence bounded in L^2_t has a subsequence convergent in L^2_s.*

Proof (The proof is due to Roe [147].) Let B be the unit ball of L^2_t. For given $\varepsilon > 0$, we can choose a closed subspace W of L^2_t of finite codimension such that $\|f\|_{L^2_s} < \varepsilon$ for every $f \in B \cap W$ as follows. Take W to be the space of f such that $\hat{f}(\xi) = 0$ if $|\xi| < N$ for a sufficiently large N. Then the unit sphere of L^2_t/W is precompact, and can be covered by finitely many balls of radius ε. Then B can be covered by finitely many balls of radius 2ε in the L^2_s-norm. Since ε is arbitrary, B is precompact in the L^2_s-norm. □

Define the several spaces of series as follows:

$$\ell^2_{-\infty}(n) = \left\{(a_\xi)_{\xi\in\mathbb{Z}^n} \;\middle|\; a_\xi \in \mathbb{C}\right\},$$

$$\ell^2_s(n) = \left\{ (a_\xi)_{\xi\in\mathbb{Z}^n} \in \ell^2_{-\infty}(n) \;\middle|\; \|(a_\xi)\|_{\ell^2_s} := \sum_{\xi\in\mathbb{Z}^n} (1+|\xi|^2)^s |a_\xi|^2 < \infty \right\},$$

$$\ell^2_\infty(n) = \bigcap_{s\geq 0} \ell^2_s(n),$$

$$\ell^2_f(n) = \left\{(a_\xi)_{\xi\in\mathbb{Z}^n} \in \ell^2_{-\infty}(n) \;\middle|\; a_\xi = 0 \text{ except for finitely many } \xi \right\}.$$

Then $\ell^2_f(n) \subset \ell^2_\infty(n) \subset \ell^2_s(n)$ and $\ell^2_f(n)$ is dense in each $\ell^2_s(n)$. Note that $f \in L^2_s(T^n)$ if and only if $\left(\hat{f}(\xi)\right) \in \ell^2_s(n)$.

Lemma 4.86 *Let $L\colon \ell^2_\infty(n) \to \ell^2_{-\infty}(n)$ be a linear operator. For $i = 0, 1$, suppose L is extended to the bounded operators $L\colon \ell^2_{s_i}(n) \to \ell^2_{t_i}(n)$ for some $s_i \geq t_i$, i.e., there are constants $C_i > 0$ such that*

$$\|Lf\|_{\ell^2_{t_i}} \leq C_i \|f\|_{\ell^2_{s_i}}, \quad f \in \ell^2_{s_i}.$$

Let $s_x = xs_1 + (1-x)s_0$ and $t_x = xt_1 + (1-x)t_0$ for $x \in [0,1]$. Then, for any $x \in [0,1]$, $L\colon \ell^2_{s_x} \to \ell^2_{t_x}$ is bounded, and the following estimate holds:

$$\|Lf\|_{\ell^2_{t_x}} \leq C_1^x C_0^{1-x} \|f\|_{s_x}, \quad f \in \ell^2_{s_x}. \tag{4.32}$$

Proof We prove the following inequality for every finite series $f = (a_\xi)$ and $g = (b_\xi)$.

$$\left|\langle Lf, g\rangle_{\ell^2_{s_x}}\right| \leq C_1^x C_0^{1-x} \|f\|_{\ell^2_{s_x}} \|g\|_{\ell^2_{t_x}}. \tag{4.33}$$

Since ℓ^2_f is dense in ℓ^2_s for all s, the inequality (4.33) proves the lemma.

By substituting $f = (a_\xi)$ and $g = (b_\xi)$ by $f = (1+|\xi|^2)^{s_x/2} a_\xi$ and $g = (1+|\xi|^2)^{t_x/2} b_\xi$ respectively, the inequality (4.33) is rewritten as

$$\sum_{\eta\in\mathbb{Z}^n} (1+|\eta|^2)^{t_x/2} \left(L\left(\left((1+|\xi|^2)^{-s_x/2} a_\xi\right)\right)\right)_\eta \overline{b_\eta} \leq C_1^x C_0^{1-x} \|f\|_{\ell^2_0} \|g\|_{\ell^2_0}. \tag{4.34}$$

For non-zero finite series $f = (a_\xi)$ and $g = (b_\xi)$, define the complex function $F(z)$ on $0 \leq \operatorname{Re}(z) \leq 1$ by

$$F(z) = \sum_{\eta \in \mathbb{Z}^n} (1+|\eta|^2)^{t_z/2} \left(L \left(\left((1+|\xi|^2)^{-s_z/2} a_\xi \right) \right) \right)_\eta \overline{b_\eta} C_1^{-z} C_0^{-(1-z)} \|f\|_{\ell_0^2}^{-1} \|g\|_{\ell_0^2}^{-1}.$$

It is easy to see that $F(z)$ is bounded. Since f, g are finite series, $F(z)$ is a finite sum of holomorphic functions, and hence $F(z)$ is holomorphic. For $\mathrm{Re}(z) = 0, 1$, $|F(z)| \leq 1$.

The lemma follows from the following claim based on Hadamard three-lines theorem.

Claim Let $F(z)$ be a bounded holomorphic function on $0 \leq \mathrm{Re}(z) \leq 1$ such that $|F(z)| \leq 1$ if $\mathrm{Re}(z) = 0, 1$. Then $|F(z)| \leq 1$ on $0 \leq \mathrm{Re}(z) \leq 1$.

We prove the claim. Since $F(z)$ is bounded, there is a constant $C > 0$ such that $|F(z)| \leq C$ for $z = x + iy$ with $0 \leq x \leq 1$. For $\varepsilon > 0$, let $G_\varepsilon(z) = F(z)e^{\varepsilon z^2}$. Then

$$|G_\varepsilon(z)| = |F(z)| e^{\varepsilon(x^2-y^2)} \leq C e^{-\varepsilon y^2 + \varepsilon}. \tag{4.35}$$

For $Y > 1$, let R_Y be the rectangle

$$R_Y = \{z \,|\, 0 \leq \mathrm{Re}(z) \leq 1, \ -Y \leq \mathrm{Im}(z) \leq Y\}.$$

By the estimate (4.35), we can find a sufficiently large $Y > 1$ such that $|G_\varepsilon(z)| \leq 1$ on the boundary of R_Y. By the maximum principle, $|G_\varepsilon(z)| \leq 1$ on the whole R_Y, and therefore $|F(z)| \leq e^{\varepsilon Y^2 - \varepsilon}$. Letting $\varepsilon \to 0$, we obtain $|F(z)| \leq 1$ on R_Y. Since we can take an arbitrary large Y, $|F(z)| \leq 1$ on $0 \leq \mathrm{Re}(z) \leq 1$. □

Corollary 4.87 *Lemmas 4.81 and 4.82 hold for any nonnegative real number s.*

Let X be a closed Riemannian smooth n-manifold, and E a complex vector bundle over X with rank k. Let $\{U_\alpha\}_{\alpha=1,\cdots,N}$ be an open cover of X such that there exists a local trivialization $\varphi_\alpha : E|_{U_\alpha} \to U_\alpha \times \mathbb{C}^k$ for each α. We assume each U_α is diffeomorphically identified with an open set in T^n. Let $\{\psi_\alpha^2\}$ be a smooth partition of unity subordinate to U_α.

Definition 4.88 For a nonnegative real number s, the *Sobolev s-inner product* on $C^\infty(E)$ is defined by

$$\langle u, v \rangle_{L_s^2} := \sum_\alpha \langle \varphi_\alpha(\psi_\alpha u), \varphi_\alpha(\psi_\alpha v) \rangle_s$$

The norm induced from this inner product is called the *Sobolev s-norm*. Let $L_s^2(E)$ be the completion of $\Gamma(E)$ in the Sobolev s-norm.

By using Corollary 4.87, we can prove the following facts.

- If we take another choice of $\{(U_\alpha, \varphi_\alpha, \psi_\alpha)\}$, then the norm associated with the new choice is equivalent to the old one. Consequently, $L_s^2(E)$ is independent of such choices.

- If s is a nonnegative integer, then the Sobolev L_s^2-norm in (4.4) is equivalent to the one in Definition 4.88.

It is easy to check the next theorem.

Theorem 4.89 *Lemma 4.83, Theorems 4.84 and 4.85 are true for arbitrary* (X, E).

In particular, Lemmas 4.10 and 4.11 are true.

Let $X = T^4$ and $\mathfrak{s}$ be the spin^c structure $\mathfrak{s}_0$ on T^4 with $c_1(\mathfrak{s}_0) = 0$. Since the tangent bundle of T^4 and the spinor bundles of $\mathfrak{s}$ are trivial, we can directly apply the theory in this subsection to $(T^4, \mathfrak{s}_0)$. In particular, the first-order operators $d^* + d^+$ and D_{A_0} extends to the operators from L_{s+1}^2 to L_s^2 for any nonnegative real number s by Lemma 4.86. The Sobolev space L_{s+1}^2 is the maximal domain of the first-order operator to L_s^2.

Lemma 4.90 *Theorem 4.17 is true for* $(X, \mathfrak{s}) = (T^4, \mathfrak{s}_0)$.

Proof The proof is parallel to that of Lemma 4.86. For instance, in order to prove the inequality of the form

$$\|\phi\|_{L_{s+1}^2} \le C_s \left(\|\phi\|_{L_s^2} + \|L\phi\|_{L_s^2}\right), \tag{4.36}$$

represent the operator L in terms of ℓ^2-series by Fourier transform. We know the inequality is true for nonnegative integers k by Proposition 4.15. Let $s_x = x(k-1) + (1-x)k$ for $x \in [0, 1]$. Then we can prove the inequality

$$\begin{aligned}
&\sum_{\eta \in \mathbb{Z}^n} (1 + |\eta|^2)^{s_x/2} \left(L\left(\left((1 + |\xi|^2)^{-s_x/2} a_\xi\right)\right)\right)_\eta \overline{b_\eta} \\
&\qquad \le C_s^{1-x} \left(\|\phi\|_{\ell_0^2} + \|L\phi\|_{\ell_0^2}\right)^{1-x} C_t^x \left(\|\phi\|_{\ell_0^2} + \|L\phi\|_{\ell_0^2}\right)^x \|\psi\|_{\ell_0^2}
\end{aligned}$$

for finite series $\phi = (a_\xi)$ and $\psi = (b_\eta)$ in a way similar to the claim in the proof of Lemma 4.86. Let $C_s = \max(C_k, C_{k+1})$. We obtain the inequality (4.36) for $s = s_x$. □

We prove Theorem 4.17.

Proof of Theorem 4.17 We prove the case of $L = d^* + d^+$. The proofs of the other cases are similar. By Lemma 4.90,

$$\begin{aligned}
\|\psi_\alpha a\|_{L_{s+1}^2} &\le C_s \left(\|\psi_\alpha a\|_{L_s^2} + \|L(\psi_\alpha a)\|_{L_s^2}\right) \\
&\le C_s \left(\|\psi_\alpha a\|_{L_s^2} + \|(L\psi_\alpha) a\|_{L_s^2} + \|\psi_\alpha (La)\|_{L_s^2}\right).
\end{aligned}$$

The support of $(L\psi_\alpha)$ is in the support of ψ_α. We can find a constant $C_\alpha > 0$ such that

$$\|(L\psi_\alpha)a\|_{L^2_s} \le C_\alpha \|\psi_\alpha a\|_{L^2_s}.$$

Hence we obtain

$$\|\psi_\alpha a\|_{L^2_{s+1}} \le C_s(1+C_\alpha)\left(\|\psi_\alpha a\|_{L^2_s} + \|\psi_\alpha(La)\|_{L^2_s}\right).$$

Since $\|a\|^2_{L^2_s} = \sum_\alpha \|\psi_\alpha a\|^2_{L^2_{s+1}}$ is a finite sum, the previous inequality proves the theorem. □

Next we prove the multiplication lemma (Lemma 4.12).

Lemma 4.91 *For $s \ge 0$, there exists a constant $C_s > 0$ such that*

$$\|uv\|_{L^2_s} \le C_s(\|u\|_{L^2_s}\|v\|_{L^\infty} + \|u\|_{L^\infty}\|v\|_{L^2_s}), \tag{4.37}$$

for $u, v \in L^2_s(T^n) \cap L^\infty(T^n)$.

Proof By using Hölder's inequality $(a+b)^p \le 2^{p-1}(a^p+b^p)$, we obtain

$$\begin{aligned}(1+|\xi_1+\xi_2|^2)^r &\le (1+2|\xi_1|^2+2|\xi_2|^2)^r \le 2^r(1+|\xi_1|^2+1+|\xi_2|^2)^r \\ &\le c\left((1+|\xi_1|^2)^r+(1+|\xi_2|^2)^r\right),\end{aligned}$$

where $c = 2^{2r-1}$. Let $\langle\xi\rangle = (1+|\xi|^2)^{1/2}$.

$$\begin{aligned}\langle\xi\rangle^s|\widehat{uv}(\xi)| &= (1+|\xi|^2)^{s/2}\left|\sum_\eta \hat{u}(\xi-\eta)\hat{v}(\eta)\right| \\ &\le c\sum_\eta\left((1+|\xi-\eta|^2)^{s/2}+(1+|\eta|^2)^{s/2}\right)|\hat{u}(\xi-\eta)||\hat{v}(\eta)| \\ &= c\sum_\eta(1+|\xi-\eta|^2)^{s/2}|\hat{u}(\xi-\eta)||\hat{v}(\eta)| \\ &\quad + c\sum_\eta(1+|\eta|^2)^{s/2}|\hat{u}(\xi-\eta)||\hat{v}(\eta)|.\end{aligned}$$

Hence,

$$\begin{aligned}\langle\xi\rangle^{2s}|\widehat{uv}(\xi)|^2 \le& 2c^2\left(\sum_\eta(1+|\xi-\eta|^2)^{s/2}|\hat{u}(\xi-\eta)||\hat{v}(\eta)|\right)^2 \\ &+ 2c^2\left(\sum_\eta(1+|\eta|^2)^{s/2}|\hat{u}(\xi-\eta)||\hat{v}(\eta)|\right)^2.\end{aligned}$$

We set

$$w(x) = \sum_{\xi} |\hat{u}(\xi)| e^{ik\cdot\xi}, \qquad W(x) = \sum_{\xi} (1 + |\xi|^2)^{s/2} |\hat{u}(\xi)| e^{ik\cdot\xi},$$

$$z(x) = \sum_{\xi} |\hat{v}(\xi)| e^{ik\cdot\xi}, \qquad Z(x) = \sum_{\xi} (1 + |\xi|^2)^{s/2} |\hat{v}(\xi)| e^{ik\cdot\xi}.$$

Then we have

$$\langle\xi\rangle^{2s} |\widehat{uv}(\xi)|^2 \leq 2c^2 \left(\left|\widehat{wZ}(\xi)\right|^2 + \left|\widehat{Wz}(\xi)\right|^2 \right).$$

Therefore we obtain

$$\begin{aligned}
\|uv\|_{L_s^2}^2 &= \sum_{\xi} \langle\xi\rangle^{2s} |\widehat{uv}(\xi)|^2 \leq 2c^2 \sum_{\xi} \left(\left|\widehat{wZ}(\xi)\right|^2 + \left|\widehat{Wz}(\xi)\right|^2 \right) \\
&= 2c^2 \left(\|wZ\|_{L^2}^2 + \|Wz\|_{L^2}^2 \right) \\
&\leq 2c^2 \left(\|w\|_{L^\infty}^2 \|Z\|_{L^2}^2 + \|W\|_{L^2}^2 \|z\|_{L^\infty}^2 \right) \\
&= 2c^2 \left(\|u\|_{L^\infty}^2 \|v\|_{L_s^2}^2 + \|u\|_{L_s^2}^2 \|v\|_{L^\infty}^2 \right).
\end{aligned}$$

This completes the proof. □

Corollary 4.92 *If $s - \frac{n}{2} > 0$, then $L_s^2(T^n)$ is an algebra.*

Proof If $s - n/2 > 0$, then $L_s^2 \subset C^0$ by Theorem 4.84. Hence, there is a constant $C_0 > 0$ such that

$$\|u\|_{L^\infty} \leq C_0 \|u\|_{L_s^2}.$$

The inequality (4.37) implies that

$$\|uv\|_{L_s^2} \leq C_s(\|u\|_{L_s^2}\|v\|_{L^\infty} + \|u\|_{L^\infty}\|v\|_{L_s^2}) \leq C_0 C_s(\|u\|_{L_s^2}\|v\|_{L_s^2} + \|u\|_{L_s^2}\|v\|_{L_s^2}).$$

This completes the proof. □

Lemma 4.93 *Let $M \colon \ell_\infty^2(n) \times \ell_\infty^2(n) \to \ell_{-\infty}^2(n)$ be the bilinear operator defined by $(u, v) \mapsto u * v$. For $i = 0, 1$, suppose M is extended to the bounded operators $M \colon \ell_{s_i}^2(n) \times \ell_{t_i}^2(n) \to \ell_{r_i}^2(n)$ for some $s_i, t_i \geq r_i$, i.e., there are constants $C_i > 0$ such that*

$$\|u * v\|_{\ell_{r_i}^2} \leq C_i \|u\|_{\ell_{s_i}^2} \|v\|_{\ell_{t_i}^2}, \quad u \in \ell_{s_i}^2, v \in \ell_{t_i}^2.$$

Let $s_x = xs_1 + (1-x)s_0$, $t_x = xt_1 + (1-x)t_0$ $r_x = xr_1 + (1-x)r_0$ for $x \in [0, 1]$. Then, for any $x \in [0, 1]$, $M: \ell^2_{s_x} \times \ell^2_{t_x} \to \ell^2_{r_x}$ is bounded, and the following estimate holds:

$$\|u * v\|_{\ell^2_{r_x}} \leq C_1^x C_0^{1-x} \|u\|_{\ell^2_{s_x}} \|v\|_{\ell^2_{t_x}}, \quad u \in \ell^2_{s_x}, v \in \ell^2_{t_x}. \tag{4.38}$$

Proof *Mimic the proof of Lemma 4.86.* □

Lemma 4.94 *For any $\varepsilon > 0$ and $x \in [0, 1]$, the map M extends to the bounded operator $M: \ell^2_{x\cdot\frac{n}{2}}(n) \times \ell^2_{\frac{n}{2}+\varepsilon}(n) \to \ell^2_{x\cdot\frac{n}{2}}(n)$.*

Proof Since $L^2_{\frac{n}{2}+\varepsilon}$ is embedded in C^0 by Theorem 4.84, the map M extends to $\ell^2_0(n) \times \ell^2_{\frac{n}{2}+\varepsilon}(n) \to \ell^2_0(n)$. By Corollary 4.92, the map M also extends to $\ell^2_{\frac{n}{2}+\varepsilon}(n) \times \ell^2_{\frac{n}{2}+\varepsilon}(n) \to \ell^2_{\frac{n}{2}+\varepsilon}(n)$. Use Lemma 4.93. □

Lemma 4.95 (Lemma 4.12) *Suppose real numbers s_1, s_2, s satisfy that $s_1, s_2 \geq s \geq 0$ and $s_1 + s_2 - s > \frac{n}{2}$. Then the map $(f, g) \mapsto fg$ extends to $L^2_{s_1} \times L^2_{s_2} \to L^2_s$.*

Proof Suppose $s > \frac{n}{2}$. Note that L^2_s is an algebra by Corollary 4.92 and $L^2_{s_i} \subset L^2_s$ $(i = 1, 2)$. Then we have

$$L^2_{s_1} \times L^2_{s_2} \hookrightarrow L^2_s \times L^2_s \hookrightarrow L^2_s.$$

Suppose $s \leq \frac{n}{2}$. Let $\varepsilon = s_1 + s_2 - s - \frac{n}{2} > 0$. By Lemma 4.94, we have

$$\begin{aligned} L^2_s \times L^2_{\frac{n}{2}+\varepsilon} &\hookrightarrow L^2_s, \\ L^2_{\frac{n}{2}+\varepsilon} \times L^2_s &\hookrightarrow L^2_s. \end{aligned} \tag{4.39}$$

If $s \leq s_2$, then $s_1 \leq s_1 + s_2 - s = \frac{n}{2} + \varepsilon$. Then there exists $\theta \in [0, 1]$ such that $(1-\theta)s + \theta\left(\frac{n}{2} + \varepsilon\right) = s_1$. Then $(1-\theta)\left(\frac{n}{2} + \varepsilon\right) + \theta s = s_1$. Applying Lemma 4.93 to (4.39), we obtain the lemma. □

Unique Continuation of the Dirac Equation

The purpose of this subsection is to prove the following unique continuation theorem for Dirac equations.

Theorem 4.96 *Suppose Φ is a L^2_k-solution $(k > 2)$ of the Dirac equation $D_A\Phi = 0$. If Φ vanishes on some open set, then $\Phi(x) = 0$ for all $x \in X$.*

Remark 4.97 It is known that the unique continuation property holds for the solutions of rather general elliptic PDEs of 2nd order [4].

We follow the proof due to Booss-Bavnbek and Wojciechowski [21, Chapter 8]. For $x \in X$, choose a local geodesic polar coordinate centered at x. Choose $r_0, T > 0$ such that the ball with radius r_0+T centered at x is contained in the polar coordinate. For $0 < r \leq r_0 + T$, let $S_x(r)$ and $B_x(r)$ be respectively the sphere and open ball with radius r centered at x. The polar coordinate gives the identification

$$\begin{aligned} \overline{B_x(r_0+T) \setminus B_x(r_0)} &\overset{\cong}{\to} [0, T] \times S^3, \\ (r, \theta) &\mapsto (u := r - r_0, \theta). \end{aligned} \tag{4.40}$$

The induced metric on $[0, T] \times S^3$ has a form of $du \otimes du + g_u$ where g_u is the induced metric on $S_u := \{u\} \times S^3$. We can prove the following estimate of Carleman type.

Lemma 4.98 *For a Riemannian metric, a connection A and small $r_0 > 0$, there exist constants T and C such that the following inequality holds for sufficient large R:*

$$\begin{aligned} R \int_0^T \int_{S_u} e^{R(T-u)^2} \|\phi(u, \theta)\|^2 d\theta du \\ \leq C \int_0^T \int_{S_u} e^{R(T-u)^2} \|D_A\phi(u, \theta)\|^2 d\theta du. \end{aligned} \tag{4.41}$$

The proof will be given later. Assuming Lemma 4.98, we first prove Theorem 4.96.

Proof of Theorem 4.96 Let Φ be a solution to $D_A\Phi = 0$. Suppose $\Phi(x) = 0$ for some x. We prove that, if Φ vanishes on a small neighborhood of x, then Φ also vanishes on a larger neighborhood as follows.

Claim If there exists a small $r_0 > 0$ such that $\Phi(y) = 0$ for all $y \in B_x(r_0)$, then $\Phi(y) = 0$ for all $y \in B_x(r_0 + T/2)$ for T in Lemma 4.98.

Obviously, the claim proves the theorem. We prove the claim. Fix a smooth cut-off function ρ such that $\rho(r) = 1$ for $r \leq r_0 + \frac{8T}{10}$ and $\rho(r) = 0$ for $r > r_0 + \frac{9T}{10}$. Let $\phi(r, \theta) = \rho(r)\Phi(r, \theta)$. Then

$$\begin{aligned} \operatorname{supp} \phi &\in B_x\left(r_0 + \frac{9T}{10}\right) \setminus B_x(r_0), \\ \operatorname{supp} D_A\phi &\in B_x\left(r_0 + \frac{9T}{10}\right) \setminus B_x\left(r_0 + \frac{8T}{10}\right). \end{aligned} \tag{4.42}$$

Then we have

$$e^{RT^2/4} \int_{r_0}^{r_0+\frac{T}{2}} \int_{S(r)} \|\phi(r, \theta)\|^2 d\theta dr = \int_0^{\frac{T}{2}} \int_{S_u} e^{RT^2/4} \|\phi(u, \theta)\|^2 d\theta du$$

$$\leq \int_0^T \int_{S_u} e^{R(T-u)^2} \|\phi(u,\theta)\|^2 d\theta du$$

$$\overset{(4.41)}{\leq} \frac{C}{R} \int_0^T \int_{S_u} e^{R(T-u)^2} \|D_A\phi(u,\theta)\|^2 d\theta du$$

$$\overset{(4.42)}{\leq} \frac{C}{R} e^{RT^2/25} \int_0^T \int_{S_u} \|D_A\phi(u,\theta)\|^2 d\theta du.$$

Hence we obtain

$$\int_{r_0}^{r_0+\frac{T}{2}} \int_{S(r)} \|\phi(r,\theta)\|^2 d\theta dr \leq \frac{C}{R} e^{-21RT^2/100} \int_{r_0}^{r_0+T} \int_{S(r)} \|D_A\phi(r,\theta)\|^2 d\theta dr.$$

Letting $R \to \infty$, we obtain the claim. □

Proof of Lemma 4.98 Under the identification (4.40), we can take trivializations of spinor bundles over $[0, T] \times S^3$ and a connection A_0 such that the Dirac operator associated with A_0 has a form of

$$D_{A_0} = \frac{\partial}{\partial u} + D_u,$$

where D_u is a Dirac operator on $S_u = \{u\} \times S^3$. For a general connection A, the 1-form $A - A_0$ is decomposed into

$$A - A_0 = cdu + a_\theta,$$

where cdu is a radial form and a_θ is a 1-form on S_u. Then the Dirac operator D_A is written as

$$D_A = \frac{\partial}{\partial u} + D_u + \frac{1}{2}\rho(cdu + a_\theta).$$

Note that D_u is symmetric and $\frac{\partial}{\partial u} + \frac{1}{2}\rho(cdu + a_\theta)$ is anti-symmetric.

Define ψ by

$$\phi = e^{-R(T-u)^2}\psi.$$

Then (4.41) is rewritten as

$$R \int_0^T \int_{S_u} \|\psi(u,\theta)\|^2 d\theta du$$

$$\leq C \int_0^T \int_{S_u} \left\| \frac{\partial \psi}{\partial u} + D_u \psi + B\psi + R(T-u)\psi \right\|^2 d\theta du, \qquad (4.43)$$

where $B = \frac{1}{2}\rho(cdu + a_\theta)$. Then

$$\begin{aligned}
&\int_0^T \int_{S_u} \left\| \frac{\partial \psi}{\partial u} + D_u \psi + B\psi + R(T-u)\psi \right\|^2 d\theta du \\
&= \iint \left\| \frac{\partial \psi}{\partial u} + B\psi \right\|^2 \theta du + \iint \|D_u\psi + R(T-u)\psi\|^2 d\theta du \\
&\quad +2\mathrm{Re} \iint \left\langle \frac{\partial \psi}{\partial u} + B\psi, D_u\psi + R(T-u)\psi \right\rangle d\theta du. \qquad (4.44)
\end{aligned}$$

Note that $\mathrm{Re} \iint \langle \psi, P_+\psi \rangle d\theta du = \frac{1}{2} \iint \langle \psi, P_+\psi \rangle d\theta du$ for a symmetric operator P_+, and $\mathrm{Re} \iint \langle \psi, P_-\psi \rangle d\theta du = 0$ for an anti-symmetric operator P_-. The last term of (4.44) is rearranged as follows:

$$\begin{aligned}
&2\mathrm{Re} \iint \left\langle \frac{\partial \psi}{\partial u} + B\psi, D_u\psi + R(T-u)\psi \right\rangle d\theta du \\
&= 2\mathrm{Re} \iint \left\langle \frac{\partial \psi}{\partial u}, D_u\psi + R(T-u)\psi \right\rangle d\theta du + 2\mathrm{Re} \iint \langle B\psi, D_u\psi \rangle d\theta du \\
&= -2\mathrm{Re} \iint \left\langle \psi, \left\{ \frac{\partial}{\partial u}(D_u + R(T-u)) \right\} \psi \right\rangle d\theta du - 2\mathrm{Re} \iint \langle \psi, BD_u\psi \rangle d\theta du \\
&= \iint \left\langle \psi, -\frac{\partial D_u}{\partial u}\psi + R\psi \right\rangle d\theta du + \mathrm{Re} \iint \langle \psi, [D_u, B]\psi \rangle d\theta du \\
&= R \int_0^T \|\psi\|^2_{L^2(S_u)} du + \iint \left\langle \psi, -\frac{\partial D_u}{\partial u}\psi + [D_u, B]\psi \right\rangle d\theta du, \qquad (4.45)
\end{aligned}$$

where $\|\cdot\|_{L^2(S_u)}$ is the L^2-norm on S_u. The inequality (4.43) follows from (4.44) and (4.45) if there is a constant K with $0 < K < 1$ such that

$$\begin{aligned}
&\left| \iint \left\langle \psi, -\frac{\partial D_u}{\partial u}\psi + [D_u, B]\psi \right\rangle d\theta du \right| \\
&\leq K \left(R \int_0^T \|\psi\|^2_{L^2(S_u)} du + \iint \|D_u\psi + R(T-u)\psi\|^2 d\theta du \right). \qquad (4.46)
\end{aligned}$$

Since D_u is a first-order elliptic operator, there is a constant C_1 such that the estimate holds:

$$\|\psi\|_{L^2_1} \le C_1(\|\psi\|_{L^2} + \|D_u\psi\|_{L^2}),$$

for any spinor ψ on S_u. We can also find a constant C_2 such that the estimate holds:

$$\left\| -\frac{\partial D_u}{\partial u}\psi + [D_u, B]\psi \right\|_{L^2} \le C_2\|\psi\|_{L^2_1}.$$

Note that C_1 and C_2 depend only on the Riemannian metric, the connection A and the initial choice for Sobolev norms. Below we will use the estimate

$$ab = \frac{a}{c}\cdot cb \le \frac{1}{2}\left(\frac{a^2}{c^2} + c^2b^2\right). \tag{4.47}$$

Then

$$\begin{aligned}
&\left|\iint \left\langle \psi, -\frac{\partial D_u}{\partial u}\psi + [D_u, B]\psi \right\rangle d\theta du\right| \\
&\le \int_0^T \|\psi\|_{L^2}\left\| -\frac{\partial D_u}{\partial u}\psi + [D_u, B]\psi \right\|_{L^2} du \\
&\le C_2\int_0^T \|\psi\|_{L^2}\|\psi\|_{L^2_1} du \\
&\le C_1C_2\int_0^T \|\psi\|_{L^2}(\|\psi\|_{L^2} + \|D_u\psi\|_{L^2}) du \\
&\le C_1C_2\int_0^T \|\psi\|_{L^2}\left(\|(D_u + R(T-u))\psi\|_{L^2} + (R(T-u)+1)\|\psi\|_{L^2}\right) du \\
&\le C_1C_2(RT+1)\int_0^T \|\psi\|^2_{L^2} du + C_1C_2\int_0^T \|\psi\|_{L^2}\|(D_u + R(T-u))\psi\|_{L^2} du \\
&\overset{(4.47)}{\le} C_1C_2(RT+1)\int_0^T \|\psi\|^2_{L^2} du \\
&\qquad + C_1C_2\int_0^T \frac{1}{2}\left\{C_1C_2\|\psi\|^2_{L^2} + \frac{1}{C_1C_2}\|(D_u + R(T-u))\psi\|^2_{L^2}\right\} du \\
&= C_1C_2\left(T + \frac{C_1C_2+2}{2R}\right)R\int_0^T \|\psi\|^2_{L^2} + \frac{1}{2}\int_0^T \|(D_u + R(T-u))\psi\|^2_{L^2} du.
\end{aligned}$$

By choosing sufficiently small T (say $T < \frac{1}{4C_1C_2}$), the inequality (4.46) holds for large R (say $R > 2C_1C_2(C_1C_2+2)$). □

Construction of the Determinant Line Bundles

The purpose of this subsection is to explain the construction of the determinant line bundles. Let X and Y be separable Hilbert spaces, and $\mathcal{F}(X, Y)$ be the space of Fredholm operators $D\colon X \to Y$. Let X^* and Y^* be the dual spaces of X and Y respectively. For $D \in \mathcal{F}(X, Y)$, let $D^*\colon Y^* \to X^*$ be the adjoint operator of D.

Definition 4.99 The determinant of D is the 1-dimensional real vector space

$$\det(D) = \bigwedge^{\max}(\ker D) \otimes \bigwedge^{\max}(\ker D^*).$$

The goal of this subsection is to prove that $\coprod_{D\in\mathcal{F}(X,Y)} \det(D)$ forms a locally trivial real line bundle over $\mathcal{F}(X, Y)$. (A reference for determinant line bundles is [151, Appendix A]. A general reference for functional analysis is [105, 150].)

The next lemma is well-known as the stability of Fredholm operators. (See e.g. [105].)

Lemma 4.100 *Let $D\colon X \to Y$ be a Fredholm operator.*

1. *If $K\colon X \to Y$ is a compact operator, then $D + K$ is a Fredholm operator and* $\operatorname{ind}(D + K) = \operatorname{ind} D$.
2. *There is a constant $\varepsilon > 0$ such that, if $B\colon X \to Y$ is a bounded linear operator with $\|B\| < \varepsilon$, then $D + B$ is a Fredholm operator, and* $\operatorname{ind}(D + B) = \operatorname{ind} D$.
3. *The adjoint operator $D^*\colon Y^* \to X^*$ is Fredholm and* $\operatorname{ind}(D) = -\operatorname{ind}(D^*)$.

We need several lemmas.

Lemma 4.101 *Let $D\colon X \to Y$ be a bounded operator.*

1. *If there exists a constant $c > 0$ such that*

$$\|x\|_X \le c\|Dx\|_Y, \tag{4.48}$$

   then D is injective and has a closed range.
2. *If D is surjective, then D admits a right inverse S, that is, a bounded operator $S\colon Y \to X$ such that $D \circ S = \mathrm{id}_Y$.*

Proof

1. Injectivity is obvious. Suppose y is in the closure of $\operatorname{im} D$. Then there is a sequence $\{x_n\} \subset X$ such that $\lim_{n\to\infty} Dx_n = y$. Since $\{Dx_n\}$ is a Cauchy sequence, the inequality (4.48) implies $\{x_n\}$ is also a Cauchy sequence, and hence there is a limit $x = \lim_{n\to\infty} x_n$. Then $y = Dx$.
2. Since $\ker D$ is a closed subspace in the Hilbert space X, there is a closed orthogonal complement L of $\ker D$. Let $P\colon X \to L$ be the orthogonal projection. For $y \in Y$, we can find $x \in X$ such that $Dx = y$ by the surjectivity of D. Define S by $Sy = Px$. Note that the definition of S does not depend on

the choice of x and $D \circ S = \mathrm{id}_Y$. It is easy to see that S is bounded by using the open mapping theorem.

□

For a subspace $W \subset X$, let

$$W' = \{f \in X^* \mid f(u) = 0 \text{ for } u \in W\}.$$

It is easy to see the next lemma.

Lemma 4.102 *Let $D\colon X \to Y$ be a bounded operator. Then* $\ker D^* = (\mathrm{im}\, D)'$. *Furthermore, if D has a closed range, then there is a natural isomorphism* $(Y/\,\mathrm{im}\, D)^* \cong (\mathrm{im}\, D)'$.

Corollary 4.103 *For a Fredholm map D, there is a natural isomorphism*

$$\det(D) \cong \bigwedge^{\max}(\ker D) \otimes \bigwedge^{\max}(\mathrm{coker}\, D)^*.$$

Lemma 4.104 *For a finite-dimensional vector space, there is a natural isomorphism*

$$\bigwedge^{\max} V \otimes \bigwedge^{\max} V^* \cong \mathbb{R}.$$

Hence $\left(\bigwedge^{\max} V\right)^* \cong \bigwedge^{\max} V^*$.

Proof The isomorphism $\bigwedge^{\max} V \otimes \bigwedge^{\max} V^* \to \mathbb{R}$ is induced by the map $V \otimes V^* \to \mathbb{R}$, $v \otimes f \mapsto f(v)$. The inverse is given by $1 \mapsto (v_1 \wedge \cdots \wedge v_n) \otimes (v_1^* \wedge \cdots \wedge v_n^*)$ where $(v_1, \ldots, v_n)$ is a basis of V and v_i^* is a unique dual of v_i. (This does not depend on the choice of basis.) □

Lemma 4.105 *Let V be a finite-dimensional vector space, and W be a linear subspace of V. Then there is a natural isomorphism*

$$\bigwedge^{\max} V \cong \bigwedge^{\max} W \otimes \bigwedge^{\max}(V/W).$$

Proof Choose a basis $(v_1, \ldots, v_n)$ of V such that $(v_1, \cdots, v_k)$ is a basis of W. The isomorphism is given by

$$v_1 \wedge \cdots \wedge v_n \mapsto (v_1 \wedge \cdots \wedge v_k) \otimes ([v_{k+1}] \wedge \cdots \wedge [v_n]).$$

This does not depend on the choice of basis as above. □

Theorem 4.106 *Let*

$$\det(\mathcal{F}(X, Y)) = \{(D, v) \mid D \in \mathcal{F}(X, Y),\ v \in \det(D)\}.$$

Then $\det(\mathcal{F}(X,Y)) \to \mathcal{F}(X,Y)$, $(D, v) \mapsto D$, *is a real line bundle over* $\mathcal{F}(X,Y)$.

Proof We need to show the local triviality. Suppose first that $D \in \mathcal{F}(X,Y)$ is surjective. If $B\colon X \to Y$ is a bounded operator with sufficiently small $\|B\|$, then $D + B$ is a surjective Fredholm operator. The surjectivity is proved by applying Lemma 4.101 to $(D+B)^*$. The Fredholm property is obtained from Lemma 4.100. Since D is surjective, D has a right inverse S by the assertion 2 of Lemma 4.101. Then we have an isomorphism

$$\ker(D+B) \to \ker D, \quad x \mapsto x + SBx,$$

for every B with sufficiently small $\|B\|$. Therefore $\det(D+B) = \bigwedge^{\max} \ker(D+B)$ form a locally trivial vector bundle in a neighborhood of a surjective Fredholm operator D.

The general case is reduced to the surjective case as follows. For a Fredholm operator D, we can choose an injective linear map $F\colon \mathbb{R}^k \to Y$ such that the operator

$$D \oplus F\colon X \oplus \mathbb{R}^k \to Y, \quad (x, v) \mapsto Dx + Fv,$$

is surjective. For instance, let $k = \dim \operatorname{coker} D$, and choose $y_1, \ldots, y_k$ which span a subspace W such that $\operatorname{im} D + W = Y$. Define F by $F(v) = \sum_i v_i y_i$.

Consider the exact sequence

$$0 \longrightarrow \ker D \xrightarrow{\alpha} \ker(D \oplus F) \xrightarrow{\beta} \operatorname{im} D \cap \operatorname{im} F \longrightarrow 0,$$

where α, β are defined by $\alpha(x) = (x, 0)$, $\beta(x, v) = Fv = -Dx$. Hence there is a natural isomorphism

$$\operatorname{im} D \cap \operatorname{im} F \cong \ker(D \oplus F)/\ker D.$$

By Lemma 4.105,

$$\bigwedge^{\max} \ker(D \oplus F) \cong \bigwedge^{\max} \ker D \otimes \bigwedge^{\max} (\operatorname{im} D \cap \operatorname{im} F).$$

We prove that there is a natural isomorphism $\bigwedge^{\max}(\operatorname{im} D \cap \operatorname{im} F) \cong \bigwedge^{\max} \ker D^*$. Applying Lemma 4.105 to the inclusion $\operatorname{im} D \cap \operatorname{im} F \subset \operatorname{im} F$, we obtain

$$\bigwedge^{\max} (\operatorname{im} D \cap \operatorname{im} F) \otimes \bigwedge^{\max} \left(\frac{\operatorname{im} F}{\operatorname{im} D \cap \operatorname{im} F} \right) \cong \bigwedge^{\max} \operatorname{im} F \cong \bigwedge^{\max} \mathbb{R}^k \cong \mathbb{R}.$$

Hence, by Lemma 4.104,

$$\bigwedge^{\max}(\operatorname{im} D \cap \operatorname{im} F) \cong \left(\bigwedge^{\max} \frac{\operatorname{im} F}{\operatorname{im} D \cap \operatorname{im} F}\right)^* \cong \bigwedge^{\max}\left(\frac{\operatorname{im} F}{\operatorname{im} D \cap \operatorname{im} F}\right)^*.$$

There is a natural isomorphism

$$\frac{\operatorname{im} F}{\operatorname{im} D \cap \operatorname{im} F} \cong \frac{Y}{\operatorname{im} D}.$$

By Lemma 4.102,

$$\left(\frac{Y}{\operatorname{im} D}\right)^* \cong \ker D^*.$$

Summing up these, we obtain $\bigwedge^{\max}(\operatorname{im} D \cap \operatorname{im} F) \cong \bigwedge^{\max} \ker D^*$ and

$$\det(D \oplus F) = \bigwedge^{\max} \ker(D \oplus F) \cong \bigwedge^{\max} \ker D \otimes \bigwedge^{\max} \ker D^* \cong \det(D).$$

This completes the proof. □

Problems

4.107 Verify Proposition 4.4.

4.108 Verify (4.1).

4.109 Verify Lemma 4.8.

4.110 Verify Proposition 4.20.

4.111 Verify (4.8).

4.112 Let L be a complex line bundle over $S^1 \times S^1$ equipped with a U(1)-connection A. Verify that the Chern number $\langle c_1(L), [S^1 \times S^1]\rangle$ is equal to the degree of the map

$$S^1 \ni t \mapsto \mathrm{Hol}_{S^1 \times \{t\}}(A) \in \mathrm{U}(1).$$

(Hint: The first Chern class of L is the Euler class of L. Count the zeros of a generic section of L.)

4.113 Verify that the first Chern class of the universal line bundle $\mathbb{L}$ is given by

$$c_1(\mathbb{L}) = 1 \otimes u + \sum_{i=1}^{b_1} \gamma_i^\vee \otimes \mu(\gamma_i),$$

where $u = \mu([x_0])$, $(\gamma_1, \ldots, \gamma_{b_1})$ is a basis for $H_1(X; \mathbb{Z})/\mathrm{Tor}$, and $(\gamma_1^\vee, \ldots, \gamma_{b_1}^\vee)$ is its dual basis for $H^1(X; \mathbb{Z})$.

4.114 Verify Corollary 4.40.

4.115 Complete the proof of Proposition 4.46.

4.116 Verify Proposition 4.60.

Chapter 5
Bauer–Furuta Theory

This chapter describes the fundamentals of the Bauer–Furuta theory.

- The monopole map $\boldsymbol{\mu}_{SW}$ is introduced. This is a map between certain Hilbert bundles such that the equation $\boldsymbol{\mu}_{SW} = 0$ is the Seiberg–Witten equations.
- The Bauer–Furuta invariant is defined as a stable homotopy class of a finite-dimensional approximation of the monopole map. The relationship between the Seiberg–Witten and Bauer–Furuta invariants is also discussed.
- As applications of the finite-dimensional approximation, we give the proofs of the Donaldson diagonalization theorem and Furuta's $10/8$-inequality.

5.1 Monopole Map

In this section, we introduce the monopole map $\boldsymbol{\mu}_{SW}$ which is a $\mathrm{U}(1)$-equivariant map between certain Hilbert bundles such that $\boldsymbol{\mu}_{SW}^{-1}(0)/\mathrm{U}(1)$ is identified with the Seiberg–Witten moduli space $\mathcal{M}(\mathfrak{s})$.

Choose a reference $\mathrm{U}(1)$-connection A_0 on L. We define the map

$$m\colon \sqrt{-1}\,\mathrm{Ker}(d^*\colon \Omega^1(X) \to \Omega^0(X)) \oplus \Gamma(S^+) \to \sqrt{-1}\Omega^+(X) \oplus \Gamma(S^-)$$

by

$$m(a, \phi) = (F_{A_0}^+ + d^+a - q(\phi), D_{A_0+a}\phi).$$

Note that the equation $m(a, \phi) = 0$ is the Seiberg–Witten equations.

The map m is decomposed into the sum $m = \ell_0 + c_0$, where ℓ_0 is the linear map given by $\ell_0 = (d^+, D_{A_0})$, and c is the quadratic part given by $c_0(a, \phi) = (F_{A_0}^+ - q(\phi), \frac{1}{2}\rho(a)\phi)$. We consider the Sobolev completions of the domain and the

T. Kato, N. Nakamura, *Monopole Maps and L^2-cohomology*, Springer Monographs in Mathematics, https://doi.org/10.1007/978-981-96-8472-4_5

target of m. Choose $k \geq 4$. Extend $d^*: \Omega^1(X) \to \Omega^0(X)$ to $d^*: L^2_k(X; \Lambda^1) \to L^2_{k-1}(X; \Lambda^0)$. Then m is extended to the smooth map

$$m\colon \sqrt{-1}\,\mathrm{Ker}\, d^* \oplus L^2_k(X; S^+) \to L^2_{k-1}(X; \sqrt{-1}\Lambda^+ \oplus S^-).$$

The linear part ℓ_0 is a Fredholm map of index

$$\frac{1}{4}(c_1(\mathfrak{s})^2 - \sigma(X)) - b_0(X) + b_1(X) - b^+(X),$$

and c_0 is a quadratic compact map.

We take the L^2_{k+1}-completion of the gauge transformation group $\mathcal{G}$, denoted by the same symbol $\mathcal{G}$ to simplify the notation. Then the $\mathcal{G}$-action is smooth. By Theorem 4.34, The space

$$A_0 + \sqrt{-1}\,\mathrm{Ker}(d^*: L^2_k(X; \Lambda^1) \to L^2_{k-1}(X; \Lambda^0))$$

is a global slice for the $\mathcal{G}$-action at $(A_0, 0)$. For $\eta \in \sqrt{-1}L^2_{k-1}\left(X; \Lambda^+\right)$, we have

$$m^{-1}(\eta) = \{\text{solutions to } (4.22)\} \cap \ker d^*.$$

Let K be $\mathrm{Ker}(d^* \circ \delta) \cong \mathrm{U}(1) \times H^1(X; \mathbb{Z})$ in (4.13). Since the $\mathcal{G}$-action on $\sqrt{-1}L^2_k(X; \Lambda^1)$ is given by $\sqrt{-1}\delta$, K acts on $\mathrm{Ker}\, d^*$. Let K act on $L^2_{k-1}(X; S^-)$ as usual, and trivially on $L^2_{k-1}(X; \Lambda^+)$. Then m is K-equivariant, and we have

$$m^{-1}(\eta)/K = \mathcal{M}(\mathfrak{s}, \eta).$$

Note that the decomposition $\sqrt{-1}\,\mathrm{Ker}\, d^* = \sqrt{-1}(H^1(X, \mathbb{R}) \oplus \mathrm{Im}\, d^*)$, where $H^1(X, \mathbb{R})$ is identified with the space of harmonic 1-forms. Consider the following commutative diagram:

$$\begin{array}{ccc}
\sqrt{-1}H^1(X, \mathbb{R}) \oplus \sqrt{-1}\,\mathrm{Im}\, d^* \oplus L^2_k(X; S^+) & \xrightarrow{m \oplus \varpi} & L^2_{k-1}(X; \sqrt{-1}\Lambda^+(X) \oplus S^-) \oplus H^1(X; \mathbb{R}) \\
\Big\downarrow \text{projection} & & \Big\downarrow \text{projection} \\
H^1(X; \mathbb{R}) & = & H^1(X; \mathbb{R})
\end{array}$$

Dividing the diagram by $H^1(X; \mathbb{Z}) \subset K = \mathrm{U}(1) \times H^1(X; \mathbb{Z})$, we obtain a fiber-preserving $\mathrm{U}(1)$-equivariant map $\boldsymbol{\mu}_{\mathrm{SW}}$ between Hilbert bundles,

$$\begin{array}{ccc}
\mathcal{E} & \xrightarrow{\boldsymbol{\mu}_{\mathrm{SW}}} & \mathcal{F} \\
\Big\downarrow \pi & & \Big\downarrow \\
J & = & J
\end{array} \tag{5.1}$$

where

$$\begin{aligned}
J &= H^1(X;\mathbb{R})/H^1(X;\mathbb{Z}),\\
\mathcal{E} &= \left(\sqrt{-1}H^1(X,\mathbb{R}) \oplus \sqrt{-1}\operatorname{Im} d^* \oplus L^2_k\left(X;S^+\right)\right)/H^1(X;\mathbb{Z}),\\
\mathcal{F} &= L^2_{k-1}\left(X;\sqrt{-1}\Lambda^+ \oplus S^-\right) \times_{H^1(X;\mathbb{Z})} H^1(X;\mathbb{R}).
\end{aligned}$$

The bundles $\mathcal{E}$ and $\mathcal{F}$ have the Sobolev inner products.

Definition 5.1 We call the map $\boldsymbol{\mu}_{SW}$ the *monopole map.*

Note that the monopole map $\boldsymbol{\mu}_{SW}$ is U(1)-equivariant, and $\boldsymbol{\mu}_{SW}^{-1}(\eta)/\mathrm{U}(1)$ is identified with the Seiberg–Witten moduli space $\mathcal{M}(\mathfrak{s},\eta)$.

The map $\boldsymbol{\mu}_{\mathrm{SW}}$ is decomposed into the sum $\boldsymbol{\mu}_{\mathrm{SW}} = \ell + c$ of linear and quadratic parts as follows:

$$\begin{aligned}
\ell\left(\sqrt{-1}h, b, \phi\right) &= \left(d^+ b, D_{A_0+\sqrt{-1}h}\phi\right),\\
c\left(\sqrt{-1}h, b, \phi\right) &= \left(F^+_{A_0} - q(\phi), \frac{1}{2}\rho(b)\phi\right),
\end{aligned}$$

for $(\sqrt{-1}h, b, \phi) \in \sqrt{-1}H^1(X,\mathbb{R}) \oplus \sqrt{-1}\operatorname{Im} d^* \oplus L^2_k\left(X;S^+\right)$.

Lemma 5.2 *The quadratic map c is compact, that is, c maps bounded sets to relatively compact sets.*

Proof Apply Rellich's lemma (Lemma 4.11). (Problem 5.52.) □

Remark 5.3 When we assume ℓ as a family $\ell = \{(d^+, D_{A_0+\sqrt{-1}h})\}_{[h]\in J}$ of elliptic operators parametrized by J, the index of ℓ is defined as an element of the U(1)-equivariant KO-group $KO_{\mathrm{U}(1)}(J)$.

5.2 Finite-Dimensional Approximation and Stable Cohomotopy

The final goal of this section is to give the construction of the Bauer–Furuta invariant [15, 16, 18]. Their invariant is constructed as a stable homotopy class in a certain stable cohomotopy group by taking finite-dimensional approximations of the monopole map $\boldsymbol{\mu}_{\mathrm{SW}}$.

A stable homotopy class will be constructed for a U(1)-equivariant map $\boldsymbol{\mu}\colon \mathcal{E} \to \mathcal{F}$ satisfying the following conditions:

1. $\boldsymbol{\mu}$ has a form $\boldsymbol{\mu} = \ell + c$, where ℓ is a linear Fredholm map and c is a compact map.
2. The inverse image $\boldsymbol{\mu}^{-1}(0)$ is bounded.

Remark 5.4 The monopole map $\boldsymbol{\mu}_{\mathrm{SW}}$ satisfies the above conditions 1 and 2. Especially, the compactness of the moduli space implies the second condition. Furthermore, the monopole map $\boldsymbol{\mu}_{\mathrm{SW}}$ also satisfies the following third condition:

3. The restriction of $\boldsymbol{\mu}_{\mathrm{SW}}$ to the U(1)-fixed part is identified with the inclusion

$$\begin{aligned}(\mathcal{E})^{\mathrm{U}(1)} &= J \times \sqrt{-1}\,\mathrm{im}\, d^* \hookrightarrow J \times \sqrt{-1}\,\mathrm{im}\, d^* \oplus H^+ \\ &\cong J \times \sqrt{-1}\,\mathrm{im}\, d^+ \oplus H^+ = (\mathcal{F})^{\mathrm{U}(1)},\end{aligned}$$

where $H^+ = H^+\left(X; \sqrt{-1}\mathbb{R}\right)$.

Let $\boldsymbol{\mu}$ be a map satisfying the above conditions 1 and 2. For $R > 0$, let $D_R \subset \mathcal{E}$ be the disk bundle whose fibers are disks of radius R. By the second condition, there is an $R > 0$ such that $\boldsymbol{\mu}^{-1}(0)$ is contained in the interior of D_R. Let $S_R = \partial D_R$ be the sphere bundle of D_R. Then $\boldsymbol{\mu}(v) \neq 0$ for any $v \in S_R$.

Lemma 5.5 ([59, Lemma3.2]) *There is a positive real number δ such that $\|\boldsymbol{\mu}(v)\| \geq \delta$ for any $v \in S_R$.*

Proof Suppose there is a sequence $\{v_1, v_2, \ldots\} \subset S_R$ such that $\|\boldsymbol{\mu}(v_i)\|_{L^2_{k-1}} \to 0$ as $i \to \infty$. Let $\pi\colon S_R \to J$ be the projection of the sphere bundle. Since J is compact, there is a subsequence such that $\pi v_i \to t_0$ for some $t_0 \in J$. On a small neighborhood U of t_0, choose a local trivialization $(S_R)|_U = U \times F$, where $F = \sqrt{-1}\,\mathrm{Im}\, d^* \oplus L^2_k\left(X; S^+\right)$. We may assume a subsequence is a bounded sequence in F. Since a bounded sequence in a Hilbert space has a weakly converging subsequence, we may assume $\{v_1, v_2, \ldots\}$ has a weak limit v_∞. Recall that $\boldsymbol{\mu} = \ell + c$ where ℓ is an elliptic operator and c is a compact operator on each fiber. Then the sequence $\{c(v_i)\}$ is strong convergent to $c(v_\infty)$ in L^2_{k-1}. Since $\|\boldsymbol{\mu}(v_i)\|_{L^2_{k-1}} = \|\ell(v_i) + c(v_i)\|_{L^2_{k-1}} \to 0$, $\{\ell(v_i)\}$ is also strong convergent to $-c(v_\infty)$ in L^2_{k-1}. By Rellich's lemma (Lemma 4.11), the sequence $\{v_i\}$ is *strongly convergent to v_∞ in L^2_{k-1}*. By the elliptic estimate,

$$\|v_i - v_j\|_{L^2_k} \leq \text{const.} \left(\|v_i - v_j\|_{L^2_{k-1}} + \|\ell(v_i - v_j)\|_{L^2_{k-1}}\right).$$

Hence $\{v_i\}$ is a Cauchy sequence in L^2_k and strong convergent to v_∞ in L^2_k. Thus we obtain $v_\infty \in S$ and $\boldsymbol{\mu}(v_\infty) = 0$. This is a contradiction. □

Recall that the monopole map is $\boldsymbol{\mu}\colon \mathcal{E} \to \mathcal{F}$. By the Kuiper theorem [104], we can trivialize $\mathcal{F} = J \times \mathcal{U}$, where $\mathcal{U} = L^2_{k-1}\left(X; \sqrt{-1}\Lambda^+ \oplus S^-\right)$. We consider the composite map of $\boldsymbol{\mu}$ with the projection $\mathcal{F} = J \times \mathcal{U} \to \mathcal{U}$. The composite map is also denoted by $\boldsymbol{\mu}$. Recall that $\boldsymbol{\mu}$ is decomposed into the sum $\boldsymbol{\mu} = \ell + c$ of the linear part ℓ and the quadratic part c. The linear part ℓ is assumed as a family $\ell = \{(d^+, D_{A_0+\sqrt{-1}h})\}_{[h]\in J}$ of elliptic operators over J, and the index of ℓ is defined as an element of the equivariant KO group $KO_{\mathrm{U}(1)}(J)$. Let ℓ_h be the restriction of ℓ to the fiber over $h \in J$. Then ℓ_h is a Fredholm map. Since J is compact, we have the next lemma.

Lemma 5.6 *There is a finite-dimensional subspace $U_0 \subset \mathcal{U}$ such that* $\operatorname{Im}\ell_h + U_0 = \mathcal{U}$ *for every $h \in J$.*

Let $E_U = \ell^{-1}(U)$ and $F_U = J \times U$ for a finite-dimensional subspace $U \subset \mathcal{U}$. Then the restriction of ℓ to E_U is a fiberwise-linear map from E_U to U. If $U \supset U_0$, then $\operatorname{ind}\ell = [E_U] - [F_U] \in KO_{\mathrm{U}(1)}(J)$. Let $P_U\colon \mathcal{U} \to U$ be the L^2-orthogonal projection. By definition, $\ell(E_U) = U$, but the image $c(E_U)$ is not necessarily contained in U. We define the *finite-dimensional approximation* of $\boldsymbol{\mu}$ by

$$f_U = \ell + P_U c\colon E_U \to U.$$

Lemma 5.7 ([59, Lemma 3.3]) *For $\varepsilon > 0$, there is a large finite-dimensional subspace $\hat{U}$ such that $\hat{U} \supset U_0$ and $\|(1 - P_{\hat{U}})c(v)\| < \varepsilon$ for $v \in S_R$.*

Proof Consider a sequence of finite-dimensional subspaces

$$U_0 \subset U_1 \subset \cdots \subset \mathcal{U}$$

satisfying that $\lim_{n\to\infty} P_{U_n}(u) = u$ for $u \in \mathcal{U}$. Suppose there is a sequence $v_1, v_2, \ldots$ in S_R such that $\|(1 - P_{U_n})c(v_n)\| \geq \varepsilon$. Replacing with a subsequence, we may assume the sequence is weakly convergent to v_∞ and the sequence $c(v_1), c(v_2), \ldots$ is strong convergent to $c(v_\infty)$. Since $\lim_{n\to\infty} P_{U_n}(u) = u$, there is a large n_0 such that $\|(1 - P_{U_{n_0}})c(v_{n_\infty})\| < \varepsilon/2$. Since $c(v_n) \to c(v_\infty)$ strongly, we have $\|(1 - P_{U_{n_0}})(c(v_{n_1}) - c(v_{n_\infty}))\| < \varepsilon/2$ for some $n_1 \geq n_0$. Since $U_{n_0} \subset U_{n_1}$, we have $\|(1 - P_{U_{n_1}})c(v_{n_1})\| \leq \|(1 - P_{U_{n_0}})c(v_{n_1})\|$. Then we obtain

$$\begin{aligned}\|(1 - P_{U_{n_1}})c(v_{n_1})\| &\leq \|(1 - P_{U_{n_0}})c(v_{n_1})\| \\ &\leq \|(1 - P_{U_{n_0}})(c(v_{n_1}) - c(v_{n_\infty}))\| + \|(1 - P_{U_{n_0}})c(v_{n_\infty})\| \\ &\leq \frac{\varepsilon}{2} + \frac{\varepsilon}{2} = \varepsilon.\end{aligned}$$

This is a contradiction. □

For a finite-dimensional subspace $U \subset \mathcal{U}$, let $D_U = D_R \cap E_U$ and $S_U = S_R \cap E_U = \partial D_U$. Then D_U and S_U are disk and sphere bundles of E_U.

Corollary 5.8 *For δ in Lemma 5.5, choose ε such that $0 < \varepsilon < \delta$ and set $r = \delta - \varepsilon$. Choose $\hat{U}$ in Lemma 5.7 for ε. Then $\|f_{\hat{U}}(v)\| \geq r$ for $v \in S_{\hat{U}}$. Furthermore, if $U \supset \hat{U}$, then $\|f_U(v)\| \geq r$ for $v \in S_U$.*

Let $T(E_U)$ be the Thom space of E_U and S^U be the one-point compactification of U. By Corollary 5.8, the map f_U for $U \supset \hat{U}$ induces a U(1)-equivariant based map $f_U^+ \colon T(E_U) \to S^U$ as follows. We have a U(1)-equivariant homotopy equivalence $g \colon T(E_U) \to E_U/(\overline{E_U \setminus D_U}) = D_U/S_U$. By restriction, f_U defines a relative map $(D_U, S_U) \to (U, \overline{U \setminus D_r(U)})$, where $D_r(U)$ is the disk centered at 0 with radius r in U. Collapsing S_U, we obtain a map $\bar{f}_U \colon D_U/S_U \to U/(\overline{U \setminus D_r(U)}) = D_r(U)/\partial D_r(U)$. The map

$$(D_r(U), \partial D_r(U)) \to (S^U, \infty), \quad u \mapsto \frac{1}{r - \|u\|} u$$

induces a U(1)-equivariant homotopy equivalence $e \colon D_r(U)/\partial D_r(U) \to S^U$. Define f_U^+ by $f_U^+ = e \circ \bar{f}_U \circ g$.

Proposition 5.9 *Suppose $U_1, U_2 \subset \mathcal{U}$ be subspaces such that $\hat{U} \subset U_1 \subset U_2$. Let $U_2 = W \oplus U_1$ be the L^2-orthogonal decomposition. Then $f_{U_2}^+$ is U(1)-equivariantly homotopic to $(\mathrm{id}_W)^+ \wedge f_{U_1}^+$ as based maps, where $(\mathrm{id}_W)^+ \colon S^W \to S^W$ is the extension of the identity map of W to its one-point compactification.*

Proof Note that the restriction of $\ell|_{E_W}$ to each fiber of E_W is a linear isomorphism and E_W is a trivial bundle. We may assume the identification $E_{U_2} = E_{U_1} \times W$. Let $P_{E_{U_1}} \colon E_{U_2} \to E_{U_1}$ be the fiberwise L^2-orthogonal projection. The homotopy defined by

$$\ell + ((1-t)P_{U_2} + tP_{U_1}) \circ c \circ ((1-t)\,\mathrm{id}_{E_{U_2}} + tP_{E_{U_1}})$$

induces a homotopy between $f_{U_2}^+$ and $(\mathrm{id}_W)^+ \wedge f_{U_1}^+$. □

Choose $\hat{U}$ in Corollary 5.8. Let $\mathcal{W}$ be the orthogonal complement of $\hat{U}$ in $\mathcal{U}$. For a finite-dimensional subspace $W \subset \mathcal{W}$, let $U_W = W \oplus \hat{U}$. We define the U(1)-equivariant stable cohomotopy group $\{T(\operatorname{ind} \ell), S^0\}_{\mathcal{U}}^{\mathrm{U}(1)}$ by

$$\{T(\operatorname{ind} \ell), S^0\}_{\mathcal{U}}^{\mathrm{U}(1)} := \operatorname*{colim}_{W \subset \mathcal{W}} [S^W \wedge T(E_{\hat{U}}), S^W \wedge S^{\hat{U}}]^{\mathrm{U}(1)}.$$

(For the generality of the equivariant stable cohomotopy groups, see [1, 166].) By Proposition 5.9, we obtain a class $[f_{U_W^+}] \in \{T(\operatorname{ind}\ell), S^0\}_{\mathcal{U}}^{\mathrm{U}(1)}$ from the finite-dimensional approximations f_{U_W} of $\boldsymbol{\mu}$. In particular, the finite-dimensional approximation f_{U_W} of the monopole map $\boldsymbol{\mu}_{\mathrm{SW}}$ defines the class

$$\widetilde{\mathrm{SW}}_{X,\mathfrak{s}} = [f_{U_W}^+] \in \{T(\operatorname{ind}\ell), S^0\}_{\mathcal{U}}^{\mathrm{U}(1)}.$$

Definition 5.10 We call the stable homotopy class $\widetilde{\mathrm{SW}}_{X,\mathfrak{s}}$ the *stable cohomotopy Seiberg–Witten invariant* or *Bauer–Furuta invariant*.

The next lemma is obvious from the construction.

Lemma 5.11 *Let $\mu_t : \mathcal{E}_t \to \mathcal{F}_t$ $(t \in [0, 1])$ be a continuous family such that, for each t, $\boldsymbol{\mu}_t$ satisfies conditions 1 and 2 of the beginning of this section. More precisely, suppose that there are Hilbert bundles $\tilde{\mathcal{E}}$ and $\tilde{\mathcal{F}}$ over $[0, 1] \times J$ and a fiber-preserving map $\tilde{\boldsymbol{\mu}} : \tilde{\mathcal{E}} \to \tilde{\mathcal{F}}$ such that the restriction $\boldsymbol{\mu}_t = \ell_t + c_t$ of $\tilde{\boldsymbol{\mu}} = \tilde{\ell} + \tilde{c}$ to $\{t\} \times J$ satisfies conditions 1 and 2. Then the finite-dimensional approximations f_0 and f_1 of $\boldsymbol{\mu}_0$ and $\boldsymbol{\mu}_1$ induce the same class $[f_0^+] = [f_1^+]$ via the natural isomorphisms*

$$\{T(\operatorname{ind}\ell_0), S^0\}_{\mathcal{U}}^{\mathrm{U}(1)} \cong \{T(\operatorname{ind}\tilde{\ell}), S^0\}_{\mathcal{U}}^{\mathrm{U}(1)} \cong \{T(\operatorname{ind}\ell_1), S^0\}_{\mathcal{U}}^{\mathrm{U}(1)}.$$

Theorem 5.12 *The stable homotopy class $\widetilde{\mathrm{SW}}_{X,\mathfrak{s}}$ is independent of the choice of Riemannian metrics.*

Proof Suppose we have two choices of Riemannian metrics. Connect them by a continuous path. Then we can construct a continuous family of the monopole maps over $[0, 1] \times J$ as in Lemma 5.11. Then the homotopy classes for $t = 0$ and $t = 1$ are the same. □

Remark 5.13 When $b_1(X) = 0$, a finite-dimensional approximation f has a form of

$$f : \mathbb{C}^{x+k} \oplus \mathbb{R}^y \to \mathbb{C}^x \oplus \mathbb{R}^b,$$

where $k = \operatorname{ind}_{\mathbb{C}} D$ and $b = b^+(X)$. By using the equivariant Freudenthal suspension theorem [1, §3] and [166, II.2], we can prove that

$$\{T(\operatorname{ind}\ell), S^0\}_{\mathcal{U}}^{\mathrm{U}(1)} \cong [S^{\mathbb{C}^{x+k}\oplus\mathbb{R}^y}, S^{\mathbb{C}^x\oplus\mathbb{R}^{y+b}}]^{\mathrm{U}(1)},$$

for sufficiently large x and y.

5.3 Relation to the Seiberg–Witten Invariants

In this section, we first prove that the Seiberg–Witten invariants are computed by using the finite-dimensional approximations of the monopole maps. Then we explain the Bauer–Furuta theorem which shows that the Bauer–Furuta invariant is a stable cohomotopy lift of the Seiberg–Witten invariant (Theorem 5.23).

5.3.1 Approximated Moduli Spaces

Fix $\eta \in L^2_{k-1}\left(X; \sqrt{-1}\Lambda^+\right)$. Choose $R > 0$ such that $\boldsymbol{\mu}^{-1}(\eta)$ is contained in the interior of D_R. The next lemma says that, for a large U, we can consider $(f_U)^{-1}(P_U\eta) \cap D_R$ is an "approximation" of $\boldsymbol{\mu}^{-1}(\eta)$. (Cf. [78, §4.1].)

Lemma 5.14 *Let $U_1 \subset U_2 \subset \cdots$ be a sequence of finite-dimensional subspaces of $\mathcal{U}$ satisfying that $\lim_{n\to\infty} P_{U_n}(u) = u$ for $u \in \mathcal{U}$. Let R' be a real number such that $R' \geq R$. Suppose a sequence $v_1, v_2, \ldots$ in $\mathcal{U}$ satisfies the following conditions:*

- $v_n \in U_n$ *and* $\|v_n\| \leq R'$.
- $f_{U_n}(v_n) = (\ell + P_{U_n}c)(v_n) = P_{U_n}\eta$.

Then:

- *there exists N such that $\|v_n\| < R$ for $n \geq N$;*
- *there is a subsequence of $v_1, v_2, \ldots$ which is strong convergent to some v_∞ such that $\boldsymbol{\mu}(v_\infty) = (\ell + c)(v_\infty) = \eta$.*

Proof Since $\{v_n\}$ is bounded, we may assume $\{v_n\}$ has a weak limit v_∞. Note that $P_{U_n}c$ is a compact operator for each n. By a diagonal argument, we can find a subsequence such that $P_{U_n}c(v_n)$ converge strongly to $c(v_\infty)$. The rest of the proof is similar to that of Lemma 5.5. □

Remark 5.15 Lemma 5.14 implies that, for a large n, even if there is a solution v to the equation $f_{U_n}(v) = P_{U_n}\eta$ outside D_R, it is very far from D_R.

Note that

$$\mathcal{E}^{\mathrm{U}(1)} = \{(a, \phi) \in \mathcal{E} \mid \phi \equiv 0\}.$$

Set

$$\mathcal{E}^* = \{(a, \phi) \in \mathcal{E} \mid \phi \not\equiv 0\}.$$

Then U(1) acts freely on $\mathcal{E}^*$. Define the Hilbert vector bundle $\mathcal{V}$ by

$$\mathcal{V} = \mathcal{E}^* \times_{\mathrm{U}(1)} \mathcal{U} \to \mathcal{E}^*/\,\mathrm{U}(1).$$

The monopole map $\boldsymbol{\mu}\colon \mathcal{E} \to \mathcal{U}$ induces a section $\bar{\boldsymbol{\mu}}\colon \mathcal{E}^*/\mathrm{U}(1) \to \mathcal{V}$, whose zero set is the irreducible part of the moduli space, $\mathcal{M}^*(\mathfrak{s})$. In general, $\bar{\boldsymbol{\mu}}^{-1}(\eta) = \mathcal{M}^*(\mathfrak{s}, \eta)$ for $\eta \in \mathcal{P}_{k-1}$. We assume $b_2^+(X) \geq 1$. Then, by Corollary 4.59, we can choose η so that $\boldsymbol{\mu}^{-1}(\eta)/\mathrm{U}(1) = \mathcal{M}(\mathfrak{s}, \eta)$ contains no reducible and is a compact manifold. Recall that the Seiberg–Witten invariant for $(X, \mathfrak{s})$ is defined by using the fundamental class of $\bar{\boldsymbol{\mu}}^{-1}(\eta) = \mathcal{M}(\mathfrak{s}, \eta)$.

Choose a large $U = U_n$ in Lemma 5.14 such that $n \geq N$ and $U \supset \hat{U}$. Note that the map $f_U = \ell + P_U c\colon E_U \to U$ is a U(1)-equivariant smooth map from a finite-rank vector bundle over J to a finite-dimensional vector space. Let $E_U^* = E_U \cap \mathcal{E}^*$. Define the vector bundle V_U by

$$V_U = E_U^* \times_{\mathrm{U}(1)} U \to E_U^*/\mathrm{U}(1).$$

Then V_U is a subbundle of the restriction of $\mathcal{V}$ to $E_U^*/\mathrm{U}(1)$. The finite-dimensional approximation $f_U\colon E_U \to U$ induces a section $\bar{f}_U\colon E_U^*/\mathrm{U}(1) \to V_U$. By Lemma 5.14, we may consider $\bar{f}_U^{-1}(v) \cap D_R$ as an approximation of the moduli space $\bar{\boldsymbol{\mu}}^{-1}(\eta) = \mathcal{M}^*(\mathfrak{s}, \eta)$, where $v = P_U\eta$. We shall prove that the Seiberg–Witten invariant for $(X, \mathfrak{s})$ can be obtained from $\bar{f}_U^{-1}(v)$. For this purpose, we need to perturb $\bar{f}_U$ in general. Notice the following important facts:

- Since the restriction of the monopole map $\boldsymbol{\mu}$ to the fixed point set $\mathcal{E}^{\mathrm{U}(1)}$ of the U(1)-action is $d^+ + F_{A_0}^+$ on each fiber of $\mathcal{E}^{\mathrm{U}(1)}$, the image of $\boldsymbol{\mu}|_{\mathcal{E}^{\mathrm{U}(1)}}$ is an affine subspace of $\mathcal{U}^{\mathrm{U}(1)}$ with codimension b^+.
- By the choice of U_0 in Lemma 5.6, the image of the restriction of f_U to $E_U^{\mathrm{U}(1)}$ is also an affine subspace of $U^{\mathrm{U}(1)}$ with codimension b^+.

Suppose b^+ is positive and choose $v \in U^{\mathrm{U}(1)} \backslash \mathrm{Im}\left(f_U|_{E_U^{\mathrm{U}(1)}}\right)$. Then $(f_U - v)^{-1}(0) = f_U^{-1}(v)$ contains no reducible. It is uncertain that a choice of v makes the section of V_U induced from $f_U - v$ transverse to the zero section because v is restricted to $U^{\mathrm{U}(1)}$. We will need a further perturbation. The bundle V_U is decomposed into the direct sum $V_U = \underline{U}^{\mathrm{U}(1)} \oplus V'$, where $\underline{U}^{\mathrm{U}(1)}$ is a trivial bundle with fiber $U^{\mathrm{U}(1)}$ and V' is a complex vector bundle of the spinor part.

Lemma 5.16 *For $v \in U^{\mathrm{U}(1)} \setminus \mathrm{Im}\left(f_U|_{E_U^{\mathrm{U}(1)}}\right)$ and a section s' of V', define the section s of $V_U = \underline{U}^{\mathrm{U}(1)} \oplus V'$ by*

$$s(u) = (-v, s'(u)), \tag{5.2}$$

and perturb $\bar{f}_U$ to $\bar{f}_U' := \bar{f}_U + s$. A generic choice of v and s' makes $\bar{f}_U'$ transversal to the zero section, and the zero locus $M(v) := \left(\bar{f}_U'\right)^{-1}(0)$ is a $d(\mathfrak{s})$-dimensional compact manifold away from reducibles. Moreover we can take s' such that s' vanishes near reducibles, that is, there exists a small open neighborhood N of $E_U^{\mathrm{U}(1)}$ in E_U such that $s' = 0$ on $\left(E_U^* \cap N\right)/\mathrm{U}(1)$.

Proof Let p_1 be the projection $V_U \to \underline{U}^{\mathrm{U}(1)} = \left(E_U^*/\mathrm{U}(1)\right) \times U^{\mathrm{U}(1)} \to U^{\mathrm{U}(1)}$. Choose a regular value v of $p_1 \bar{f}_U$ such that $v \in U^{\mathrm{U}(1)} \setminus \mathrm{Im}\left(f_U|_{E_U^{\mathrm{U}(1)}}\right)$. Then the inverse image $\left(p_1 \bar{f}_U\right)^{-1}(v)$ is a manifold, and $u = (a, \phi)$ is in $\left(p_1 \bar{f}_U\right)^{-1}(v)$ if and only if

$$d^+a + P_U F_{A_0}^+ - P_U q(\phi) = v.$$

This equation implies that the norm of ϕ is bounded below by a constant multiple of the distance of v and $\mathrm{Im}\left(f_U|_{E_U^{\mathrm{U}(1)}}\right)$. Then we can choose a small section s' which vanishes near reducibles and makes $\bar{f}_U'$ transversal to the zero section. □

Definition 5.17 We call $M(v) = (\bar{f}_U')^{-1}(0)$ the *approximated moduli space.*

We want to compare $M(v)$ with the genuine moduli space $\mathcal{M}(\mathfrak{s}, \eta)$. For this purpose, we embed the section $\bar{f}_U'$ of V_U in $\mathcal{V}$ as follows. Note that $E_U^*/\mathrm{U}(1)$ is a finite-dimensional submanifold of $\mathcal{E}/\mathrm{U}(1)$, and V_U is a subbundle of the restriction of $\mathcal{V}$ to $E_U^*/\mathrm{U}(1)$.

There is a U(1)-equivariant map $\tilde{s}\colon E_U^* \to U$ which induces the section s. Since s' vanishes near reducibles, $\tilde{s}$ extends over $E_U^{\mathrm{U}(1)}$ to the U(1)-equivariant map $\tilde{s}\colon E_U \to U$. Let $U^\perp$ be the orthogonal complement of U in $\mathcal{U}$, and decompose $\mathcal{E} = E_U \oplus E_{U^\perp}$ where $E_{U^\perp} = \ell^{-1}(U^\perp)$. Note that the map $(f_U + \tilde{s})\colon E_U \to U$ which induces $\bar{f}_U'$ is written as

$$(f_U + \tilde{s})(u) = \ell(u) + p_U c(u) + \tilde{s}(u).$$

Define the U(1)-equivariant map $F_U\colon E_U \oplus E_{U^\perp} \to U \oplus U^\perp$ by

$$F_U(u \oplus u') = (\ell(u) + p_U c(u \oplus u') + \tilde{s}(u), \ell(u')).$$

Lemma 5.18 *Let $\bar{F}_U$ be the section of $\mathcal{V}$ induced from F_U. Then $\bar{F}_U^{-1}(0) = M(v)$. If $\bar{f}_U'$ is transverse to the zero-section, then $\bar{F}_U$ is also transverse to the zero-section.*

Proof Since the restriction of ℓ to each fiber of $E_{U^\perp}$ is a linear isomorphism, $\ell(u') = 0$ for $u' \in E_{U^\perp}$ implies that $u' = 0$. Then $F_U(u \oplus u') = 0$ if and only if $u' = 0$ and $(f_U + \tilde{s})(u) = \ell(u) + p_U c(u) + \tilde{s}(u) = 0$. Thus $\bar{F}_U^{-1}(0) = (\bar{f}_U')^{-1}(0) = M(v)$. Now the statement for transversality is obvious. □

Lemma 5.19 *Suppose $b^+(X) \geq 2$. Choose a generic $\eta_1 \in \mathcal{P}_{k-1}$ so that $\mathcal{M}(\eta_1, \mathfrak{s})$ contains no reducible and the transversality holds. Then $M(v) = \bar{F}_U^{-1}(0)$ is cobordant to $\bar{\mu}^{-1}(\eta_1) = \mathcal{M}(\eta_1, \mathfrak{s})$ in $\mathcal{E}^*/\mathrm{U}(1) \times [0, 1]$. Moreover, when $b^+(X) = 1$, the same statement holds if v and η_1 are in the same chamber.*

Proof Let $\tilde{\mathcal{P}}$ be the space of L_1^2 maps

$$\eta\colon [0,1] \to \mathcal{P}_{k-1} = L_{k-1}^2\left(X; \sqrt{-1}\Lambda^+\right),$$

satisfying $\eta(0) = 0$ and $\eta(1) = \eta_1$. Note that such paths are continuous. We define the map

$$\tilde{F}\colon E_U \oplus E_{U^\perp} \times [0,1] \times \tilde{\mathcal{P}} \to U \oplus U^\perp$$

by

$$\tilde{F}(u \oplus u', t, \eta) = \ell(u \oplus u') + ((1-t)P_U + t)\, c(u \oplus u') - \eta(t) + (1-t)\tilde{s}(u).$$

Note that $\tilde{F}|_{t=0} = F_U$ and $\tilde{F}|_{t=1} = \boldsymbol{\mu} - \eta_1$. Define $\tilde{\mathcal{P}}^*$ as the subspace of $\tilde{\mathcal{P}}$ such that $\tilde{F}(w,t,\eta) = 0$ has no reducible solution.

We claim that $\tilde{\mathcal{P}}^*$ is non-empty for arbitrary choice of v and η_1 if $b^+(X) \geq 2$, and for v and η_1 in the same chamber if $b^+(X) = 1$. In fact, a reducible $w = (a, 0)$ is a solution of $\tilde{F}(w,t,\eta) = 0$ if and only if

$$d^+ a + ((1-t)P_U + t)\, F_{A_0}^+ - \eta(t) + (1-t)v = 0.$$

We can choose a path η such that the above equation has no solution under the condition of the claim.

We consider the restriction of $\tilde{F}$ to $\tilde{\mathcal{P}}^*$. We claim that the differential of $\tilde{F}$ is surjective at every point in $\tilde{F}^{-1}(0)$. For $t = 0, 1$, this is true by the assumption. For $0 < t < 1$, the claim is verified by the same argument in the proof of Lemma 4.53. Then $\tilde{F}^{-1}(0)$ is a smooth manifold with boundary, and the projection $\pi : \tilde{F}^{-1}(0) \to \tilde{\mathcal{P}}^*$ is a smooth map. By the Sard–Smale theorem (Theorem 4.55), the quotient space $\pi^{-1}(\eta)/\operatorname{U}(1)$ for a generic $\eta \in \tilde{\mathcal{P}}^*$ is a smooth manifold whose boundary is the disjoint union of $\bar{F}_U^{-1}(0)$ and $\mathcal{M}(\eta_1, \mathfrak{s})$. □

The space $E_U^*/\operatorname{U}(1)$ is a fiber bundle over J whose fiber is isomorphic to

$$\mathbb{C}P^N \times (0, \infty) \times V,$$

where N is some integer and V is a finite-dimensional real vector space.

Lemma 5.20 *The following isomorphisms hold:*

$$\begin{aligned} H^*(E_U^*/\operatorname{U}(1); \mathbb{Z}) &\cong H^*(\mathbb{C}P^N; \mathbb{Z}) \otimes H^*(T^{b_1}; \mathbb{Z}) \\ &\cong \mathbb{Z}[u]/(u^{N+1}) \otimes \Lambda^* H^1(T^{b_1}; \mathbb{Z}), \end{aligned}$$

where u is a generator of degree 2.

Proof This follows from the Leray–Hirsch theorem. (See [72, §4.D], Problem 5.53.) □

We introduce the μ-map $\mu_f : H_j(X;\mathbb{Z}) \to H^{2-j}(E_U^*/\mathrm{U}(1);\mathbb{Z})$ $(j = 0, 1)$ for a finite-dimensional approximation f_U as follows: For a generator $[x_0]$ of $H^0(X;\mathbb{Z})$, define $\mu_f([x_0])$ by

$$\mu_f([x_0]) = u.$$

For a class $\gamma \in H_1(X;\mathbb{Z})$, the Kronecker pairing induces a map

$$\tilde{\gamma} : J = H^1(X;\mathbb{R})/H^1(X;\mathbb{Z}) \to \mathbb{R}/\mathbb{Z}, \quad c \mapsto \langle c, \gamma\rangle \mod \mathbb{Z}.$$

Since $H^1(J;\mathbb{Z}) \cong [J, S^1]$, $\tilde{\gamma}$ determines a class $[\tilde{\gamma}]$ in $H^1(J;\mathbb{Z})$, where $[J, S^1]$ is the set of free homotopy classes of maps from J to S^1. Define $\mu_f(\gamma)$ by

$$\mu_f(\gamma) = \pi^*[\tilde{\gamma}],$$

where $\pi : E_U^*/\mathrm{U}(1) \to J$ is the projection.

Corollary 5.21 *If $b^+(X) \geq 2$, then the Seiberg–Witten invariant can be calculated by using an approximated moduli space $M(v)$ as follows:*

$$\mathrm{SW}_{X,\mathfrak{s},O}(u^m \otimes (\gamma_1 \wedge \cdots \wedge \gamma_k)) = \langle u^m \otimes (\mu_f(\gamma_1) \wedge \cdots \wedge \mu_f(\gamma_k)), [M(v)]\rangle.$$

If $b^+(X) = 1$, then $\mathrm{SW}^{\pm}_{X,\mathfrak{s},\mathfrak{h},O_1}$ can be calculated by using an approximated moduli space $M(v)$ for v in $\mathscr{C}^{\pm}(\mathfrak{h})$.

We complete the proofs of the well-definedness of the Seiberg–Witten invariants. First, we give the proof of the case when $b^+(X) \geq 2$.

Proof of Theorem 4.67, 1 When $b^+(X) \geq 2$, the moduli space $\mathcal{M}(\mathfrak{s}, \eta)$ for any η is cobordant to an approximated moduli space $M(v)$ by Lemmas 5.19 and 5.18. By Corollary 5.21, $\mathrm{SW}_{X,\mathfrak{s},O}$ does not depend on η. Suppose two Riemannian metrics g_0 and g_1 on X are given. Choosing a continuous path g_t $(t \in [0, 1])$ of Riemannian metrics connecting g_0 and g_1, we obtain a continuous family of monopole maps $\{(\boldsymbol{\mu}_{\mathrm{SW}})_{g_t}\}$. We can also construct a continuous family of finite-dimensional approximations as follows. The family of monopole maps forms a U(1)-equivariant map

$$\tilde{\boldsymbol{\mu}}_{\mathrm{SW}} = \bigcup_t (\boldsymbol{\mu}_{\mathrm{SW}})_{g_t} : \tilde{\mathcal{E}} \to \tilde{\mathcal{U}},$$

where $\tilde{\mathcal{E}}$ is a bundle over $[0, 1] \times J$ and $\tilde{\mathcal{U}}$ is a bundle over $[0, 1]$ such that the restriction of $\tilde{\mu}_{\mathrm{SW}}$ to $\{t\} \times J$ is $\left(\mu_{\mathrm{SW}}\right)_{g_t}$. Trivializing $\tilde{\mathcal{U}} = [0, 1] \times \mathcal{U}$ and choosing a large $U \subset \mathcal{U}$, we obtain a finite-dimensional approximation $\tilde{f}_U$ of $\tilde{\mu}_{\mathrm{SW}}$ which forms a family of finite-dimensional approximations. Perturbing $\tilde{f}_U$, we finally get a cobordism between the approximated moduli spaces for g_0 and g_1. □

Next we give the proof of the case when $b^+(X) = 1$.

Proof of Theorem 4.76, 1 The proof is similar to the case when $b^+(X) \geq 2$. In this case, we can construct a cobordism between moduli spaces if their parameters are in the same chamber. □

Remark 5.22 We can directly prove assertions 1 of Theorems 4.67 and 4.76 in the infinite-dimensional setting of Chap. 4 by an argument similar to the proof of Lemma 5.19. See e.g. [131, 165].

We also give the proof of a special case of the wall-crossing formula (Corollary 4.78).

Proof of Corollary 4.78 Let $a = \mathrm{ind}_{\mathbb{C}} D_{A_0}$. Then $0 \leq d(\mathfrak{s}) = 2\,\mathrm{ind}_{\mathbb{C}} D_{A_0} - (1 + b^+) = 2a - 2$. Fix a Riemannian metric g on X. We choose $\eta_\pm$ as

$$\eta_\pm = \frac{2\pi}{\sqrt{-1}}(c_{\mathfrak{s}} \pm \omega_g(\mathfrak{h})),$$

where $c_{\mathfrak{s}}$ is the harmonic representative of $c_1(\mathfrak{s})$. Then $(g, \eta_\pm) \in \mathscr{C}^\pm(\mathfrak{h})$. In order to compare the moduli spaces $\mathcal{M}(\mathfrak{s}, \eta_+)$ and $\mathcal{M}(\mathfrak{s}, \eta_-)$, we compare their approximated moduli spaces. Choose a finite-dimensional approximation $f = f_U : E_U \to U$. Since $b_1 = 0$, E_U is a finite-dimensional vector space. Let

$$v = P_U \frac{2\pi}{\sqrt{-1}} \omega_g(\mathfrak{h}).$$

We consider the 1-parameter family $f_t = f - tv$ $(-1 \leq t \leq 1)$. Then, perturbing $f_{\pm 1}$, we obtain the approximated moduli spaces $M_\pm := M(\pm v)$ of $\mathcal{M}(\mathfrak{s}, \eta_\pm)$ respectively. Consider the $\mathrm{U}(1)$-equivariant map

$$F : I \times E_U \to U, \quad F(t, u) = f_t(u),$$

where $I = [-1, 1]$. The map F induces a section $\bar{F}$ of the bundle $I \times V_U$ over $I \times (E_U^*/\mathrm{U}(1))$. We will perturb F so that the zero locus $W = \bar{F}^{-1}(0)/\mathrm{U}(1)$ gives a cobordism between $M_\pm$ with a contribution from the reducible. More precisely, we verify that W has the boundary $\partial W = M_+ \cup -M_- \cup -\mathbb{C}P^{a-1}$.

First, we examine the situation around $(0, 0) \in I \times E_U$. The local structure of F around $(0, 0)$ is described by its differential $DF_{(0,0)} : T_{(0,0)}(I \times E_U) = \mathbb{R} \times E_U \to T_0 U = U$ of F at $(0, 0)$ of the form

$$DF_{(0,0)}\colon \mathbb{R}\times(\mathbb{C}^{x+a}\times\mathbb{R}^{y})\to\mathbb{C}^{x}\times\mathbb{R}^{y+b},$$

where $a=\mathrm{ind}_{\mathbb{C}} D_{A_0}$, $b=b^+=1$, x, y are some integers and $\mathrm{U}(1)$ acts on $\mathbb{C}$ by multiplication and on $\mathbb{R}$ trivially. By construction, the restriction of $DF_{(0,0)}$ to the $\mathrm{U}(1)$-fixed part $T(I\times E_U)^{\mathrm{U}(1)}=\mathbb{R}\times\{0\}\times\mathbb{R}^{y}$ is an isomorphism. However, $DF_{(0,0)}$ may have a nontrivial cokernel in the complex part. Choosing a $\mathrm{U}(1)$-equivariant linear surjection $L\colon\mathbb{C}^{x+a}\to\mathbb{C}^{x}$ and a cut-off function around $(0,0)$, we perturb F locally around $(0,0)$ by L. Then the zero locus of F around $(0,0)$ is isomorphic to $\mathbb{C}^a$. Hence the quotient space $F^{-1}(0)/\mathrm{U}(1)$ has a form of a cone of $\mathbb{C}P^{a-1}=\mathbb{C}P^{d(\mathfrak{s})/2}$ near $(0,0)$.

Choose a small $\mathrm{U}(1)$-invariant open neighborhood N of $(0,0)$ in $I\times E_U$ so that the induced section $\bar{F}$ is transversal to the zero section on $\bar{N}^*=((I\times E_U^*)\cap N)/\mathrm{U}(1)$ and $\bar{F}^{-1}(0)\cap\bar{N}^*\cong(0,\varepsilon)\times\mathbb{C}P^{d(\mathfrak{s})/2}$ for a small $\varepsilon>0$. Then it is easy to perturb $\bar{F}$ outside $\bar{N}^*$ to attain the transversality. Removing a small neighborhood of $(0,0)$ from the zero locus $\bar{F}^{-1}(0)$, we obtain the cobordism W with $\partial W=M_+\cup -M_-\cup -\mathbb{C}P^{d(\mathfrak{s})/2}$.

Then we have the relation of the fundamental classes

$$[M_+]-[M_-]-[\mathbb{C}P^{d(\mathfrak{s})/2}]=0,$$

which implies the wall-crossing formula. □

5.3.2 *The Relation Between the Seiberg–Witten and Bauer–Furuta Invariants*

The relation between the Seiberg–Witten and Bauer–Furuta invariants is given by Bauer and Furuta [18].

Theorem 5.23 (Bauer and Furuta [18, Proposition 3.3]) *Suppose $b^+-b_1\geq 2$. The choice of a homology orientation O of* (4.28) *determines a group homomorphism*

$$\tau\colon\{T(\mathrm{ind}\,\ell),S^0\}^{\mathrm{U}(1)}_{\mathcal{U}}\to\mathrm{Hom}(\mathbb{Z}[u]\otimes\Lambda H_1(X;\mathbb{Z})/\mathrm{Tor},\mathbb{Z})$$

which maps $\widetilde{\mathrm{SW}}_{X,\mathfrak{s}}$ to $\mathrm{SW}_{X,\mathfrak{s},O}$.

Proof of Corollary 4.78 Each element in $\{T(\mathrm{ind}\,\ell),S^0\}^{\mathrm{U}(1)}_{\mathcal{U}}$ is represented by a pointed $\mathrm{U}(1)$-map $f\colon T(E_U)\to S^U$. Perturb f so that the following are satisfied:

- The image of the U(1)-fixed point set $(T(E_U))^{\mathrm{U}(1)}$ by f is in a subspace W of $(S^U)^{\mathrm{U}(1)}$ of codimension at least $b^+ - b_1 \geq 2$. (Note that, if f is a finite-dimensional approximation of the monopole map, we can take a subspace of codimension b^+.)
- f is smooth outside of a small neighborhood of $(T(E_U))^{\mathrm{U}(1)}$.

By the construction of the approximated moduli spaces, we obtain a section $\bar{f}'$ of V_U transversal to the zero section. Then the space $M = \bar{f}^{-1}(0)$ is a $d(\mathfrak{s})$-dimensional compact submanifold in $E_U^*/\mathrm{U}(1) \subset \mathcal{E}^*/\mathrm{U}(1)$. The correspondence

$$\{T(\operatorname{ind}\ell), S^0\}_{\mathcal{U}}^{\mathrm{U}(1)} \ni [f] \mapsto [M] \in H_*(\mathcal{E}^*/\mathrm{U}(1)) \cong H_*(\mathbb{C}P^\infty \times T^{b_1})$$

gives a well-defined group homomorphism (Problem 5.54). The theorem follows from this. □

When $b_1(X) = 0$, the equivariant stable cohomotopy group $\{T(\operatorname{ind}\ell), S^0\}_{\mathcal{U}}^{\mathrm{U}(1)}$ can be identified with a non-equivariant stable cohomotopy group as follows.

Proposition 5.24 ([18, Proposition 3.4]) *If $b_1(X) = 0$ and $b^+(X) \geq 2$, then the equivariant stable cohomotopy group $\{T(\operatorname{ind}\ell), S^0\}_{\mathcal{U}}^{\mathrm{U}(1)}$ is isomorphic to the non-equivariant stable cohomotopy group $\pi^{b-1}(\mathbb{C}P^{k-1})$, where $b = b^+(X)$ and $k = \operatorname{ind}_{\mathbb{C}} D$.*

Proof of Corollary 4.78 (For the generality of equivariant stable homotopy, see [1], [166], [16, §3].) Let $D_k = D(\mathbb{C}^k)$ be the unit disk in $\mathbb{C}^k$ and $S_k = S(\mathbb{C}^k) = \partial D_k$. We have a pointed U(1)-equivariant cofiber sequence

$$S_k^+ \xrightarrow{i} D_k^+ \to Ci,$$

where $D_k^+ = D_k \sqcup *$, $S_k^+ = S_k \sqcup *$, the map i is the inclusion, and Ci is the cone of i. Associating with this cofiber sequence, we have a long exact sequence

$$\cdots \to \{S^1 \wedge S_k^+, S^b\}^{\mathrm{U}(1)} \to \{Ci, S^b\}^{\mathrm{U}(1)} \to \{D_k^+, S^b\}^{\mathrm{U}(1)} \to \{S_k^+, S^b\}^{\mathrm{U}(1)} \to \cdots.$$

Note that Ci is U(1)-homotopic to $S^{\mathbb{C}^k}$. Since U(1) acts on S_k freely, there is an isomorphism $\{S_k^+, S^a\}^{\mathrm{U}(1)} \cong \pi^a(\mathbb{C}P^{k-1})$. (See [1, Theorem 5.3].) Since D_k^+ is U(1)-homotopic to S^0, $\{D_k^+, S^a\}^{\mathrm{U}(1)}$ is trivial if $a \geq 1$. Under the assumption that $b_1(X) = 0$ and $b = b^+(X) \geq 2$, we obtain

$$\begin{aligned}\{T(\operatorname{ind}\ell), S^0\}^{\mathrm{U}(1)} &\cong \{S^{\mathbb{C}^k}, S^{\mathbb{R}^b}\}^{\mathrm{U}(1)} \cong \{Ci, S^b\}^{\mathrm{U}(1)} \cong \{S^1 \wedge S_k^+, S^b\}^{\mathrm{U}(1)} \\ &\cong \pi^{b-1}(\mathbb{C}P^{k-1}).\end{aligned}$$

This completes the proof. □

When $b_1(X) = 0$ and $d(\mathfrak{s}) = 0$, the Bauer–Furuta and Seiberg–Witten invariants give the same information as follows.

Corollary 5.25 *If $b_1(X) = 0$, $b^+(X) \geq 2$ and $d(\mathfrak{s}) = 0$, then we have an identification $\{T(\operatorname{ind}\ell), S^0\}_{\mathcal{U}}^{U(1)} \cong \mathbb{Z}$ such that*

$$\tau\left(\widetilde{\mathrm{SW}}_{X,\mathfrak{s}}\right) = \mathrm{SW}_{X,\mathfrak{s}}(1).$$

Proof of Corollary 4.78 When $d(\mathfrak{s}) = 2k - 1 - b = 0$, Proposition 5.24 implies that

$$\{T(\operatorname{ind}\ell), S^0\}_{\mathcal{U}}^{U(1)} \cong \pi^{b-1}(\mathbb{C}P^{k-1}) \cong \pi^{2k-2}(\mathbb{C}P^{k-1}) \cong \mathbb{Z}.$$

Then, the map τ in Theorem 5.23 induces an isomorphism $\{T(\operatorname{ind}\ell), S^0\}_{\mathcal{U}}^{U(1)} \to \mathbb{Z}$ such that $\tau\left(\widetilde{\mathrm{SW}}_{X,\mathfrak{s}}\right) = \mathrm{SW}_{X,\mathfrak{s}}(1)$. □

After learning this corollary, the readers may ask the question: Are there differences between $\mathrm{SW}_{X,\mathfrak{s}}$ and $\widetilde{\mathrm{SW}}_{X,\mathfrak{s}}$? In fact, a difference of $\mathrm{SW}_{X,\mathfrak{s}}$ and $\widetilde{\mathrm{SW}}_{X,\mathfrak{s}}$ is observed in the case of connected sums as follows.

Theorem 5.26 (Witten [169]) *If X_1 and X_2 are closed connected oriented 4-manifold with $b^+(X_i) \geq 1$ for $i = 1, 2$, then $\mathrm{SW}_{X_1\#X_2,\tilde{\mathfrak{s}}}$ is trivial for every spin^c structure $\tilde{\mathfrak{s}}$.*

Detailed proofs of Theorem 5.26 are given in [53, 139, 151].

Theorem 5.27 (Bauer [15]) *Let $X_1, \dots, X_n$ be closed connected oriented 4-manifolds with $b_1(X_i) = 0$ and $b^+(X_i) \geq 2$ for $i = 1, \dots, n$, and X be the connected sum $X_1\#\cdots\#X_n$. Then $\widetilde{\mathrm{SW}}_{X,\mathfrak{s}}$ is nontrivial if and only if the following conditions are satisfied:*

- $n \leq 4$.
- $b^+(X_i) \equiv 3 \bmod 4$ *for each* i.
- *There is a spin^c structure $\mathfrak{s}_i$ for each i such that $\mathrm{SW}_{X_i,\mathfrak{s}_i}(1)$ is odd, and $\mathfrak{s} = \mathfrak{s}_1\#\cdots\#\mathfrak{s}_n$.*
- $b^+(X) \equiv 4 \bmod 8$ *if* $n = 4$.

Note that $\mathrm{SW}_{X_i,\mathfrak{s}_i}(1)$ can be non-zero only if $d(\mathfrak{s}_i) = 0$. Since $d(\mathfrak{s}_i\#\mathfrak{s}_j) = d(\mathfrak{s}_i) + d(\mathfrak{s}_j) + 1$, we have $d(\mathfrak{s}_1\#\cdots\#\mathfrak{s}_n) = n - 1$ which is positive if $n \geq 2$. Note that the $K3$ surface with canonical spin^c structure $\mathfrak{s}_0$ has a nontrivial invariant $\mathrm{SW}_{K3,\mathfrak{s}_0}(1) = \pm 1$ by Theorem 4.73. Then the connected sums X of up to four $K3$ surfaces satisfy the conditions of Theorem 5.27, and therefore have nontrivial $\widetilde{\mathrm{SW}}_{X,\mathfrak{s}}$. However, if $n \geq 2$, then $\mathrm{SW}_{X,\mathfrak{s}} \equiv 0$ by Theorem 5.26. This shows that the Bauer–Furuta invariants are truly stronger than the ordinary Seiberg–Witten invariants.

As mentioned in Sect. 2.8.4, there is a conjecture called the *simple type conjecture* on the Seiberg–Witten invariants with positive $d(\mathfrak{s})$. A closed oriented 4-manifold X with $b^+(X) \geq 2$ is said to be of *Seiberg–Witten simple type* if $\mathrm{SW}_{X,\mathfrak{s}}(u^k) = 0$ for every $k > 0$ and every spin^c structure $\mathfrak{s}$ on X.

Conjecture 5.28 (Simple Type Conjecture) All closed oriented 4-manifolds with $b^+ \geq 2$ are of Seiberg–Witten simple type.

Remark 5.29 The conjecture is still open. The mod 2 version of the Seiberg–Witten simple type is studied in [89].

Remark 5.30 There are examples of $(X, \mathfrak{s})$ satisfying $b^+(X) \geq 2$, $d(\mathfrak{s}) > 0$ and $\mathrm{SW}_{X,\mathfrak{s}}(\beta) \neq 0$ for $\beta \neq u^k$. For instance, if Y satisfies that $b^+(Y) \geq 2$ and $\mathrm{SW}_{Y,\mathfrak{s}}(1) \neq 0$ for some $\mathfrak{s}$, then the connected sum $X = Y\#k(S^1 \times S^3)$ has the invariant $\mathrm{SW}_{X,\tilde{\mathfrak{s}}}(\mu(\gamma_1) \wedge \cdots \wedge \mu(\gamma_k)) \neq 0$ where γ_i is the class of $S^1 \times pt$ in the ith component of $S^1 \times S^3$ and $\tilde{\mathfrak{s}}$ is a unique extension of $\mathfrak{s}$.

Remark 5.31 The Bauer–Furuta theory is now applied in many studies. We list some of the subsequent developments of the Bauer–Furuta invariants.

- Theorem 5.23 requires the condition that $b^+ - b_1 \geq 2$. In particular, it seems difficult to recover the wall-crossing formula through $\{T(\operatorname{ind}\ell), S^0\}^{\mathrm{U}(1)}_{\mathcal{U}}$. (For instance, if $b^+ = 1$ and $b_1 = 0$, then $\{T(\operatorname{ind}\ell), S^0\}^{\mathrm{U}(1)}_{\mathcal{U}}$ is trivial.) Bauer [16] gives an alternative definition of the Bauer–Furuta invariant taking into account of the third property of the monopole map in Remark 5.4. Then the new version of the Bauer–Furuta invariant recovers the Seiberg–Witten invariant in the range of $b^+ \geq 2$, and encodes the information of the wall-crossing if $b^+ = 1$.
- Manolescu [118] uses the idea of finite-dimensional approximation to define the Seiberg–Witten Floer homotopy type for rational homology 3-spheres, which is a kind of spectrum whose homology is the Seiberg–Witten Floer homology. (We refer the readers to Kronheimer–Mrowka's book [102] for the Seiberg–Witten Floer homology.) The Bauer–Furuta invariant is extended to the relative Bauer–Furuta invariant for 4-manifolds with boundary. The connected sum formula is also extended to a more general gluing formula for manifolds glued along rational homology spheres [119].
- Furuta et al. [61] proved a non-vanishing theorem of the Bauer–Furuta invariant of rational cohomology $K3\#K3$, which can be considered as a higher version of the non-vanishing theorems of the Seiberg–Witten invariants of homotopy $K3$ and homology tori due to Morgan and Szabó [132] and Ruberman and Strle [149]. T.-J. Li [109, 110], Bauer [17] also generalized the results of [132, 149] by using the framework of the Bauer–Furuta theory, and obtained interesting applications. Kato et al. [88] proved a similar non-vanishing theorem for the family Seiberg–Witten invariant, which is applied to construct non-smoothable topological families of 4-manifolds.
- The computation of the Bauer–Furuta invariants is particularly difficult when $b_1(X) > 0$ and $d(\mathfrak{s}) > 0$. (To our knowledge, no concrete computation in this case is known.) Sasahira [153] and Ishida and Sasahira [80, 81] introduced a

spin-bordism version of the Seiberg–Witten invariant which is stronger than the ordinary Seiberg–Witten invariant and simpler than the Bauer–Furuta invariant. In fact, if the spin-bordism invariant is nontrivial, then the Bauer–Furuta invariant also is. They proved a connected-sum formula and non-vanishing results of the spin-bordism invariants. As a consequence, there is a series of 4-manifolds with $b_1(X) > 0$ which has nontrivial Bauer–Furuta invariants with $d(\mathfrak{s}) > 0$.

- As mentioned in [16] and [58], the Bauer–Furuta invariant can be straightforwardly extended to more advanced settings such as 4-manifolds with group actions and families of 4-manifolds. The Bauer–Furuta invariant for group actions is studied by Szymik [160], Nakamura [136] and Sung [157, 158]. Recently, there has been a lot of research on gauge theory on families of 4-manifolds. The Bauer–Furuta invariant for families is also studied by many authors, and a great number of remarkable results have been obtained: See Szymik [159], Baraglia and Konno [13], Kronheimer and Mrowka [103], J. Lin [113], and Baraglia [12].
- Iida [78] introduced a Bauer–Furuta type refinement of Kronheimer–Mrowka's invariant [101] for 4-manifolds with contact boundary.

5.4 Donaldson's Diagonalization Theorem for Definite 4-Manifolds

The purpose of this section is to prove Donaldson's diagonalization theorem (Theorem 2.36). Let X be a closed connected oriented smooth 4-manifold with $n := b_2(X)$. Let

$$Q_X\colon H_2(X;\mathbb{Z})/\mathrm{Tor} \times H_2(X;\mathbb{Z})/\mathrm{Tor} \to \mathbb{Z}$$

be the intersection form of X, formally defined by

$$Q_X(\alpha,\beta) = \langle PD(\alpha) \cup PD(\beta), [X]\rangle,$$

where $PD(\alpha)$ is the Poincaré dual of α and $[X] \in H_4(X;\mathbb{Z})$ is the fundamental class of X.

Theorem 5.32 (Donaldson [35, 38]) *If the intersection form Q_X of X is positive or negative definite, then Q_X is diagonalizable, that is,*

$$Q_X \cong \pm(\underbrace{(1) \oplus \cdots \oplus (1)}_{n}).$$

Remark 5.33 Donaldson first proved Theorem 5.32 for simply-connected 4-manifolds by using the Yang–Mills theory [35]. Furuta [57] and Fintushel and Stern [46] extended it to the case of 4-manifolds with more general fundamental groups. Finally, Donaldson [38] removed any assumption on the fundamental groups. We will give an alternative proof by using the finite-dimensional approximation of the monopole map.

First, we recall the definition of the characteristic elements in a lattice.

Definition 5.34 An element w in a lattice L is called *characteristic* if $w \cdot v \equiv v \cdot v$ mod 2 for any $v \in L$.

A characterization of the diagonal form is given by N. Elkies [42].

Theorem 5.35 (Elkies [42]) *Let L be a definite unimodular form over $\mathbb{Z}$. If every characteristic element $w \in L$ satisfies that $|w \cdot w| \geq \operatorname{rank} L = n$, then L is isomorphic to the standard form $\pm n(1)$.*

Theorem 5.32 is proved by using Elkies' theorem together with the following lemma.

Lemma 5.36 *Let X be a closed connected oriented smooth 4-manifold with $b^+(X) = 0$. For every cohomology class $C \in H^2(X;\mathbb{Z})$ which is an integral lift of $w_2(X)$, the inequality $|C \cdot C| \geq b_2(X)$ holds.*

Assuming Lemma 5.36, we prove Theorem 5.32.

Proof of Theorem 5.32 By reversing the orientation of X if necessary, we may assume $b^+(X) = 0$. Note that an integral lift of $w_2(X)$ is a characteristic element by Wu's formula (e.g. [123, Chapter 23]). Then Lemma 5.36 implies that every characteristic element C satisfies that $|C \cdot C| \geq b_2(X)$. The theorem follows from Theorem 5.35. □

The rest of this section is devoted to the proof of Lemma 5.36.

5.4.1 Preliminaries on Equivariant Cohomology Theory

Before proving Lemma 5.36, we give some preliminaries on the equivariant cohomology theory.

Let G be a compact topological group. Let BG be the classifying space of G, and $EG \to BG$ be the universal bundle which is a principal G-bundle with contractible EG.

For a space M with G-action, $M_G = EG \times_G M$ is called the *homotopy quotient of* M. For a pair (M, A) of G-spaces, the *equivariant cohomology (Borel cohomology)* of M is given by

$$H_G^*(M, A) := H^*(M_G, A_G).$$

In particular, $H_G^*(pt) = H^*(BG)$. If G acts on M freely, then $H_G^*(M) = H^*(M/G)$. If G acts on M trivially, then $H_G^*(M) = H_G^*(pt) \otimes H^*(M)$.

Let V be a (real) n-dimensional orthogonal representation of G, and $D(V)$ (resp. $S(V)$) be the unit disk (resp. unit sphere) in V. Assume V is oriented and the G-action preserves the orientation. Consider the oriented vector bundle $V_G := EG \times_G V \to BG$. Then the Thom isomorphism of $V_G \to BG$ is

$$H^{*+n}(D(V_G), S(V_G); \mathbb{Z}) \cong H^*(BG; \mathbb{Z}),$$

where $D(V_G)$ (resp. $S(V_G)$) is the unit disk (resp. unit sphere) bundle in V_G. In terms of equivariant cohomology, this is rewritten as

$$H_G^{*+n}(D(V), S(V); \mathbb{Z}) \cong H_G^*(pt; \mathbb{Z}),$$

which gives a special case of the *Thom isomorphism* for equivariant cohomology. Now the Thom class $\tau(V_G) \in H^{*+n}(D(V_G), S(V_G); \mathbb{Z})$ of the bundle V_G is assumed as a generator $\tau_G(V)$ of $H_G^{*+n}(D(V), S(V); \mathbb{Z})$, which is called the *equivariant Thom class*. The restriction of $\tau_G(V)$ to the origin O of V is called the *equivariant Euler class* $e_G(V)$ which is the ordinary Euler class $e(V_G)$ of the bundle V_G. For oriented orthogonal representations V and W, the Euler classes of the bundles V_G and W_G satisfy the relation $e(V_G \oplus W_G) = e(V_G)e(W_G)$. This is rewritten as

$$e_G(V \oplus W) = e_G(V)e_G(W). \tag{5.3}$$

Remark 5.37 We refer the readers to [166] for the generality of the Thom isomorphism for the equivariant cohomology.

Let $G = \mathrm{U}(1)$. Then we can take $BG = \mathbb{C}P^\infty$ and $EG = S^\infty$.

- $H_G^*(\mathrm{pt}; \mathbb{Z}) = H^*(BG; \mathbb{Z}) = \mathbb{Z}[u]$ where u is a generator of degree 2.
- Let G act on $\mathbb{C}^n$ by multiplication. Then

$$H_G^*(D(\mathbb{C}^n); \mathbb{Z}) \cong H_G^*(O; \mathbb{Z}) = H^*(BG; \mathbb{Z}) = \mathbb{Z}[u],$$
$$H_G^*(S(\mathbb{C}^n); \mathbb{Z}) = H^*(\mathbb{C}P^{n-1}; \mathbb{Z}) = \mathbb{Z}[u]/(u^n),$$
$$H_G^*(D(\mathbb{C}^n), S(\mathbb{C}^n); \mathbb{Z}) = \mathbb{Z}[u]u^n. \tag{5.4}$$

(continued)

The proof is an exercise (Exercise 5.55).

- The equivariant Thom isomorphism

$$H_G^*(O;\mathbb{Z}) \xrightarrow{\cong} H_G^{*+2n}(D(\mathbb{C}^n), S(\mathbb{C}^n);\mathbb{Z})$$

is given by multiplication of the equivariant Thom class $\tau_G(\mathbb{C}^n) = u^n$. The equivariant Euler class is $e_G(\mathbb{C}^n) = u^n \in H_G^*(O;\mathbb{Z}) = \mathbb{Z}[u]$.

- Let G act on $\mathbb{R}^m$ trivially. Then

$$H_G^*(D(\mathbb{R}^m), S(\mathbb{R}^m);\mathbb{Z}) = H_G^*(pt;\mathbb{Z}) \otimes H^*(D(\mathbb{R}^m), S(\mathbb{R}^m);\mathbb{Z}).$$

A generator $\tau_G(\mathbb{R}^m)$ of $H^m(D(\mathbb{R}^m), S(\mathbb{R}^m);\mathbb{Z})$ is specified by choosing an orientation of $\mathbb{R}^m$. Then the Thom isomorphism is

$$H_G^*(pt) = \mathbb{Z}[u] \xrightarrow{\cong} H_G^{*+m}(D(\mathbb{R}^m), S(\mathbb{R}^m);\mathbb{Z}) = \mathbb{Z}[u]\tau_G(\mathbb{R}^m).$$

5.4.2 *Proof of Lemma 5.36*

We give the proof of Lemma 5.36.

Proof of Lemma 5.36 Since $c_1(\mathfrak{s})$ is an integral lift of $w_2(X)$ for a spin^c structure $\mathfrak{s}$ on X, $c_1(\mathfrak{s})$ is a characteristic element, and the set of all characteristic elements is written as

$$c_1(\mathfrak{s}) + 2H^2(X;\mathbb{Z}).$$

This is equal to the set of first Chern classes of all spin^c structures. Therefore it suffices to prove that $-c_1(\mathfrak{s})^2 \geq b_2(X)$ for every spin^c structure $\mathfrak{s}$. Take a finite-dimensional approximation $f_U : E_U \to U$ of the monopole map for $(X, \mathfrak{s})$. Choose a point $p \in J$, and consider the restriction f of f_U to the fiber of $E_U \to J$ over p. Then f is a U(1)-equivariant proper map, and assumed as a map of the following form:

$$f : \mathbb{C}^{x+k} \oplus \mathbb{R}^y \to \mathbb{C}^x \oplus \mathbb{R}^{y+b},$$

where $k = \operatorname{ind} D_{A_0}$, $b = b^+(X)$, x, y are some positive integers and U(1) acts on $\mathbb{C}$ by multiplication and on $\mathbb{R}$ trivially.

Since $b^+(X) = 0$, the restriction of f to the fixed point set of the $\mathrm{U}(1)$-action is a linear isomorphism,

$$f^{\mathrm{U}(1)} \colon \{0\} \oplus \mathbb{R}^y \xrightarrow{\cong} \{0\} \oplus \mathbb{R}^y .$$

Composing a suitable homotopy, we may assume f induces a map between pairs,

$$\bar{f} \colon (D(\mathbb{C}^{x+k} \oplus \mathbb{R}^y), S(\mathbb{C}^{x+k} \oplus \mathbb{R}^y)) \to (D(\mathbb{C}^{x} \oplus \mathbb{R}^y), S(\mathbb{C}^{x} \oplus \mathbb{R}^y)).$$

We apply the functor H^*_G to the following commutative diagram:

$$\begin{array}{ccc}
(D(\mathbb{C}^{x+k} \oplus \mathbb{R}^y), S(\mathbb{C}^{x+k} \oplus \mathbb{R}^y)) & \xrightarrow{\ \bar{f}\ } & (D(\mathbb{C}^{x} \oplus \mathbb{R}^y), S(\mathbb{C}^{x} \oplus \mathbb{R}^y)) \\
{\scriptstyle i_1}\uparrow & & \uparrow{\scriptstyle i_2} \\
(D(\{0\} \oplus \mathbb{R}^y), S(\{0\} \oplus \mathbb{R}^y)) & \underset{\cong}{\xrightarrow{\ f^{\mathrm{U}(1)}\ }} & (D(\{0\} \oplus \mathbb{R}^y), S(\{0\} \oplus \mathbb{R}^y)) .
\end{array}$$

Then it follows from the Thom isomorphisms that we obtain the following commutative diagram:

$$\begin{array}{ccc}
\mathbb{Z}[u]\tau_G(\mathbb{C}^{x+k})\tau_G(\mathbb{R}^y) & \xleftarrow{\ \bar{f}^*\ } & \mathbb{Z}[u]\tau_G(\mathbb{C}^{x})\tau_G(\mathbb{R}^y) \\
\downarrow{\scriptstyle i_1^*} & & \downarrow{\scriptstyle i_2^*} \\
\mathbb{Z}[u]\tau_G(\mathbb{R}^y) & \underset{\cong}{\xleftarrow{\ (f^{\mathrm{U}(1)})^*\ }} & \mathbb{Z}[u]\tau_G(\mathbb{R}^y).
\end{array}$$

Note that there exists $\alpha \in \mathbb{Z}[u]$ such that $\bar{f}^*(\tau_G(\mathbb{C}^x)\tau_G(\mathbb{R}^y)) = \alpha\tau_G(\mathbb{C}^{x+k})\tau_G(\mathbb{R}^y)$. By the commutativity of the diagram, we obtain

$$i_1^*(\alpha\tau_G(\mathbb{C}^{x+k})\tau_G(\mathbb{R}^y)) = i_2^*(\tau_G(\mathbb{C}^x)\tau_G(\mathbb{R}^y)).$$

Since $i_1^*\tau_G(\mathbb{C}^{x+k}) = e_G(\mathbb{C}^{x+k}) = u^{x+k}$ and $i_2^*\tau_G(\mathbb{C}^x) = e_G(\mathbb{C}^x) = u^x$, the equality

$$\alpha u^{x+k} = u^x$$

holds in $\mathbb{Z}[u]$, which implies that

$$0 \geq k = \operatorname{ind} D_{A_0} = \frac{1}{8}(c_1(\mathfrak{s})^2 - \operatorname{sign}(X)) = \frac{1}{8}(c_1(\mathfrak{s})^2 + b_2(X)).$$

Hence we obtain $-c_1(\mathfrak{s})^2 \geq b_2(X)$. $\square$

Remark 5.38

- Frøyshov [50–52, 54] constructed and studied additive homology cobordism invariants for homology 3-spheres by using both Yang–Mills and Seiberg–Witten theories, which are now called the Frøyshov invariants. By using these invariants, Frøyshov obtained constraints on the intersection forms of definite 4-manifolds with boundary, which can be considered as extensions of Donaldson's theorem (Theorem 5.32).
- Frøyshov [55] proved a version of the diagonalization theorem for the intersection forms with local coefficients by using SO(3)-Yang–Mills theory. Nakamura [137] proved another local coefficient version of the diagonalization theorem by using Pin(2)-Seiberg–Witten theory.

5.5 10/8-Inequality

In this section, we prove Furuta's theorem (Theorem 1.1).

Theorem 5.39 (10/8-Inequality, Furuta [59]) *Let X be a closed oriented smooth spin 4-manifold. If the intersection form of X is not definite, then the following inequality holds:*

$$b_2(X) \geq \frac{10}{8}|\sigma(X)| + 2. \tag{5.5}$$

Remark 5.40 As mentioned in Sect. 2.8.3, Theorem 2.40, the inequality (5.5) is rewritten as $\ell \geq 2|k| + 1$ for the intersection form $kE_8 \oplus \ell H$. Donaldson [36, 38] proved a partial result that, if $\ell = 1, 2$, then $k = 0$, by using the Yang–Mills theory. This partial result is often referred to as Theorem B and C, while the diagonalization theorem (Theorem 5.32) is referred to as Theorem A.

The proof of Theorem 5.39 requires two ingredients: the j-symmetry on spin structures and the equivariant Thom isomorphism for equivariant K-theory, which will be explained in the subsequent subsections.

5.5.1 *j*-Symmetry on Spin Structure

If $\mathfrak{s}$ is a spin^c structure associated with a spin structure, then the Seiberg–Witten equations have an additional symmetry. In the spin case, the spinor bundles have quaternionic structures. Especially the j-action on spinors is defined. By extending the j-action to connections suitably, the Seiberg–Witten equations also have a j-symmetry. Note that the determinant line bundle L of spin $\mathfrak{s}$ is trivial. Let A_0 be a trivial flat connection on L. Then a U(1) connection A on L is written as $A_0 + a$, $a \in L^2_k(X; \sqrt{-1}\Lambda^1)$. The j-action is defined by

$$j \cdot (A_0 + a) = A_0 - a.$$

Then a j-action is induced on the curvature $F_{A_0} + da = da$ by

$$j \cdot (da) = -da.$$

Let $\tilde{\mathcal{G}}_{k+1} = \langle L^2_{k+1}(X; \mathrm{U}(1)), j\rangle = L^2_{k+1}(X; \mathrm{Pin}(2))$. (See Proposition 4.21.) Now it is easy to see that the Seiberg–Witten equations (4.2) are $\tilde{\mathcal{G}}$-equivariant.

By choosing the trivial flat connection A_0 as a reference connection, the monopole map $\boldsymbol{\mu}\colon \mathcal{E} \to \mathcal{F}$ (5.1) is $\langle \mathrm{U}(1), j\rangle = \mathrm{Pin}(2)$-equivariant. Let $\varepsilon\colon \mathrm{Pin}(2) \to \mathrm{Pin}(2)/\mathrm{U}(1) = \{\pm 1\}$ be the projection, and Pin(2) act on $H^1(X; \mathbb{R})$ by multiplication of the value of ε. Then this induces the Pin(2)-action on $J = H^1(X; \mathbb{R})/H^1(X; \mathbb{Z})$, and the projection $\pi : \mathcal{E} \to J$ is also Pin(2)-equivariant. The finite-dimensional approximations of the monopole map $\boldsymbol{\mu}$ are also Pin(2)-equivariant.

5.5.2 Preliminaries on K-Theory

As mentioned before, the proof of Theorem 5.39 requires the equivariant K-theory, especially, the equivariant Thom isomorphism. In this subsection, we give a quick review of nonequivariant K-theory. The next subsection gives a summary of several results in the equivariant K-theory and a proof of a special case of the equivariant Thom isomorphism. General references are [5, 73, 154].

First, we define K-groups.

Definition 5.41 For a compact Hausdorff space X, let $\mathrm{Vect}(X)$ be the set of isomorphism classes of complex vector bundles over X. Then $\mathscr{A} = \mathrm{Vect}(X)$ has the structure of an abelian semigroup, where the additive structure is given by direct sum $\oplus$. Let $F(\mathscr{A})$ be the free abelian group generated by the elements of $\mathscr{A}$, and $S(\mathscr{A})$ be the subgroup of $F(\mathscr{A})$ generated by the elements of the form $a+b-(a\oplus b)$. Define the *K-group* of X as

$$K(X) = F(\mathscr{A})/S(\mathscr{A}).$$

That is, $K(X)$ is the *Grothendieck group* of the abelian semigroup $\mathrm{Vect}(X)$.

Let ε^n be the trivial bundle over X of rank n. The class of $E \in \mathrm{Vect}(X)$ in $K(X)$ is also denoted by E. Note that, for $E, F \in \mathrm{Vect}(X)$,

$$E + F = E \oplus F, \quad -E - F = -(E \oplus F).$$

Then every element in $K(X)$ is represented by $E_0 - E_1$ for some $E_0, E_1 \in \mathrm{Vect}(X)$, and $E_0 - E_1 = F_0 - F_1$ if and only if there exists $P \in \mathrm{Vect}(X)$ such that $E_0 \oplus F_1 \oplus P \cong F_0 \oplus E_1 \oplus P$. It is easy to see that the class of ε^0 is the zero of $K(X)$.

A multiplication is defined from the tensor product of vector bundles as follows:

$$(E_0 - E_1)(F_0 - F_1) = E_0 \otimes F_0 - E_0 \otimes F_1 - E_1 \otimes F_0 + E_1 \otimes F_1.$$

This makes $K(X)$ a commutative ring with the identity ε^1. In general, it is convenient to write ε^n as n. For example, $\varepsilon^n \otimes E = nE$.

Lemma 5.42 *For each $E \in \mathrm{Vect}(X)$, there exists $E' \in \mathrm{Vect}(X)$ such that $E \oplus E' \cong \varepsilon^n$.*

Proof of Theorem 5.32 We construct an embedding of E into a trivial bundle as a subbundle. Then a required bundle E' is obtained as a complement of E in the trivial bundle. For simplicity, assume X is connected. By using Urysohn's lemma, we can find a finite cover $\{U_j\}$ of X and a map $\varphi_j\colon E \to X \times \mathbb{R}^k$ for each j such that $E|_{U_j}$ is trivial and the restriction of φ_j to $E|_{U_j}$ is an isomorphism onto $U_j \times \mathbb{R}^k$. Then the direct sum $\oplus_j \varphi_j$ is a required embedding of E into $X \times (\mathbb{R}^k \oplus \cdots \oplus \mathbb{R}^k)$. □

Then every element in $K(X)$ is represented by $E - n$ for some $E \in \mathrm{Vect}(X)$ and n. For $E_0 - E_1 \in K(X)$, choose E_1' such that $E_1 \oplus E_1' \cong \varepsilon^n$ by Lemma 5.42. Then $E_0 - E_1 = E_0 + E_1' - (E_1 + E_1') = E_0 \oplus E_1' - n$.

A map $f\colon X \to Y$ induces a ring homomorphism $f^*\colon K(Y) \to K(X)$ given by $E_0 - E_1 \mapsto f^*E_0 - f^*E_1$. Then it holds that $(fg)^* = g^*f^*$. If f is homotopic to g, then $f^* = g^*$.

An exterior product $K(X) \otimes K(Y) \to K(X \times Y)$ is defined by

$$(E, F) \mapsto p_1^*E \otimes p_2^*F,$$

where p_1 and p_2 are the projections of $X \times Y$ to X and Y. This is also a ring homomorphism.

Choose a base point $x_0 \in X$. Let $\iota\colon \{x_0\} \to X$ be the inclusion. The *reduced K-group* $\tilde{K}(X)$ is defined by

$$\tilde{K}(X) := \ker(\iota^*\colon K(X) \to K(x_0)).$$

In general, $\tilde{K}(X)$ is an ideal of $K(X)$, and therefore a ring which may not have the identity. If $f\colon X \to Y$ is a based map, then the induced map $f^*\colon \tilde{K}(Y) \to \tilde{K}(X)$ is also a ring homomorphism.

For a pair (X, A) of compact Hausdorff spaces, we define the relative K-group $K(X, A)$ by $K(X, A) := \tilde{K}(X/A)$ with base point A/A in X/A.

Proposition 5.43 *The inclusion and quotient maps* $A \xrightarrow{i} X \xrightarrow{q} X/A$ *induce an exact sequence* $K(X, Y) = \tilde{K}(X/A) \to \tilde{K}(X) \to \tilde{K}(A)$.

Proof of Theorem 5.32 Since qi is the composition $A \to A/A \hookrightarrow X/A$ and $\tilde{K}(A/A) = 0$, we have $i^*q^* = 0$, and therefore $\operatorname{im} q^* \subset \ker i^*$.

We prove that $\ker i^* \subset \operatorname{im} q^*$. Suppose $E - n \in \tilde{K}(X)$ is in $\ker i^*$. Adding some trivial bundle to E, we may assume E is trivial over A. Choosing a frame of E over A, we can identify all the fibers over A to a single fiber over a point in A. Let $\bar{E}$ be the quotient space by this identification of fibers. Then we have a surjective map $\bar{E} \to X/A$. We need to prove the local triviality near the point A/A. Note that the frame over A can be extended to a frame over a neighborhood of A by using the Tietze extension theorem. Then the extended frame induces a local trivialization of $\bar{E}$ over a neighborhood of A/A. By construction, we see that $E = q^*(\bar{E})$. □

It is easy to verify the next lemma.

Lemma 5.44 *If A is contractible, then* $q^*\colon \operatorname{Vect}(X/A) \to \operatorname{Vect}(X)$ *is bijective.*

Let CX and SX be the cone and suspension of X. Consider the following sequence of inclusions:

$$\begin{array}{ccccccccc} A & \hookrightarrow & X & \hookrightarrow & X \cup CA & \hookrightarrow & (X \cup CA) \cup CX & \hookrightarrow & ((X \cup CA) \cup CX) \cup C(X \cup CA). \\ & & & & \downarrow & & \downarrow & & \downarrow \\ & & & & X/A & & SA & & SX \end{array}$$

Since the vertical maps of the above sequence are contractions, the following long exact sequence is obtained from Proposition 5.43 and Lemma 5.44:

$$\cdots \to \tilde{K}(SX) \to \tilde{K}(SX) \to \tilde{K}(X/A) \to \tilde{K}(X) \to \tilde{K}(A). \tag{5.6}$$

Remark 5.45 The group $\tilde{K}(S^n X)$ is denoted by $\tilde{K}^{-n}(X)$. The Bott periodicity theorem (see [5, 6, 73]) says that the isomorphism $\tilde{K}^{-n}(X) \cong \tilde{K}^{-n-2}(X)$ holds for every n. Then the exact sequence (5.6) is folded into the following six-term exact sequence:

$$\begin{array}{ccccc} K^0(X, A) & \longrightarrow & \tilde{K}^0(X) & \longrightarrow & \tilde{K}^0(A) \\ \uparrow & & & & \downarrow \\ \tilde{K}^1(A) & \longleftarrow & \tilde{K}^1(X) & \longleftarrow & K^1(X, A) \end{array}$$

There is the unreduced version of the exact sequence. Let X_+ be the disjoint union of X and a base point '+'. Then $K(X) = \tilde{K}(X_+)$, and $\tilde{K}(S(X_+)) = \tilde{K}(S(X) \vee S^1) = \tilde{K}(S(X)) \oplus \tilde{K}(S^1) = \tilde{K}(S(X))$ since $\tilde{K}(S^1) = 0$. If we set $K^0(X) = K(X)$ and $K^1(X) = \tilde{K}^1(X)$, then the unreduced version is obtained from the above sequence.

Consider the product $X \times Y$ of based spaces X and Y with base points x_0 and y_0. The smash product of X and Y is $X \wedge Y = X \times Y / X \vee Y$ where $X \vee Y = X \times \{y_0\} \cup \{x_0\} \times Y$. We construct the exterior product $\tilde{K}(X) \otimes \tilde{K}(Y) \to \tilde{K}(X \wedge Y)$ as follows. The long exact sequence for the pair $(X \times Y, X \vee Y)$ is

$$\begin{array}{ccccc} \tilde{K}(S(X \times Y)) \to & \tilde{K}(S(X \vee Y)) & \to \tilde{K}(X \wedge Y) \to \tilde{K}(X \times Y) \to & \tilde{K}(X \vee Y) \\ & \| & & \| \\ & \tilde{K}(SX) \oplus \tilde{K}(SY) & & \tilde{K}(X) \oplus \tilde{K}(Y) \end{array}$$

The last horizontal map has a splitting $\tilde{K}(X) \oplus \tilde{K}(Y) \to \tilde{K}(X \times Y)$ given by $a \oplus b \mapsto p_1^* a + p_2^* b$. Similarly, the first horizontal map has a splitting. Therefore we obtain

$$\tilde{K}(X \times Y) \cong \tilde{K}(X \wedge Y) \oplus \left(\tilde{K}(X) \oplus \tilde{K}(Y)\right),$$

$$\tilde{K}(X \wedge Y) = \operatorname{Ker}\left(\tilde{K}(X \times Y) \to \tilde{K}(X) \oplus \tilde{K}(Y)\right).$$

For $a \in \tilde{K}(X) = \operatorname{Ker}(K(X) \to K(x_0))$ and $b \in \tilde{K}(Y) = \operatorname{Ker}(K(Y) \to K(y_0))$, the restrictions of $p_1^* a$ and $p_2^* b$ respectively to $K(\{x_0\} \times Y)$ and $K(X \times \{y_0\})$ are zero. Hence $p_1^* a p_2^* b$ is in the kernel of $\tilde{K}(X \times Y) \to \tilde{K}(X) \oplus \tilde{K}(Y)$, and therefore defines a unique element in $\tilde{K}(X \wedge Y)$. Hence the correspondence $a \otimes b \mapsto p_1^* a p_2^* b$ defines the exterior product $\tilde{K}(X) \otimes \tilde{K}(Y) \to \tilde{K}(X \wedge Y)$.

An element of $K(X, A)$ is represented by $F_0 - F_1$ where $F_0, F_1 \in \operatorname{Vect}(X/A)$ such that $F_0|_{(A/A)} \cong F_1|_{(A/A)}$. Then the pull-back bundles $q^* F_0$ and $q^* F_1$ via the quotient map $q \colon X \to X/A$ have the properties that the restrictions $q^* F_0|_A$ and $q^* F_1|_A$ are trivial and $q^* F_0|_A \cong q^* F_1|_A$. Conversely, the bundles $E_0, E_1 \in \operatorname{Vect}(X)$ such that $E_0|_A \cong E_1|_A$ (but $E_0|_A$ and $E_1|_A$ may be nontrivial) determine an element in $K(X, A)$ as follows.

Lemma 5.46 *For $E_0, E_1 \in \operatorname{Vect}(X)$, suppose there is an isomorphism $\alpha \colon E_1|_A \xrightarrow{\cong} E_0|_A$. Then the triple (E_0, E_1, α) defines an element in $K(X, A)$.*

Proof of Theorem 5.32 Let X_0 and X_1 be copies of X, and let $Y = X_0 \cup_A X_1$ be the space identifying the points in A. Identifying $E_0|_A$ and $E_1|_A$ by α, we obtain a bundle $E_0 \cup_\alpha E_1$ over Y, hence $E_0 \cup_\alpha E_1 \in K(Y)$. Note that the identification $X_0 = X_1$ induces a retraction $r \colon Y \to X_1$ which gives a splitting of the exact sequence (5.6) for (Y, X_1). Hence we obtain $K(Y) = K(Y, X_1) \oplus K(X_1)$. Then $E_0 \cup_\alpha E_1 - r^* E_1$ can be assumed as an element of $K(Y, X_1) \cong K(X, A)$. □

Let V be a finite-dimensional vector space and O be the origin. Let $D(V)$ and $S(V)$ be the unit disk and sphere of V, and $\pi \colon D(V) \to O$ be the projection. Assume $\Lambda^* V$ as a vector bundle over O, and consider the pull-back $\pi^* \Lambda^* V$ over $D(V)$. For $v \neq 0$ in V, we have an isomorphism

$$\gamma = v \wedge \cdot + v \lrcorner \cdot \colon \Lambda^{\text{even}} V \to \Lambda^{\text{odd}} V,$$

where $\Lambda^{\text{even}} V = \sum_i \left(\Lambda^{2i} V\right)$ and $\Lambda^{\text{odd}} V = \sum_i \left(\Lambda^{2i+1} V\right)$. Then $(\Lambda^{\text{even}} V, \Lambda^{\text{odd}} V, \gamma)$ defines an element λ_V in $K(D(V), S(V))$. Let V^* be the dual of V. Similarly $(\Lambda^{\text{even}} V^*, \Lambda^{\text{odd}} V^*, \gamma^*)$ defines an element λ_V^* in $K_G(D(V), S(V))$.

5.5.3 *Equivariant K-Theory and Equivariant Thom Isomorphism*

Let G be a compact topological group.

A vector bundle $p\colon E \to X$ is a *G-vector bundle* if E and X are G-spaces and p is G-equivariant, and for any $g \in G$ and $x \in X$, the action $g\colon E_x \to E_{gx}$ is a linear isomorphism of vector spaces.

We assume X is a compact Hausdorff G-space. Let $\mathrm{Vect}^G(X)$ be the set of isomorphism classes of G-equivariant complex vector bundles over X. We assume every complex G-bundle is equipped with a G-invariant Hermitian metric. The equivariant K-group $K_G(X)$ is defined as the Grothendieck group of the abelian semigroup $\mathrm{Vect}^G(X)$. Every construction in Sect. 5.5.2 is straightforwardly extended to K_G and almost everything in Sect. 5.5.2 holds for K_G. For the generality of the equivariant K-theory, see [5, 154].

We summarize several facts.

- If X is a point, then $K_G(\mathrm{pt})$ is the complex representation ring $R(G)$ of G.
- $K_G(X)$ is an algebra over $R(G)$.
- A G-map $f\colon X \to Y$ induces a homomorphism $f^*\colon K_G(Y) \to K_G(X)$. Furthermore, G-equivariantly homotopic maps induce the same homomorphism.
- If G acts freely on X, then the map $K(X/G) \to K_G(X)$ defined by pull-back is a ring isomorphism.
- If G acts trivially on X, then tensoring a representation with a vector bundle induces the map $R(G) \otimes K(X) \to K_G(X)$, which is an isomorphism of $R(G)$-algebras.
- For every $E \in \mathrm{Vect}^G(X)$, there exist $E' \in \mathrm{Vect}^G(X)$ and a G-module M such that $E \oplus E'$ is isomorphic to the trivial G-bundle $X \times M$. (For the proof, the Peter–Weyl theorem is used. See [154, Proposition 2.4].)
- For a based G-space X with base point x_0, the reduced K_G-group is defined as $\tilde{K}_G(X) = \ker(K_G(X) \to K_G(x_0))$. For a based pair (X, A) of G-spaces, the relative K_G-group is defined as $K_G(X, A) = \tilde{K}_G(X/A)$.
- We have the following long exact sequence:

$$\cdots \to \tilde{K}_G(SX) \to \tilde{K}_G(SX) \to \tilde{K}_G(X/A) \to \tilde{K}_G(X) \to \tilde{K}_G(A).$$

(continued)

- The exterior product $\tilde{K}_G(X) \otimes \tilde{K}_G(Y) \to \tilde{K}_G(X \wedge Y)$ is defined. In particular, we have the exterior product

$$K_G(D(V), S(V)) \otimes K_G(D(W), S(W)) \to K_G(D(V \oplus W), S(V \oplus W))$$

for representations V and W of G.
- Let V be a representation of G. Then $(\Lambda^{\text{even}} V, \Lambda^{\text{odd}} V, \gamma)$ defines an element λ_V in $K_G(D(V), S(V))$. Let V^* be the dual of V. Similarly $(\Lambda^{\text{even}} V^*, \Lambda^{\text{odd}} V^*, \gamma^*)$ defines an element λ_V^* in $K_G(D(V), S(V))$.
- The restriction of λ_V to the origin O of V is

$$\Lambda^{\text{even}} V - \Lambda^{\text{odd}} V = \sum_i (-1)^i \left(\Lambda^i V\right) \in K_G(O) \cong R(G).$$

In the proof of Theorem 5.39, we need the following theorem which is a special case of the G-equivariant Thom isomorphism.

Theorem 5.47 (Atiyah [6]) *Let $\tau_G(V) = \lambda_V^* \in K_G(D(V), S(V))$. Then $\tau_G(V)$ is the G-equivariant Thom class such that the Thom isomorphism (equivariant Bott periodicity) holds,*

$$K_G(D(V), S(V)) \cong R(G)\tau_G(V).$$

The restriction of $\tau_G(V)$ to the origin O is the K_G-theoretic Euler class which is given by

$$e_G(V) := \lambda_{-1}(V^*) = \sum_i (-1)^i [\Lambda^i V^*] \in R(G).$$

The rest of this subsection is devoted to the proof of Theorem 5.47. Define the homomorphism $\beta\colon R(G) \to K_G(D(V), S(V))$ by multiplication $x \mapsto x\lambda_V^*$. We construct the inverse $\alpha\colon K_G(D(V), S(V)) \to R(G)$ such that $\alpha\beta = 1$ and $\beta\alpha = 1$. The construction of α uses the indices of elliptic operators.

Let $P = P(V \oplus 1)$ be the complex projective space of $V \oplus \varepsilon^1$. Let $H^* \to P$ be the tautological line bundle, and H be its dual. Then $H^* = H^{-1}$. Let $\pi\colon P \to O$ be the projection to the origin O. Consider the bundle $\Lambda^*(H^{-1} \otimes \pi^* V^*)$ over P. Since $P(V \oplus 1) \setminus P(V) \cong V$, and the restriction of H^{-1} to V is trivial, we obtain $\Lambda^*(H^{-1} \otimes \pi^* V^*)|_V = \lambda_V^*$. This shows that the image of λ_V^* by the homomorphism

$$j\colon K(D(V), S(V)) = K(P(V \oplus 1), P(V)) \to K(P(V \oplus 1))$$

is $\Lambda^*(H^{-1} \otimes \pi^* V^*) = \sum_i (-1)^i H^{-i} \left(\Lambda^i V^*\right)$. We consider the G-equivariant elliptic operator

$$D = \bar{\partial} + \bar{\partial}^* \colon \Omega^{0,\text{even}}(P(V \oplus 1)) \to \Omega^{0,\text{odd}}(P(V \oplus 1)).$$

For a holomorphic G-vector bundle Q over $P = P(V \oplus 1)$, D extends to the operator

$$D_Q \colon \Omega^{0,\text{even}}(Q) \to \Omega^{0,\text{odd}}(Q).$$

Then there are G-module isomorphisms

$$\operatorname{Ker} D_Q = \sum_k H^{2k}(P; \mathscr{O}(Q)), \quad \operatorname{Coker} D_Q = \sum_k H^{2k+1}(P; \mathscr{O}(Q)).$$

Thus we have $\operatorname{ind} D_Q = \sum_i (-1)^i H^i(P; \mathscr{O}(Q)) \in R(G)$. By the Kodaira–Nakano vanishing theorem (see [65]), we have $H^i(P; \mathscr{O}(H^{-r})) = 0$ for $r > 0$ and $i \leq n$ where $n = \dim V = \dim P$. Note also that

$$H^i(P; \mathscr{O}) = \begin{cases} 0 & (i \geq 1) \\ \mathbb{C} & (i = 0). \end{cases}$$

Then we have $\operatorname{ind} D_Q = 1$ for $Q = j(\lambda_V^*)$. We define the map α as the composition

$$K_G(D(V), S(V)) \xrightarrow{j} K_G(P(V \oplus 1)) \xrightarrow{\operatorname{ind} D_\cdot} R(G).$$

Then $\alpha\beta = 1$ by definition. We prove that $\beta\alpha = 1$. The next lemma is easy to check.

Lemma 5.48 *The map α extends to the map $\tilde{\alpha} \colon K_G(D(V \oplus V), S(V \oplus V)) \to K_G(D(V), S(V))$ such that the following diagram commutes:*

$$\begin{array}{ccc} K_G(D(V), S(V)) \otimes K_G(D(V), S(V)) & \xrightarrow{\alpha \times 1} & R(G) \otimes K_G(D(V), S(V)) \\ \downarrow & & \downarrow \\ K_G(D(V \oplus V), S(V \oplus V)) & \xrightarrow{\tilde{\alpha}} & K_G(D(V), S(V)) \end{array}$$

where the vertical arrows are the exterior products.

Since the maps $V \times V \to V \times V$ given by $(u, v) \mapsto (v, -u)$ and $(u, v) \mapsto (-v, u)$ are G-equivariantly homotopic to the identity by rotation, these induce the identity on $K_G(D(V \oplus V), S(V \oplus V))$. The map $u \mapsto -u$ on V induces an involution $x \mapsto \hat{x}$ on $K_G(D(V), S(V))$. Then we have

$$xy = \hat{y}x = y\hat{x}.$$

Set $b = \lambda_V^*$. Then $\alpha(b) = 1$. By Lemma 5.48, we obtain

$$\beta(\alpha(y)) = \alpha(y)b = \tilde{\alpha}(yb) = \tilde{\alpha}(b\hat{y}) = \alpha(b)\hat{y} = \hat{y}.$$

Since $\alpha\beta = 1$, we have $y = \beta\alpha\beta\alpha(y) = \beta(\alpha\beta)\alpha(y) = \beta\alpha(y) = \hat{y}$. Hence $\beta\alpha = 1$. This completes the proof of Theorem 5.47.

5.5.4 Proof of Theorem 5.39

In this subsection, we prove Theorem 5.39. We use the following representations of G.

- The trivial representation $\mathbb{R}$.
- The 1-dimensional representation $\tilde{\mathbb{R}}$ on which U(1) acts trivially, and j acts by multiplication of -1. Let $\tilde{\mathbb{C}} = \tilde{\mathbb{R}} \otimes_{\mathbb{R}} \mathbb{C}$.
- The space of quaternions $\mathbb{H}$ on which G acts by left multiplication.

Note that $h^* = h$ and $c^* = c$.

Lemma 5.49 *Let* $c = [\tilde{\mathbb{C}}]$ *and* $h = [\mathbb{H}]$. *Then the complex representation ring* $R(G)$ *is isomorphic to*

$$\mathbb{Z}[c, h]/(c^2 - 1, ch - h).$$

Proof of Theorem 5.32 It is straightforward to see that $c^2 = 1$ and $ch = h$. For $n \in \mathbb{Z}$, let h_n be the representation induced from the 1-dimensional representation of $U(1) \subset \mathrm{Pin}(2)$ given by the multiplication of z^n. Note that $h_1 = h$ and $h_0 = 1 + c$. By Frobenius reciprocity, we have $h_{-n} = h_n$, $h_m h_n = h_{m+n} + h_{|m-n|}$. The irreducible representations are 1, c and h_n ($n \geq 1$). In $R(G)$, h_n is represented by 1, c, h. □

We will use the following generators:

$$w = \lambda_{-1}(\tilde{\mathbb{C}}) = 1 - c, \quad z = \lambda_{-1}(\mathbb{H}) = 2 - h.$$

Then we have

$$R(G) = \mathbb{Z}[w, z]/(w^2 - 2w, zw - 2w).$$

Now we prove Theorem 5.39.

Proof of Theorem 5.39 (Cf. [25, 59, 120].) Let $k = -\sigma(X)/16$ which is an integer by Rokhlin's theorem. By reversing the orientation of X if necessary, we may assume $k \geq 0$. Then the inequality (5.5) is equivalent to $b^+(X) \geq 2k + 1$.

Take a finite-dimensional approximation $f_U : E_U \to U$. Note that the origin O of J is a fixed point of the G-action. The restriction f of f_U to the fiber of $E_U \to J$ over O is a G-equivariant map of the following form:

$$f : \mathbb{H}^{x+k} \oplus \tilde{\mathbb{R}}^y \to \mathbb{H}^x \oplus \tilde{\mathbb{R}}^{y+b}, \tag{5.7}$$

where $k = \frac{1}{2} \operatorname{ind}_{\mathbb{C}} D_{A_0} = -\sigma(X)/16$, $b = b^+(X)$, x, y are some positive integers. Take the complexification of f as

$$f : \mathbb{H}^{2x+2k} \oplus \tilde{\mathbb{C}}^y \to \mathbb{H}^{2x} \oplus \tilde{\mathbb{C}}^{y+b},$$

$$f\left(u \otimes 1 + v \otimes \sqrt{-1}\right) = f(u) \otimes 1 + f(v) \otimes \sqrt{-1}.$$

(Note that $\mathbb{H} \otimes_{\mathbb{R}} \mathbb{C} \cong \mathbb{H}^2$.) The restriction of f to the $\mathrm{U}(1)$-fixed point set is a linear inclusion

$$f^{\mathrm{U}(1)} : \{0\} \oplus \tilde{\mathbb{C}}^y \to \{0\} \oplus \tilde{\mathbb{C}}^{y+b}.$$

As in the proof of Lemma 5.36, we have the commutative diagram of G-equivariant maps:

$$\begin{array}{ccc}
(D(\mathbb{H}^{2x+2k} \oplus \tilde{\mathbb{C}}^y), S(\mathbb{H}^{2x+2k} \oplus \tilde{\mathbb{C}}^y)) & \xrightarrow{\bar{f}} & (D(\mathbb{H}^{2x} \oplus \tilde{\mathbb{C}}^y), S(\mathbb{H}^{2x} \oplus \tilde{\mathbb{C}}^{y+b})) \\
{\scriptstyle i_1}\uparrow & & \uparrow{\scriptstyle i_2} \\
(D(\{0\} \oplus \tilde{\mathbb{C}}^y), S(\{0\} \oplus \tilde{\mathbb{C}}^y)) & \xhookrightarrow{\bar{f}^{\mathrm{U}(1)}} & (D(\{0\} \oplus \tilde{\mathbb{C}}^y), S(\{0\} \oplus \tilde{\mathbb{C}}^{y+b})).
\end{array}$$

Applying the K_G-functor to the diagram and using the Thom isomorphism, we obtain

$$\begin{array}{ccc}
R(G)\tau_G(\mathbb{H}^{2x+2k})\tau_G(\tilde{\mathbb{C}}^y) & \xleftarrow{\bar{f}^*} & R(G)\tau_G(\mathbb{H}^{2x})\tau_G(\tilde{\mathbb{C}}^{y+b}) \\
\downarrow{\scriptstyle i_1^*} & & \downarrow{\scriptstyle i_2^*} \\
R(G)\tau_G(\tilde{\mathbb{C}}^y) & \xleftarrow{(\bar{f}^{\mathrm{U}(1)})^*} & R(G)\tau_G(\tilde{\mathbb{C}}^{y+b}).
\end{array}$$

There exists an element $\alpha \in R(G)$ such that

$$\bar{f}^*(\tau_G(\mathbb{H}^{2x})\tau_G(\tilde{\mathbb{C}}^{y+b})) = \alpha\tau_G(\mathbb{H}^{2x+2k})\tau_G(\tilde{\mathbb{C}}^y).$$

It follows from the commutativity of the diagram that

$$(i_1^* \circ \bar{f}^*)(\tau_G(\mathbb{H}^{2x})\tau_G(\tilde{\mathbb{C}}^{y+b})) = (\bar{f}^{\mathrm{U}(1)} \circ i_2^*)(\tau_G(\mathbb{H}^{2x})\tau_G(\tilde{\mathbb{C}}^{y+b})).$$

Since $i_k^* \tau_G(\mathbb{H}) = e_G(\mathbb{H}) = z$ and $(\bar{f}^{\mathrm{U}(1)})^* \tau_G(\tilde{\mathbb{C}}) = e_G(\tilde{\mathbb{C}}) = w$, both sides are rewritten as

$$(i_1^* \circ \bar{f}^*)(\tau_G(\mathbb{H}^{2x})\tau_G(\tilde{\mathbb{C}}^{y+b})) = \alpha z^{2x+2k} \tau_G(\tilde{\mathbb{C}}^y),$$
$$(\bar{f}^{\mathrm{U}(1)} \circ i_2^*)(\tau_G(\mathbb{H}^{2x})\tau_G(\tilde{\mathbb{C}}^{y+b})) = z^{2x} w^b \tau_G(\tilde{\mathbb{C}}^y).$$

Using the relations $w^2 = 2w$ and $zw = 2w$, we obtain

$$\alpha z^{2x+2k} = z^{2x} w^b = 2^{2x+b-1} w. \tag{5.8}$$

Note that α has a form of $\alpha = Cw + P(z)$, where $C \in \mathbb{Z}$ and $P(z)$ is a polynomial in z. Then (5.8) is rewritten as

$$(Cw + P(z))z^{2x+2k} = C2^{2x+2k} w + z^{2x+2k} P(z) = 2^{2x+b-1} w.$$

This implies $P(z) = 0$ and $C = 2^{b-2k-1}$. Since C is an integer, we obtain $b \geq 2k + 1$. □

Remark 5.50 The proofs of Theorems 5.32 and 5.39 presented in this book are based on the same principle: *the divisibility of Euler classes implies a constraint.* The former uses the cohomological Euler class, and the latter uses the K-theoretic one. This principle is applied to families of 4-manifolds by Baraglia [11] and Konno and Nakamura [97] to obtain constraints on smooth families. Kametani [83] is remarkable in that it is observed that, for the purpose of proving the divisibility of Euler classes, there is no need to use a finite-dimensional approximation; a local model called the Kuranishi model around a reducible is sufficient. (For the Kuranishi model, see e.g. [151, Appendix B.2] and [165, §6.7].) See also Problems 5.58 and 5.62.

Remark 5.51 We list subsequent developments of the 10/8-inequality and related topics.

- Furuta and Kametani [60] sharpened the 10/8-inequality by using KO-theory. Recently, a further refinement was given by Hopkins et al. [77] by using more elaborated homotopy theory.
- Bryan [25] and Fang [43] proved the inequalities stronger than the 10/8-inequality for spin 4-manifolds with $\mathbb{Z}/2^p$-actions. Similar refinements under the actions of symmetric and alternating groups are proved in Kiyono and Liu [95], Liu [114] and the works due to X. Liu and his collaborators. Nakamura [137] considered free involutions on spin 4-manifolds which induce complex conjugation on spin structures, and obtained a local coefficient version of the 10/8-inequality for their quotient manifolds. Y. Kato [86] generalized Nakamura's work [137] to

more general involutions and $\mathbb{Z}_2 \times \mathbb{Z}_2$-actions and obtained several versions of 10/8-type inequalities.

- Manolescu [120] extends the 10/8-inequality to the case of 4-manifolds with boundary by applying K-theory to the framework of the relative Bauer–Furuta invariant. Furuta and Li also developed the 10/8- type inequality for 4-manifolds with boundary by using K-theory with local coefficient (Generalized cohomology theories and Pin(2) Seiberg–Witten Floer functors on homology 3-spheres, unpublished). J. Lin [112] refined Manolescu's inequality by KO-theory.
- For closed oriented smooth non-spin 4-manifolds X with even intersection forms, there is a conjecture called the $5/4$-*conjecture* which states

$$b_2(X) \geq \frac{5}{4}|\sigma(X)|.$$

Bohr [20] and Lee and Li [108] verified the conjecture for some classes of 4-manifolds by applying the 10/8-inequality to the spin coverings of non-spin 4-manifolds.

Problems

5.52 Verify Lemma 5.2.

5.53 Verify Lemma 5.20.

5.54 Let X and Y be closed oriented smooth manifolds. The homotopy set $[S^2 \wedge X, Y]$ of pointed maps $f\colon S^2 \wedge X \to Y$ has a natural structure of an abelian group. Perturbing f if necessary, we may assume the inverse image $M_y = f^{-1}(y)$ of a point $y \in Y$ has a fundamental class $[M_y]$ in $H_*(S^2 \wedge X; \mathbb{Z})$. Prove that the correspondence $[S^2 \wedge X, Y] \ni [f] \mapsto [M_y] \in H_*(S^2 \wedge X; \mathbb{Z})$ is a group homomorphism.

5.55 Prove the isomorphisms in (5.4).

5.56 Verify the following.

1. $BC_2 = \mathbb{R}P^\infty$ for the cyclic group C_2 of order 2.
2. $B\mathrm{SU}(2) = \mathbb{H}P^\infty$, where $\mathbb{H}P^\infty$ is the infinite-dimensional quaternionic projective space.

5.57 Let $G = \{\pm 1\}$ act on $\mathbb{R}^n$ by multiplication. Let $D(\mathbb{R}^n)$ be the unit ball in $\mathbb{R}^n$, and $S(\mathbb{R}^n) = \partial D(\mathbb{R}^n)$. Let $\mathbb{F}_2 = \mathbb{Z}/2$. Verify the following isomorphisms:

$$H^*_G(D(\mathbb{R}^n); \mathbb{F}_2) \cong H^*_G(O; \mathbb{F}_2) = H^*(BG; \mathbb{F}_2) = \mathbb{F}_2[q],$$

$$H^*_G(S(\mathbb{R}^n); \mathbb{F}_2) = H^*(\mathbb{R}P^{n-1}; \mathbb{F}_2) = \mathbb{F}_2[q]/(q^n),$$

$$H^*_G(D(\mathbb{R}^n), S(\mathbb{R}^n); \mathbb{F}_2) = \mathbb{F}_2[q]q^n,$$

where O is the origin of $\mathbb{R}^n$ and q is a generator of degree 1.

5.58 Prove the Borsuk–Ulam theorem (Proposition 2.51) in the following steps. Let the cyclic group C_2 of order 2 act on S^m and S^n antipodally. Suppose we have a C_2-equivariant map $f: S^m \to S^n$. By taking cones of S^m and S^n, the map f extends to a C_2-equivariant map $\tilde{f}: D(\mathbb{R}^{m+1}) \to D(\mathbb{R}^{n+1})$ such that the restriction of $\tilde{f}$ to the boundary $\partial D(\mathbb{R}^{m+1})$ is f. Then prove that $m \leq n$ holds by mimicking the proof of Lemma 5.36.

5.59 Prove Theorem 5.32 when $b_1(X) = 0$ by proving Lemma 5.36 by using Seiberg–Witten moduli spaces in the following steps. For a spin^c structure $\mathfrak{s}$, let k be the Dirac index of $\mathfrak{s}$. We want to prove $k \leq 0$. Suppose to the contrary that $k > 0$. Then the dimension of the moduli space $\mathcal{M}(\mathfrak{s})$ is $2k - 1 > 0$. If $b^+(X) = 0$ and $b_1(X) = 0$ then the moduli space $\mathcal{M}(\mathfrak{s})$ contains a unique reducible point by Proposition 4.57 and Corollary 4.58. Perturbing the equations and removing a neighborhood of the reducible point from $\mathcal{M}(\mathfrak{s})$, we obtain a compact manifold $\mathcal{M}'$ with boundary embedded in $\mathcal{B}^*$.

1. What is the boundary of $\mathcal{M}'$?
2. Prove that such a compact manifold $\mathcal{M}'$ cannot exist. (Note that the fundamental class of the boundary of $\mathcal{M}'$ is nontrivial in $H_*(\mathcal{B}^*; \mathbb{Z})$.)

(One can also use the approximated moduli space for the proof of the nonpositivity of k instead of $\mathcal{M}(\mathfrak{s})$.)

5.60 Let $G = \mathrm{Pin}(2)$. We have a fibration

$$G \to \mathrm{SU}(2) \to \mathbb{R}P^2,$$

where the second map is the composition of the Hopf fibration $\mathrm{SU}(2) \cong S^3 \to S^2$ with the quotient map $S^2 \to \mathbb{R}P^2$. This induces a fibration

$$\mathbb{R}P^2 \to BG \to B\mathrm{SU}(2) = \mathbb{H}P^\infty.$$

By considering the Leray–Serre spectral sequence for this fibration, prove

$$H^*(BG; \mathbb{F}_2) \cong \mathbb{F}_2[q, v]/(q^3),$$

where the degree of q is 1 and the degree of v is 4.

5.61 Let $G = \mathrm{Pin}(2)$. Let $\mathbb{H}$ and $\tilde{\mathbb{R}}$ be the representations of G in Sect. 5.5.3. For a representation V of G, let $e_G(V)$ be the equivariant Euler class in the equivariant cohomology theory $H^*_G(\cdot; \mathbb{F}_2)$. Prove that $e_G(\mathbb{H}^k) = v^k$ and $e_G(\tilde{\mathbb{R}}^m) = q^m$, where v and q are the generators of $H^*(BG; \mathbb{F}_2)$ in Problem 5.60.

5.62 (Donaldson's Theorem B and C (Remark 5.40)) Let $G = \mathrm{Pin}(2)$. Consider the following diagram:

$$\begin{array}{ccc}
\mathbb{H}^{x+k} \oplus \tilde{\mathbb{R}}^{y} & \xrightarrow{\ f\ } & \mathbb{H}^{x} \oplus \tilde{\mathbb{R}}^{y+b} \\
\uparrow & & \uparrow \\
\{0\} \oplus \tilde{\mathbb{R}}^{y} & \xhookrightarrow[f^{\mathrm{U}(1)}]{} & \{0\} \oplus \tilde{\mathbb{R}}^{y+b}
\end{array}$$

where f is the G-equivariant map of (5.7) and $f^{\mathrm{U}(1)}$ is the restriction of f to $\{0\} \oplus \tilde{\mathbb{R}}^{y}$. By using the above diagram with G-equivariant cohomology $H_G^*(\cdot; \mathbb{F}_2)$, prove that, if $b \leq 2$, then $k = 0$. (Observe that no constraint on b and k can be obtained if $b \geq 3$.)

Chapter 6
L^p-Cohomology

In this chapter, we introduce the L^p-cohomology groups on a complete Riemannian manifold. There are two L^p-cohomology theories, reduced or unreduced versions. In particular, the reduced L^2-cohomology group is canonically isomorphic to the space of L^2-harmonic forms that is the kernel of the Laplace–de Rham operator. We see that difference between the reduced and the unreduced L^2-cohomology groups are reflected by the spectral behavior of the Laplace–de Rham operator near zero. We take up the following two basic subjects:

- We present a detailed proof of Brooks' theorem which gives a characterization of amenability of the fundamental group of a compact manifold by the spectral distribution at zero of the Laplace–Beltrami operator (the Laplace–de Rham operator acting on the space of L^2-functions) of the universal covering space. We also describe basic subjects of amenability and residual finiteness of groups.
- We introduce the simplicial L^2-cohomology groups, and present a proof of the L^2-de Rham theorem by Dodziuk, which gives an isomorphism between the reduced L^2-de Rham cohomology group and the reduced simplicial L^2-cohomology group.

6.1 L^p-Cohomology Groups

Let (X, g) be a complete Riemannian manifold. Throughout this book, we always assume that a Riemannian metric is complete. Let $\Omega_c^m(X)$ be the space of smooth m-forms on X with compact support. For $1 \leq p \leq \infty$, we introduce the norm:

$$\|\alpha\|_{L^p} := \left(\int_X |\alpha(x)|^p \mathrm{vol} \right)^{\frac{1}{p}}.$$

T. Kato, N. Nakamura, *Monopole Maps and L^2-cohomology*, Springer Monographs in Mathematics, https://doi.org/10.1007/978-981-96-8472-4_6

The space of the completion of $\Omega_c^m(X)$ with respect to the norm is called the L^p-Sobolev space, and we denote the space by $L^p(X; \Lambda^m)$. We just say the Sobolev space when $p = 2$. By using the metric g and the Levi-Civita connection, we can define the L^p-Sobolev k-spaces that include l-th derivatives for $0 \leq l \leq k$. They admit the structure of the Hilbert spaces when $p = 2$. For the other cases, they admit the structure of the Banach spaces.

Let us introduce L^p-cohomology groups. There are two variants of the groups, when the base space X is non-compact.

Definition 6.1

- The unreduced L^p-cohomology group is defined by

$$H^m_{(p)}(X) := \frac{\operatorname{Ker} d : L^p_1(X; \Lambda^m) \to L^p(X; \Lambda^{m+1})}{d(\, L^p_1(X; \Lambda^{m-1})\,)}.$$

- The reduced L^p-cohomology group is defined by

$$\bar{H}^m_{(p)}(X) := \frac{\operatorname{Ker} d : L^p_1(X; \Lambda^m) \to L^p(X; \Lambda^{m+1})}{\overline{d(\, L^p_1(X; \Lambda^{m-1})\,)}},$$

where the overline on the right-hand side means the closure.

The differential d has closed range at the degree m, if the equality $H^m_{(p)}(X) = \bar{H}^m_{(p)}(X)$ holds.

Remark 6.2

1. In general, there is a canonical map $H^m_{(p)}(X) \to \bar{H}^m_{(p)}(X)$.
2. $\bar{H}^m_{(2)}(X)$ admits a Hilbert structure induced from the Sobolev norms.

Example The differential of the 0-th degree does not have closed range over the real line (Exercise Problem 6.59).

Definition 6.3 An L^p-form α on (X, g) is called L^p-harmonic, if

$$d\alpha = d^*\alpha = 0$$

holds.

Lemma 6.4 *An L^2-form α is L^2-harmonic, if and only if $\Delta\alpha = 0$, where $\Delta = dd^* + d^*d$ is the Laplace–de Rham operator.*

Proof This is well known if X is compact.

Suppose X is non-compact, and take α with $\|\alpha\|_{L^2} = 1$. It is obvious that $\Delta\alpha = 0$ holds, if $d\alpha = d^*\alpha = 0$. Conversely, suppose that $\Delta\alpha = 0$ holds. We verify the equalities $d\alpha = d^*\alpha = 0$. Let us choose an exhaustion $K_\epsilon \subset X$, and compactly supported smooth functions $a_\epsilon : X \to [0, 1]$ such that $\cup_{\epsilon>0} K_\epsilon = X$, $a_\epsilon|_{K_\epsilon} \equiv 1$ and $\|da_\epsilon\|_{L^\infty} \leq \epsilon$ hold. Then,

$$0 = \langle \Delta\alpha, a_\epsilon\alpha\rangle_{L^2} = \langle d\alpha, d(a_\epsilon\alpha)\rangle_{L^2} + \langle d^*\alpha, d^*(a_\epsilon\alpha)\rangle_{L^2}.$$

Consider the first term on the right-hand side:

$$\langle d\alpha, d(a_\epsilon\alpha)\rangle_{L^2} = \int_X a_\epsilon |d\alpha|^2 \operatorname{vol} + \langle d\alpha, da_\epsilon \wedge \alpha\rangle_{L^2}.$$

By the Cauchy–Schwarz inequality, we have the estimates

$$|\langle d\alpha, da_\epsilon \wedge \alpha\rangle_{L^2}| \leq \|da_\epsilon\|_{L^\infty} \cdot \|d\alpha\|_{L^2} \cdot \|\alpha\|_{L^2} \leq \epsilon\|d\alpha\|_{L^2}.$$

Note that the convergence $\lim_{\epsilon\to 0} \int_X a_\epsilon |d\alpha|^2 \mathrm{vol} = \|d\alpha\|^2_{L^2}$ holds.

By estimating the second term similarly, we have the estimate:

$$\int_X a_\epsilon |d\alpha|^2 \operatorname{vol} + \int_X a_\epsilon |d^*\alpha|^2 \operatorname{vol} \leq \epsilon\|d\alpha\|_{L^2} + \epsilon\|d^*\alpha\|_{L^2}.$$

This implies that for any $\epsilon > 0$, the estimate

$$\|d\alpha\|^2_{L^2} + \|d^*\alpha\|^2_{L^2} \leq \epsilon \cdot (\|d\alpha\|_{L^2} + \|d^*\alpha\|_{L^2})$$

holds. Because the left-hand side is bounded by $\frac{1}{4}(\|d\alpha\|_{L^2} + \|d^*\alpha\|_{L^2})^2$ from below, we obtain the estimate

$$\|d\alpha\|_{L^2} + \|d^*\alpha\|_{L^2} \leq 4\epsilon.$$

This implies $d\alpha = d^*\alpha = 0$. □

Let $\mathcal{H}^m(X)$ be the space of L^2-harmonic forms of degree m.

Corollary 6.5 *The decomposition*

$$L^2(X;\Lambda^m) \cong \mathcal{H}^m(X) \oplus \overline{d(L_1^2(X;\Lambda^{m-1}))} \oplus \overline{d^*(L_1^2(X;\Lambda^{m+1}))} \tag{6.1}$$

holds.

Proof It is easy to see the following orthogonality:

$$\mathcal{H}^m(X) \perp \overline{d(L_1^2(X;\Lambda^{m-1}))} \perp \overline{d^*(L_1^2(X;\Lambda^{m+1}))}.$$

Take an L^2-form α that is orthogonal to $\mathcal{H}^m(X) \oplus \overline{d(L_1^2(X;\Lambda^{m-1}))} \oplus \overline{d^*(L_1^2(X;\Lambda^{m+1}))}$. Following a similar idea to the proof of Lemma 6.4, we can check that $d\alpha = d^*\alpha = 0$ vanishes. Then $\alpha \in \mathcal{H}^m(X)$ must hold. Hence, $\alpha = 0$. □

Corollary 6.6 *If an L^2-harmonic form α satisfies $\alpha = d\eta$ for some L^2-form η, then $\alpha = 0$ holds.*

Proof

$$\begin{aligned}\langle \alpha, \alpha\rangle_{L^2} = \langle \alpha, d\eta\rangle_{L^2} &= \lim_{\epsilon\to 0} \langle \alpha, a_\epsilon \cdot d\eta\rangle_{L^2} \\ &\le \langle \alpha, d(a_\epsilon \cdot \eta)\rangle_{L^2} + \|da_\epsilon\|_{L^\infty} \cdot \|\alpha\|_{L^2} \cdot \|\eta\|_{L^2}.\end{aligned}$$

The first term in the right-hand side vanishes, because of the equalities

$$\langle \alpha, d(a_\epsilon \cdot \eta)\rangle_{L^2} = \langle d^*\alpha, a_\epsilon \cdot \eta\rangle_{L^2} = 0.$$

The second term in the right-hand side converges to zero. □

Corollary 6.7 *There is a canonical isomorphism*

$$\mathcal{H}^m(X) \cong \overline{H}^m_{(2)}(X)$$

for each $1 \le m \le \dim X$.

Proof Take an element $u \in \mathcal{H}^m(X)$. The differential $du = 0$ vanishes by definition. Let us check that this assignment gives a linear isomorphism. This is injective by (6.1).

Take an element $[u] \in \overline{H}^m_{(2)}(X)$. Then, there is a decomposition $u = u^0 + a + b \in \mathcal{H}^m(X) \oplus \overline{d(L^2_1(X; \Lambda^{m-1})} \oplus \overline{d^*(L^2_1(X; \Lambda^{m+1})}$. We claim that $b = 0$. Then, by replacing the representative from u to $u - a$, we obtain surjectivity. Let us check $b = 0$. Suppose it is non-zero $b \neq 0$. Then, there is a sequence $d^*\mu_i \in d^*(L^2_1(X; \Lambda^{m+1})$ that converges to b in L^2. Then,

$$\lim_{i \to \infty} \langle u, d^*\mu_i \rangle_{L^2} = \langle u, b \rangle \neq 0.$$

This is a contradiction, because $\langle u, d^*\mu_i \rangle_{L^2} = \langle du, \mu_i \rangle_{L^2} = 0$ must hold. □

Suppose $X := \tilde{M}$ is the universal covering space of a compact manifold, and use the Riemannian metric on X by a lift of a metric on M. Then, the fundamental group $\Gamma := \pi_1(M)$ acts on both $\mathcal{H}^m(X)$ and $\overline{H}^m_{(2)}(X)$. The above assignment is clearly Γ-equivariant. Hence, we have the following:

Corollary 6.8 *The isomorphism*

$$\mathcal{H}^m(X) \cong \overline{H}^m_{(2)}(X)$$

is Γ-equivariant, when $X := \tilde{M}$ is the universal covering space of a manifold.

6.1.1 Spectral Decomposition

Let (X, g) be a complete Riemannian manifold.

There is a spectral density function E^m_λ of the Laplace–de Rham operator $\Delta_m := d \circ d^* + d^* \circ d$ on L^2-forms of degree m such that the formula

$$\omega = \int_0^\infty \lambda \cdot dE^m_\lambda(\omega)$$

holds for any $\omega \in L^2(X, \Lambda^m)$ (e.g. see [150]).

Then, the spectral distribution

$$\operatorname{spec}(\Delta_m) \subset [0, \infty)$$

is given for each $0 \leq m \leq \dim X$.

Lemma 6.9 *The spectral distribution* $\operatorname{spec}(\Delta_m)$ *is isolated at zero for any* $0 \le m \le \dim X$, *if and only if the equality*

$$\overline{H}^m_{(2)}(X) = H^m_{(2)}(X)$$

holds for any $0 \le m \le \dim X$.

Proof Note that there is a canonical map $q : H^m_{(2)}(X) \to \overline{H}^m_{(2)}(X)$. The map q is surjective, by definition. There is the orthogonal projection $P : L^2(X; \Lambda^m) \to \mathcal{H}^m(X)$ by Corollary 6.5.

Suppose that the spectral distribution $\operatorname{spec}(\Delta_m)$ is isolated at zero. Then the restriction

$$\Delta_m : (1 - P)(L^2_2(X; \Lambda^m)) \cong (1 - P)(L^2(X; \Lambda^m))$$

is isomorphic (see Problem 6.61). Note that the equality

$$(1 - P)(L^2(X; \Lambda^m)) \;=\; \overline{d(L^2_1(X; \Lambda^{m-1}))} \oplus \overline{d^*(L^2_1(X; \Lambda^{m+1}))}$$

holds by Corollary 6.5. Hence, the embedding

$$\overline{d(L^2_1(X; \Lambda^{m-1}))} \oplus \overline{d^*(L^2_1(X; \Lambda^{m+1}))} \;\subset\; d(L^2_1(X; \Lambda^{m-1})) \oplus d^*(L^2_1(X; \Lambda^{m+1}))$$

holds, because the image of Δ_m consists of the images of d and d^*. Because both the equality

$$\left(d(L^2_1(X; \Lambda^{m-1})) \oplus d^*(L^2_1(X; \Lambda^{m+1}))\right) \cap \operatorname{Ker} d \;=\; d(L^2_1(X; \Lambda^{m-1}))$$

and the embedding $\operatorname{Ker} d \supset \overline{d(L^2_1(X; \Lambda^{m-1}))}$ hold, the equality

$$\overline{d(L^2_1(X; \Lambda^{m-1}))} \;=\; d(L^2_1(X; \Lambda^{m-1}))$$

holds. Hence, q is isomorphic.

Conversely, suppose that the map q is isomorphic. Then we have the equality $d(L^2_1(X; \Lambda^{m-1})) \;=\; \overline{d(L^2_1(X; \Lambda^{m-1}))}$, and hence, the differential d has closed range. For the formally self-adjoint operator d^*, we have

$$\begin{aligned}\overline{d^*(L^2_1(X; \Lambda^{m+1}))} &= \overline{*d*(L^2_1(X; \Lambda^{m+1}))} = \overline{*d(L^2_1(X; \Lambda^{n-m-1}))} \\ &= *d(L^2_1(X; \Lambda^{n-m-1})) = *d*(L^2_1(X; \Lambda^{m+1})) \\ &= d^*(L^2_1(X; \Lambda^{m+1})).\end{aligned}$$

Hence, d^* also has closed range.

If the spectral distribution $\text{spec}(\Delta_m)$ could not be isolated at zero, then there is a sequence $u_i \in L^2(X; \Lambda^m)$ such that the three conditions $\|u_i\|_{L^2} = 1$, $P(u_i) = 0$, and $\lim_{i\to\infty}\langle \Delta_m(u_i), u_i\rangle_{L^2} = 0$ hold.

By (6.1), each u_i is decomposed as

$$\begin{aligned} u_i = v_i + w_i &\in \overline{d(L_1^2(X; \Lambda^{m-1}))} \oplus \overline{d^*(L_1^2(X; \Lambda^{m+1}))} \\ &= d(L_1^2(X; \Lambda^{m-1})) \oplus d^*(L_1^2(X; \Lambda^{m+1})). \end{aligned}$$

Hence, we have $v_i = df_i$ and $w_i = d^* g_i$ for some $f_i \in L_1^2(X; \Lambda^{m-1})$ and $g_i \in L_1^2(X; \Lambda^{m+1})$. We may assume that both the sets $\{f_i\}_i$ and $\{g_i\}_i$ are uniformly bounded in L_1^2 as

$$\|f_i\|_{L^2} \le C\|v_i\|_{L^2}, \qquad \|g_i\|_{L^2} \le C\|w_i\|_{L^2}.$$

(Why? See Problem 6.62.) Similarly, we have the estimates

$$\|v_i\|_{L^2}^2 \le C\|d^*(v_i)\|_{L^2}^2 = C\langle \Delta_m(v_i), v_i\rangle_{L^2} \le C\|\Delta_m(v_i)\|_{L^2} \cdot \|v_i\|_{L^2}.$$

We also have $\|w_i\|_{L^2} \le C\|\Delta_m(w_i)\|_{L^2} \cdot \|w_i\|_{L^2}$.

On the other hand, we have $\lim_i \Delta_m(v_i) = \lim_i \Delta_m(w_i) = 0$. This is a contradiction, because $\|v_i\|_{L^2}^2 + \|w_i\|_{L^2}^2 = \|u_i\|_{L^2}^2 = 1$ must hold. □

6.2 Amenability and Residual Finiteness

6.2.1 Banach–Tarski Paradox

The Banach–Tarski paradox can be stated in the following way:

Theorem 6.10 ([10]) *The three-dimensional ball in $\mathbb{R}^3$ can be decomposed into finitely many disjoint subsets such that they can be rearranged to form two balls of the same size.*

In particular, there is no probability measure on $\mathbb{R}^3$ that is invariant under isometries and is defined on all subsets in $\mathbb{R}^3$ with a positive measure on the ball.

Definition 6.11 Suppose a group Γ acts on a set X.

- A pair of two subsets $A, B \subset X$ is called Γ-quasi-decomposable, if they can be partitioned into finitely many subsets as $A = A_1 \sqcup \cdots \sqcup A_n$ and $B = B_1 \sqcup \cdots \sqcup B_n$ such that $B_i = A_i \cdot g_i$ hold for all $1 \le i \le n$ for some $g_1, \cdots, g_n \in \Gamma$. We let $A \sim B$.

- X is called Γ-paradoxical, if there exist proper subsets $A, B \subset X$ with $A \cap B = \phi$ and $A \sim X \sim B$.

The condition $A \sim B$ is an equivalent relation. Namely, if $A \sim B$ and $B \sim C$, then $A \sim C$ holds (Problem 6.63).

Lemma 6.12 *The free group F_2 of rank 2 is F_2-paradoxical by the left action on itself.*

Proof Let $\{a, b\}$ be a generating set of F_2, and $\{a^{\pm 1}, b^{\pm 1}\}$ be the generating set of F_2. Recall that a reduced word consists of a sequence of elements in the generating set such that $a \cdot a^{-1}$, $a^{-1} \cdot a$ or $b \cdot b^{-1}$, $b^{-1} \cdot b$ do not appear successively in the sequence.

Let us denote by $W(x)$ as the set of reduced words starting with $x \in \{a^{\pm 1}, b^{\pm 1}\}$. Then, we can decompose it as

$$F_2 = W(a) \sqcup a \cdot W(a^{-1}) = W(b) \sqcup b \cdot W(b^{-1}).$$

The proper subsets $A := W(a) \sqcup W(a^{-1})$ and $B := W(b) \sqcup W(b^{-1})$ satisfy the relations $A \sim F_2 \sim B$. □

6.2.2 Amenable Groups

Let Γ be a finitely generated group.

Definition 6.13 A finitely generated group Γ is *amenable*, if it admits a left invariant finitely additive probability measure on the power set $\mathcal{P}(\Gamma)$ of Γ.

Lemma 6.14 *If a group Γ is Γ-paradoxical by the left action on itself, then it cannot be amenable.*

Proof Suppose Γ could be amenable and μ be a left invariant probability measure.

Let $A, B \subset \Gamma$ be two subsets with $A \cap B = \phi$ and $A \sim \Gamma \sim B$. Decompose $A = A_1 \sqcup \cdots \sqcup A_n$ and $\Gamma = \Gamma_1 \sqcup \cdots \sqcup \Gamma_n$ with $\Gamma_i = g_i \cdot A_i$ for some $g_i \in \Gamma$.

By left invariance, we have the equalities

$$\begin{aligned}\mu(A) &= \sum_{i=1}^{n} \mu(A_i) = \sum_{i=1}^{n} \mu(g_i \cdot A_i) \\ &= \mu(\sqcup_{i=1}^{n} g_i \cdot A_i) = \mu(\Gamma) = 1.\end{aligned}$$

Similarly, we have $\mu(B) = \mu(\Gamma) = 1$.

On the other hand, we should have the estimate

$$1 \geq \mu(A \sqcup B) = \mu(A) + \mu(B) = 1 + 1 = 2.$$

This is a contradiction. □

Hence, the group F_2 is non-amenable by Lemma 6.12.

Lemma 6.15 *A subgroup of an amenable groups is still amenable.*

Proof Let $H \subset G$ be a subgroup of an amenable group G. Let μ be a left invariant finitely additive measure on G. Then by the axiom of choice, there is a right transverse $R \subset G$ such that R contains exactly one element of each right coset of H. Then we set ν on H by $\nu(A) := \mu(A \cdot R)$, that gives a left invariant finitely additive probability measure on H. □

In particular, we have the following:

Corollary 6.16 *The free group F_n of rank $n \geq 2$ is non-amenable.*

Remark 6.17 It is well known as the von Neumann conjecture to claim that a group is non-amenable, if and only if it contains a free subgroup with two generators. The "if" part can be checked. Hence, the "only if" part was conjectured. It was disproved first by A. Ol'shanskii, and then by S. Adian and by M. Sapir with Ol'shanskii [143, 144].

Examples

- Finite groups are amenable.
- Finitely generated nilpotent groups are amenable.
- The property of being amenable is closed under taking subgroups, quotient groups and directed unions (see Problem 6.64).
- Finitely generated abelian groups are amenable (see Proposition 6.24).
- The group $SO(3)$ is non-amenable, because it contains F_2 as a subgroup (see [167]). (Recall Theorem 6.10.)

Proposition 6.18 ([152]) *The group $SL_2(\mathbb{Z})$ contains a free subgroup of rank 2.*

To see this, we use the so-called ping-pong lemma:

Lemma 6.19 *Suppose a group G acts on a set X. Let $a, b \in G$ be two elements and $A, B \subset X$ be two subsets such that $A \cap B = \phi$ is empty. Moreover, the inclusions*

$$B \cdot a^n \subset A, \quad A \cdot b^n \subset B$$

for all $n \neq 0$. Then the subgroup generated by a and b is free.

Before giving a proof of this fact, we present a proof of Proposition 6.18.

Proof We verify that the subgroup H generated by two generators

$$a := \begin{pmatrix} 1 & 0 \\ 2 & 1 \end{pmatrix}, \quad b := \begin{pmatrix} 1 & 2 \\ 0 & 1 \end{pmatrix}$$

is free.

Note that the group $SL_2(\mathbb{Z})$ acts on $\mathbb{Z}^2$. Choose subsets

$$A := \{ (x, y) \in \mathbb{Z}^2 : |y| < |x| \}, \quad B := \{ (x, y) \in \mathbb{Z}^2 : |y| > |x| \}.$$

Then, $(x, y) \cdot b^n = (x, 2nx + y)$, and $|2nx + y| > |x|$ is satisfied, if $|y| < |x|$ holds. Similarly, $(x, y) \cdot a^n = (x + 2ny, y)$, and $|y| < |x + 2ny|$ is also satisfied, if $|y| > |x|$ holds. □

Let us verify Lemma 6.19.

Proof Let $G_1 := \langle a^n : n \in \mathbb{Z} \rangle$ and $G_2 := \langle b^n : n \in \mathbb{Z} \rangle$ be the infinite cyclic subgroups of G. Take a reduced word $w = \prod_{i=1}^m w_i$, where $w_i \in G_{[i]}$ with $[i] = i$ mod 2. By the condition, we have the inclusions

$$B \cdot w = B \cdot \prod_{i=1}^m w_i \subset A \cdot \prod_{i=2}^m w_i \subset B \cdot \prod_{i=3}^m w_i \subset \cdots$$

Suppose m is odd. Then, the inclusion $B \cdot w \subset A$ holds. The word w represents a nontrivial element in G, because $A \cap B = \phi$.

Next, if m is even, let $w_1 = a^n$. Take the conjugate by $h := a^{-n+1}$, if $n \neq 1$ and by $h := a^{-2}$ if $n = 1$. The conjugate element $h \cdot w \cdot h^{-1} = a^{\pm 1} \cdot w_2 \cdots w_m \cdot a^*$ is a reduced word, where $* \neq 0$. Hence, it follows from the case of odd m that $h \cdot w \cdot h^{-1}$ is nontrivial. Because the conjugate element is nontrivial, w itself is also nontrivial. □

Corollary 6.20 *The groups $SL_n(\mathbb{Z})$ are all non-amenable for $n \geq 2$.*

Proof It follows from Proposition 6.18 that these groups contain F_2, the free group of rank 2. If $SL_n(\mathbb{Z})$ could be amenable, then it should be the case for its subgroup by Lemma 6.15. However, F_2 is non-amenable. □

6.2.3 Følner's Criterion

For a set X and two subsets $A, B \subset X$, define the symmetric difference by $A \triangle B := (A \cup B) \backslash (A \cap B)$.

Definition 6.21 A group Γ satisfies *Følner's criterion*, if for any finite subset $A \subset \Gamma$ and every $\epsilon > 0$, there is a non-empty finite subset $F \subset \Gamma$ such that for each $a \in A$, the following estimate holds:

$$\sharp\{\, (a \cdot F) \triangle F \,\} \leq \epsilon \cdot \sharp\{F\}.$$

A *Følner sequence* is a sequence of finite subsets $\{F_i\}_i$ of Γ such that the convergence

$$\lim_{i \to \infty} \frac{\sharp\{\, (a \cdot F_i) \triangle F_i \,\}}{\sharp\{F_i\}} = 0$$

holds for any $a \in \Gamma$.

Remark 6.22 By definition, the following properties follow immediately.

- A group admits Følner's criterion if and only if it admits a Følner sequence.
- A group Γ satisfies Følner's criterion if and only if, for any finite subset $A \subset \Gamma$ and every $0 < k < 1$, there is a non-empty finite subset $F \subset \Gamma$ such that, for each $a \in A$, the following estimate holds:

$$\sharp\{\, (a \cdot F) \cap F \,\} > k \cdot \sharp\{F\}.$$

Notations

- $$l^p(\Gamma) := \left\{ \sum_{\gamma \in \Gamma} a_\gamma \cdot \gamma \;:\; a_\gamma \in \mathbb{C},\ \sum_{\gamma \in \Gamma} |a_\gamma|^p < \infty \right\}$$
$$= \{f : \Gamma \to \mathbb{C} : \sum_{\gamma \in \Gamma} |f(\gamma)|^p < \infty\}.$$
- $a^* f(x) := f(a^{-1} \cdot x)$ for $f \in l^1(\Gamma)$. Then for $f = \sum_{\gamma \in \Gamma} b_\gamma \cdot \gamma$,

(continued)

$$a^* f = \sum_{\gamma \in \Gamma} b_{a^{-1} \cdot \gamma} \cdot \gamma = \sum_{\gamma \in \Gamma} b_\gamma \cdot (a \cdot \gamma)$$

•

$$\delta_y(x) := \begin{cases} 1 & x = y, \\ 0 & x \neq y. \end{cases}$$

- For two functions $f, g \in l^1(\Gamma)$, we define $f * g \in l^1(\Gamma)$ by

$$f * g(x) := \sum_{z \in \Gamma} f(z) \cdot g(z^{-1} \cdot x).$$

In particular, $\delta_a * \delta_y(x) = \sum_{z \in \Gamma} \delta_a(z) \cdot \delta_y(z^{-1} \cdot x) = \delta_y(a^{-1} \cdot x) = a^* \delta_y(x)$.

Lemma 6.23 *Let Γ be a finitely generated group. The set of finitely additive probability measures on Γ is compact.*

Proof The set of finitely additive probability measures is a closed subset in $[0, 1]^{\mathcal{P}(\Gamma)}$, where $\mathcal{P}(\Gamma)$ is the power set of Γ. The space $[0, 1]^{\mathcal{P}(\Gamma)}$ is compact by Tychonoff's theorem, and hence its closed subspace is also compact. □

Proposition 6.24 *A group Γ is amenable, if and only if it admits Følner sequence.*

Proof Suppose that Γ admits a Følner sequence $\{F_i\}_i$. Let μ_i be a family of finitely additive probability measures on Γ by

$$\mu_i(B) := \frac{\sharp\{ B \cap F_i \}}{\sharp\{ F_i \}}.$$

Then, a subsequence of $\{\mu_i\}_i$ converges by Lemma 6.23. Then, the limit μ_∞ is the left invariant finitely additive probability measure, because the estimate

$$|\mu_i(B) - \mu_i(g \cdot B)| \leq \frac{\sharp\{ F_i \triangle (g \cdot F_i) \}}{\sharp\{ F_i \}}$$

holds for each i. Because the left-hand side converges to zero, this implies that $\lim_{i\to\infty} \mu_i$ is left invariant.

Conversely, suppose that Γ is amenable. Let us set

$$\Phi := \{f \in l^1(\Gamma) : f \geq 0, \text{ finitely supported, and } \|f\|_{l^1} = 1\}.$$

We claim that for any finite $A \subset \Gamma$, any $\epsilon > 0$ and any $a \in A$, there exists $f \in \Phi$ such that $\|f - a^* f\|_{l^1} \leq \epsilon$. Suppose the contrary. Then, there exist $A \subset \Gamma$, $\epsilon > 0$ and $a \in A$ such that $\|f - a^* f\|_{l^1} > \epsilon$ holds for any $f \in \Phi$. Note that the set $\{f - a^* f : f \in \Phi\}$ is a convex subset of $l^1(\Gamma)$ away from 0. Recall that $l^\infty(\Gamma) = (l^1(\Gamma))^*$ by

$$\langle f, \beta \rangle = \sum_{x\in\Gamma} \beta(x) \cdot f(x)$$

for $\beta \in l^\infty(\Gamma)$ and $f \in l^1(\Gamma)$. Then by the Hahn–Banach separation theorem, there exists $\beta \in l^\infty(\Gamma)$ such that $\langle f - a^* f, \beta \rangle \geq 1$ holds for any $f \in \Phi$. In particular, for $f = \delta_y$, we have

$$1 \leq \langle \delta_y - \delta_a * \delta_y, \beta \rangle = \sum_{x\in\Gamma} \left(\delta_y(x)\beta(x) - \delta_y(a^{-1}x)\beta(x)\right) = \left(\beta - (a^{-1})^* \beta\right)(y).$$

In particular, for any probability measure m, the estimate $\langle m, \beta - (a^{-1})^* \beta \rangle \geq 1$ holds. On the other hand, there exists a left invariant probability measure $M : l^\infty(\Gamma) \to \mathbb{R}$, because Γ is assumed to be amenable. Apply M to $\beta - (a^{-1})^* \beta$, and we should have $\langle M, \beta - (a^{-1})^* \beta \rangle = 0$, which is a contradiction.

Now for $A \subset \Gamma$, $\epsilon > 0$ and $a \in A$, there is some $f \in \Phi$ such that $\|f - a^* \cdot f\|_{l^1} \leq \frac{\epsilon}{\sharp\{A\}}$ holds. Decompose as $f = \sum_{i=1}^n c_i \mathbf{1}_{F_i}$ for $c_i \geq 0$ and $F_1 \supset \cdots \supset F_n$. We have $\sum_{i=1}^n c_i \cdot \sharp\{F_i\} = 1$. This is called the layer cake representation of f. (How to decide c_i and F_i? Problem 6.65.)

For $g \in (a \cdot F_i)\Delta F_i$ and $f \in \Phi$, we have $|f(g) - (a^* f)(g)| \geq c_i$. Hence,

$$\sum_{i=1}^n c_i \cdot \sharp\{(a \cdot F_i)\,\Delta F_i\} \leq \|f - a^* f\|_{l^1} \leq \frac{\epsilon}{\sharp\{A\}} \sum_{i=1}^n c_i \cdot \sharp\{F_i\}$$

holds for each $a \in A$. Then, we have

$$\sum_{i=1}^n \sum_{a\in A} c_i \cdot \sharp\{(a \cdot F_i)\,\Delta F_i\} \leq \epsilon \sum_{i=1}^n c_i \cdot \sharp\{F_i\}.$$

By the pigeonhole principle, there is some i such that $\sum_{a\in A} \sharp\{(a \cdot F_i)\Delta F_i\} \leq \epsilon \cdot \sharp\{F_i\}$. Hence, we have

$$\sharp\{(a \cdot F_i) \triangle F_i\} \leq \epsilon \cdot \sharp\{F_i\}$$

for any $a \in A$. □

The argument above depends on [62, 138].

Remark 6.25 The set Φ in the proof of Proposition 6.24 is a subset of the set of finitely additive probability measures on Γ; however, when Γ is infinite, then there is a difference between $l^1(\Gamma)$ and the set of finitely additive probability measures. This comes from a basic fact in functional analysis that $l^\infty(\Gamma) = \left(l^1(\Gamma)\right)^*$, where the right-hand side is the dual Banach space, and the canonical linear map $l^1(\Gamma) \hookrightarrow \left(l^1(\Gamma)\right)^{**} = l^\infty(\Gamma)^*$ is isometrically injective, but not surjective. The latter is identified with the set of finitely additive finite signed measures that are absolutely continuous with respect to the standard point measure on Γ.

If a Banach space B satisfies a property that $B^{**} = B$ through the canonical map, then it is called a *reflective* Banach space. Hence, $l^1(\Gamma)$ and also $l^\infty(\Gamma)$ are non-reflective when Γ is infinite. On the other hand, it is known that $l^p(\Gamma)$ are all reflective for $1 < p < \infty$.

Example Free abelian groups are amenable. In fact, for a finite set $A \subset \mathbb{Z}^n$, there is a constant $C = C(A) > 0$ such that $\sharp\{(a \cdot F_i) \triangle F_i\} \leq C \cdot \sharp\{\partial F_i\} = C \cdot (2n) \cdot (2i)^{n-1}$, where $F_i := \{a = (a_1, \cdots, a_n) \in \mathbb{Z}^n : -i \leq a_k \leq i \text{ for } 1 \leq k \leq n\}$ and $\partial F_i := \{a \in F_i : |a_k| = i \text{ for some } 1 \leq k \leq n\}$. Hence,

$$\frac{\sharp\{(a \cdot F_i) \triangle F_i\}}{\sharp\{F_i\}} \leq \frac{C \cdot (2n) \cdot (2i)^{n-1}}{i^n} = \frac{C \cdot (2n) \cdot 2^{n-1}}{i} \to 0$$

as $i \to \infty$.

6.2.4 *Residually Finite Groups*

Definition 6.26 A discrete group Γ is residually finite, if there is a tower

$$\Gamma \supset \Gamma_1 \supset \Gamma_2 \supset \cdots$$

by normal subgroups of finite indices with $\bigcap_i \Gamma_i = 1$.

Lemma 6.27 *Finitely generated free abelian groups are residually finite.*

Proof Let us set $\mathbb{Z}(i) := 2^i \cdot \mathbb{Z}$. Then, we have a tower of normal subgroups $\mathbb{Z} \supset \mathbb{Z}(1) \supset \mathbb{Z}(2) \supset \cdots$. This satisfies the required property $\cap_i \ \mathbb{Z}(i) = 1$. □

Consider the following properties $(*)$ that Γ admits:

- a family of finite groups $\{G_i\}_i$, homomorphisms $\pi_i : G_i \to G_{i-1}$,
- surjective homomorphisms $\rho_i : \Gamma \twoheadrightarrow G_i$ with $\rho_i = \pi_{i+1} \circ \rho_{i+1}$, and
- for any $\gamma \neq e \in \Gamma$, $\rho_i(\gamma) \neq e$ holds for any $i \geq i_0$ and some $i_0 = i_0(\gamma)$.

Note that a residually finite group Γ admits $(*)$ by letting $G_i := \Gamma/\Gamma_i$ and both π_i and ρ_i are the quotient homomorphisms.

Lemma 6.28 Γ *is residually finite, if and only if it admits the above properties* $(*)$.

Proof We have to verify the "if" part only. Suppose that Γ admits $(*)$. We set $\Gamma_i := \rho_i^{-1}(e)$. Because of the commutativity, we have the inclusions $\Gamma \supset \Gamma_i \supset \Gamma_{i+1}$. The group Γ/Γ_i is finite, because the union of the orbits $\Gamma = \sqcup_{g \in G_i} \ g \cdot \Gamma_i$ holds and G_i is finite. An element $\gamma \in \cap_i \ \Gamma_i$ satisfies the property of $\rho_i(\gamma) = e$ for any i. Hence, it implies $\gamma = e$ by the condition. □

Examples

- Finite groups are residually finite by definition.
- Finitely generated nilpotent groups are residually finite.
- Fundamental groups of compact Riemann surfaces are residually finite. (See [74].)
- (Malcev) Finitely generated linear groups are residually finite.
- The fundamental groups of hyperbolic manifolds are residually finite.
- Some Baumslag–Solitar groups are not residually finite (see [32], III 18–21).

Question
Find non-residually finite Gromov word hyperbolic groups.

Let us pick up the case of $SL_n(\mathbb{Z})$, because it is important.

Lemma 6.29 $SL_n(\mathbb{Z})$ *is residually finite.*

Proof Let $\mathbb{Z}_i := \mathbb{Z}/\ 2^i \cdot \mathbb{Z}$. We set the quotient homomorphisms by $\rho_i : SL_n(\mathbb{Z}) \to SL_n(\mathbb{Z}_i)$. It is enough to check that ρ_i is surjective by Lemma 6.28. For this, it is enough to check that both groups $SL_n(\mathbb{Z})$ and $SL_n(\mathbb{Z}_i)$ are generated by elementary matrices of the form $\{1 + e_{ij}\}_{1 \leq i \neq j \leq n}$, where 1 is the identity, and the components of e_{ij} is zero except for the (i, j)-entry.

Note that the common divisor of elements in any row $(a_{i1}, a_{i2}, \cdots, a_{in})$ must be 1 for an element $(a_{ij}) \in SL_n(\mathbb{Z})$, because its determinant is a multiple of the common divisor. Hence, an element in $SL_n(\mathbb{Z})$ can be transformed to a diagonal matrix by the multiplication by elementary matrix transformations of the form $1 + e_{ij}$. We may assume that the diagonal matrix has the form $\mathrm{diag}(\pm 1, \cdots, \pm 1)$ because we can transform a diagonal matrix $\mathrm{diag}(a, b)$ to $\mathrm{diag}(1, a \cdot b)$ by

$$\begin{pmatrix} a & 0 \\ 0 & b \end{pmatrix} \to \begin{pmatrix} a & a \\ 0 & b \end{pmatrix} \to \begin{pmatrix} 1 & a \\ (a^{-1} - 1) \cdot b & b \end{pmatrix} \to \begin{pmatrix} 1 & a \\ 0 & a \cdot b \end{pmatrix} \to \begin{pmatrix} 1 & 0 \\ 0 & a \cdot b \end{pmatrix}$$

through elementary operations. In total, an element in $SL_n(\mathbb{Z})$ can be obtained by a multiple of elementary matrices.

For the case of $SL_n(\mathbb{Z}_i)$, note that $\mathbb{Z}_i$ is a local ring. In general, a local ring R has the unique maximal ideal M, and it is given by $M = R \backslash R^{\times}$. For any element in $SL_n(\mathbb{Z}_i)$, there is at least one entry in the first row that is not in M. Otherwise, the determinant of the element should be in M, but $1 \notin M$. In particular, the entry is invertible in $\mathbb{Z}_i$. Then, one can follow a parallel argument to the case of $SL_n(\mathbb{Z})$. □

Corollary 6.30 *The groups $SL_n(\mathbb{Z})$ are residually finite and non-amenable for all $n \geq 2$.*

Proof This is obtained by Corollary 6.20 with Lemma 6.29. □

6.3 Brooks' Theorem

In this book, we always assume that group actions on spaces are from the right hand side.

Theorem 6.31 ([24]) *Let M be a compact manifold and $X := \tilde{M}$ be the universal covering space. Then, the fundamental group $\pi_1(M)$ is amenable, if and only if* $\mathrm{spec}(\Delta_0)$ *of X contains* 0.

Any finite group is amenable. It is obvious that $\mathrm{spec}(\Delta_0)$ of X contains 0, if the fundamental group is finite. Hence, below we assume that the fundamental group is infinite. Note that there are no non-zero L^2-harmonic functions on X, if the fundamental group is infinite. In particular, in the case where zero is not the eigenvalue, even if $\mathrm{spec}(\Delta_0)$ contains 0.

In this subsection, we verify that the fundamental group $\pi_1(M)$ is amenable only if $\mathrm{spec}(\Delta_0)$ of X contains 0. The rest of the proof will be treated in Sect. 6.3.1 below.

Before providing the proof, we introduce a notion of *fundamental domain*. Let M be a topological space and X be its universal covering space. Hence, the fundamental groups $\pi_1(M)$ acts on X as the deck transformation group.

Definition 6.32 A fundamental domain of the universal covering space X is a closed subset K such that:

- the union $\cup_{g\in\pi_1(M)} \ (K \cdot g) \ = \ X$ covers, and
- the interior $\mathring{K}$ satisfies $\left(\mathring{K} \cdot g\right) \cap \mathring{K} = \phi$ for any $g \neq e \in \pi_1(M)$.

Remark 6.33

- A fundamental domain always exists.
- A fundamental domain is far from unique. For example, let $B \subset K$ be any closed subset, and set

$$K' := \left(\overline{K \backslash B}\right) \cup (B \cdot g)$$

for any $g \in \pi_1(M)$. Then, K' is also a fundamental domain of the universal covering space X. Note that we do not necessarily assume that a fundamental domain is connected.
- In the case when M is a manifold, we always choose a fundamental domain by a manifold possibly with corners.

To verify Theorem 6.31, we prepare the following proposition.

Proposition 6.34 *Let M be a compact manifold, and let X be its universal covering space. Then, $\pi_1(M)$ is amenable, if and only if for any fundamental domain $K \subset X$ and any $\epsilon > 0$, there is a finite subset $F \subset \pi_1(M)$ such that the isoperimetric inequality*

$$\frac{\mathrm{area}(\partial H)}{\mathrm{vol}(H)} < \epsilon$$

holds, where

$$H := \cup_{g\in F} \ (K \cdot g).$$

Assuming Proposition 6.34, we finish providing the proof of the "only if" part of Theorem 6.31 as follows.

Proof It follows from Proposition 6.24 and Remark 6.22 that it is enough to verify that $\inf \operatorname{spec}(\Delta_0) = 0$ if and only if $\pi_1(M)$ satisfies Følner's criterion.

Note that we have fixed a fundamental domain $K \subset X$. We also fix a Riemannian metric on M arbitrarily and lift it on X.

Given $\epsilon > 0$ and F, and let H be as in Proposition 6.34.

Let χ_H be the characteristic function such that

$$\chi_H(x) = \begin{cases} 1 & x \in H, \\ 0 & x \notin H. \end{cases}$$

Of course it is not continuous, and hence, we modify it slightly as follows. Take sufficiently small $0 < \epsilon_1 < \epsilon_2 < 1$, and $\psi : [0, 1] \to [0, 1]$ be a smooth function such that

$$\psi(t) = \begin{cases} 1 & t > \epsilon_2 \\ 0 & t < \epsilon_1. \end{cases}$$

Then, we have a compactly supported continuous function f_F on X by

$$f_F(x) := \psi(\operatorname{dist}(x, H^c)) \cdot \chi_H(x).$$

This is not yet smooth; however, one can easily make it smooth by a minor modification. Note that $\nabla(f_F)$ and $\Delta(f_F)$ are both uniformly bounded in L^∞, independently of the choice of F (recall that we have fixed a fundamental domain K).

Then, there are positive constants $C_1, C_2 > 0$ independent of the choice of F, such that two estimates

$$\int_X f_F \cdot \Delta(f_F) \leq C_1 \cdot \operatorname{area}(\partial H),$$

$$\int_X f_F^2 \geq C_2 \cdot \operatorname{vol}(H),$$

hold. Combining these estimates, we obtain

$$\frac{\int_X f_F \cdot \Delta(f_F)}{\int_X f_F^2} \leq \frac{C_1}{C_2} \cdot \frac{\operatorname{area}(\partial H)}{\operatorname{vol}(H)}.$$

Note the equalities

$$\inf \operatorname{spec}(\Delta_0) = \inf_{f \in C_c^\infty(X)} \frac{\int_X f \cdot \Delta(f)}{\int_X f^2} = \inf_{f \in C_c^\infty(X)} \frac{\int_X |\nabla(f)|^2}{\int_X f^2}.$$

Hence, combining the above estimates with Proposition 6.34, we finish the proof of the "only if" part of Theorem 6.31. □

Finally, we verify Proposition 6.34 below.

Proof There is a finite set $A := \{a_1, \cdots a_n\} \subset \pi_1(M)$ such that $(K \cdot a_i) \cap K \neq \phi$, and the set $\{a_1, \cdots, a_n\}$ generates $\pi_1(M)$.

First, assume that $\pi_1(M)$ is amenable. For $\epsilon > 0$ and A, there is a finite set $F \subset \pi_1(M)$ as in Følner's criterion. Then, we have the estimates

$$\begin{aligned}\text{area}(\partial H) &\leq \text{area}(\partial K) \cdot \sharp\{g \in F : a_i \cdot g \notin F \text{ for some } i\} \\ &\leq \text{area}(\partial K) \cdot n \cdot \epsilon \cdot \sharp(F),\end{aligned}$$

where we used Følner's criterion only for the second inequality. Note the equality $\text{vol}(H) = \text{vol}(K) \cdot \sharp(F)$. Then, we have the inequality

$$\frac{\text{area}(\partial H)}{\text{vol}(H)} \leq \frac{\text{area}(\partial K)}{\text{vol}(K)} \cdot n \cdot \epsilon.$$

Let us verify the converse, and assume the condition of Proposition 6.34. Note that if Følner's criterion holds for (A', ϵ), then it also does for (A, ϵ) for any subset $A \subset A'$. Let $A := \{a_1, \cdots, a_n\}$ and $\epsilon > 0$ be as in Følner's criterion. We can rechoose and fix the fundamental domain K such that $(K \cdot a_i) \cap K \neq \phi$ for any $1 \leq i \leq n$ (see Remark 6.33). Moreover, we may extend A to the set A' such that it contains any element $g \in \pi_1(M)$ with $(K \cdot g) \cap K \neq \phi$.

For $\epsilon > 0$ and K, let F be a finite subset as in Proposition 6.34. Then, there is a constant $C_3 > 0$ such that the estimate

$$\text{area}((K \cdot g) \cap \partial H) \geq C_3$$

holds, when $(K \cdot g) \cap \partial H \neq \phi$ and $g \in F$. The constant C_3 can be chosen independent of the choice of $\epsilon > 0$ and F. Then, the estimate holds:

$$\sharp\{g \in F : (K \cdot g) \cap \partial H \neq \phi\} \leq \frac{\text{area}(\partial H)}{C_3}.$$

Note that the set $\{g \in F : (K \cdot g) \cap \partial H \neq \phi\}$ contains $F \setminus (a_i \cdot F)$ for each i, because $(K \cdot a_i \cdot g) \cap (K \cdot g) \neq \phi$ is non empty. Then, taking into account the equality $\sharp\{F\} \cdot \text{vol}(K) = \text{vol}(H)$, we obtain the estimate

$$\begin{aligned}\frac{\sharp\{(a_i \cdot F) \cap F\}}{\sharp\{F\}} &\geq 1 - \frac{\sharp\{g \in F : (K \cdot g) \cap \partial H \neq \phi\}}{\sharp\{F\}} \\ &\geq 1 - \frac{\text{area}(\partial H)}{\text{vol}(H)} \cdot \frac{\text{vol}(K)}{C_3} \\ &\geq 1 - \epsilon \cdot \frac{\text{vol}(K)}{C_3}.\end{aligned}$$

Because $\epsilon > 0$ can be arbitrarily small, it follows from Remark 6.22 that $\pi_1(M)$ satisfies Følner's criterion. □

6.3.1 Cheeger's Inequality

In this subsection, we verify Cheeger's inequality. This will be used in the proof of Theorem 6.31 in Sect. 6.3.3.

Let Y be a compact Riemannian manifold possibly with boundary, and $S \subset Y$ be a codimension 0 submanifold.

Definition 6.35 Cheeger's constant is given by

$$h(S) := \frac{\text{Area}(\partial S)}{\text{Vol}(S)}.$$

Let $f : Y \to \mathbb{R}$ be a smooth function. Let $\{f \geq t_0\}$ denote the subset $\{x \in Y : f(x) \geq t_0\} \subset Y$. Note that the subset $\{f \geq t_0\}$ is a compact manifold with boundary for a generic choice of t_0.

We introduce the following quantities:

- $R_1(f) := \left(\int_Y |\nabla f|(x)dx\right) \cdot \left(\int_Y |f(x)|dx\right)^{-1}$.
- $R(f) := \left(\int_Y |\nabla f|^2(x)dx\right) \cdot \left(\int_Y f(x)^2 dx\right)^{-1}$.

Theorem 6.36 *Let $f : Y \to \mathbb{R}$ be a smooth function. Then there is another map $g : Y \to \mathbb{R}$ such that the inequality*

$$\sqrt{R(f)} \geq \frac{h(\{g \geq t_0\})}{2}$$

holds, where g is constructed explicitly from f.

First, we have the coarea formula.

Lemma 6.37 *The coarea formula holds:*

$$\int_Y |\nabla f|(x)dx = \int_0^\infty \text{Area}(\{f = t\})dt.$$

Proof For another function $k : Y \to \mathbb{R}_{\geq 0}$, we verify the formula

$$\int_Y k(x)dx = \int_0^\infty \left\{ \int_{f=t} k \cdot |\nabla f|^{-1} d\sigma_t \right\} dt,$$

where $d\sigma_t$ is the area form on $f = t$. Then, the proof is finished by applying $k(x) = |\nabla f|(x)$.

Let s be the outward normal variable to the tangent $T\{f = t\}$. Then, the volume form is given by $\mathrm{Vol} = |\nabla f|^{-1} ds d\sigma_t$ as follows. In a local coordinate around a point on $\{f = t\}$, we set $x_1 = s$ and $\{x_2, \cdots, x_n\}$ is orthogonal to x_1. Then, we have the following pointwise equalities:

- $\langle \frac{\partial}{\partial x_1}, \nabla f \rangle = \frac{\partial f}{\partial x_1} = 1.$
- $\langle \frac{\partial}{\partial x_i}, \nabla f \rangle = 0$ for $2 \leq i \leq n$.

Hence, we have $\nabla f = g^{11} \cdot \frac{\partial}{\partial x_1}$ and $g_{1i} = 0$ for $i \geq 2$, we have

$$
\begin{aligned}
\mathrm{Vol} &= \sqrt{\det(g_{ij})_{i,j\geq 1}} \cdot dx_1 \cdots dx_n == \sqrt{g_{11}} \cdot \sqrt{\det(g_{ij})_{i,j\geq 2}} \cdot dx_1 \cdots dx_n \\
&= |\nabla f|^{-1} dx_1 \sqrt{\det(g_{ij})_{i,j\geq 2}} \cdot dx_2 \cdots dx_n \\
&= |\nabla f|^{-1} ds d\sigma_t .
\end{aligned}
$$

This gives the coarea formula. □

Lemma 6.38 *If $f \geq 0$, then there is some t_0 such that the inequality $h(\{f \geq t_0\}) \leq R_1(f)$ holds.*

Proof Set $t_0 := \inf_t h(\{f \geq t\})$. Note that the equality

$$
\int_Y f(x)dx = \int_0^\infty \mathrm{Vol}(\{f \geq t\})dt
$$

holds. Then Lemma 6.37 implies that

$$
\begin{aligned}
\int_Y |\nabla f|(x)dx &= \int_0^\infty \mathrm{Area}(\partial\{f \geq t\})dt \\
&= \int_0^\infty h(\{f \geq t\}) \cdot \mathrm{Vol}(\{f \geq t\})dt \\
&\geq h(\{f \geq t_0\}) \cdot \int_0^\infty \mathrm{Vol}(\{f \geq t\})dt = h(\{f \geq t_0\}) \cdot \int_Y f(x)dx.
\end{aligned}
$$

This finishes the proof. □

Recall that $m \in \mathbb{R}$ is the *median* of f, if both the estimates

$$
\mathrm{Vol}(\{f \geq m\}), \ \mathrm{Vol}(\{f \leq m\}) \ = \ \frac{1}{2} \mathrm{Vol}(Y)
$$

hold. Let $\bar{f} := f - m : Y \to \mathbb{R}$.

Lemma 6.39 *Suppose $\int_Y f dx = 0$. Then for $f : Y \to \mathbb{R}$, there is $g : Y \to \mathbb{R}_{\geq 0}$ such that* $\mathrm{Vol}(g \geq 0) \leq \frac{1}{2} \mathrm{Vol}(M)$ *and the estimate*

$$R(g) \leq R(f)$$

hold.

Proof We claim that the inequality $R(\bar{f}) \leq R(f)$ holds. In fact, it follows from the two properties

$$\begin{aligned}\int_Y \bar{f}^2 dx &= \int_Y (f^2 - 2mf + m^2) dx \\ &= \int_Y (f^2 + m^2) dx \geq \int_Y f^2 dx,\end{aligned}$$

and the equality $\nabla \bar{f} = \nabla f$.

Let us set

$$\bar{f}^+ := \max(0, \bar{f}), \quad \bar{f}^- := \max(0, -\bar{f}).$$

Then, we have $\bar{f} = f^+ - f^-$.

Finally, we claim the inequality

$$\min(R(f^+), R(f^-)) \leq R(\bar{f}).$$

This finishes the proof by letting g be one of $\bar{f}^\pm$ which has smaller value of R.

Note that $\bar{f}^+$ and $\bar{f}^-$ have the disjoint support mutually. The same thing holds for $\nabla \bar{f}^\pm$, because $\bar{f}^\pm(x) = 0$ implies $\nabla \bar{f}^+(x) = 0$. Hence, we have the equalities $|\nabla(\bar{f})|^2 = |\nabla(\bar{f}^+)|^2 + |\nabla(\bar{f}^-)|^2$.

By integrating, we obtain the two equalities

$$\begin{aligned}&\int_Y |\nabla(\bar{f})|^2 dx = \int_Y |\nabla(\bar{f}^+)|^2 dx + \int_Y |\nabla(\bar{f}^-)|^2 dx, \\ &\int_Y \bar{f}^2 dx = \int_Y (\bar{f}^+)^2 + \int_Y (\bar{f}^-)^2.\end{aligned}$$

Finally, the claim follows by a general estimate

$$\min\left\{\frac{A'}{A}, \frac{B'}{B}\right\} \leq \frac{A' + B'}{A + B}$$

for nonnegative values. □

Let us finish the proof of Theorem 6.36.

Proof By Lemma 6.38, $h(\{(\bar{f}^\pm)^2 \geq t_0\}) \leq R_1((\bar{f}^\pm)^2)$ holds.

For $f \geq 0$, let us verify the inequality $R_1(f^2) \leq 2\sqrt{R(f)}$. By the Cauchy–Schwarz inequality, we have the estimate

$$\int_Y |\nabla(f^2)|dx = \int_Y |2f\nabla f|dx \le 2\sqrt{\int_Y f^2dx}\cdot\sqrt{\int_Y |\nabla f|^2dx}.$$

By dividing both sides by $\int_Y f^2dx$, we have the desired inequality.

Combining these estimates, we have

$$h(\{(\bar{f}^{\pm})^2 \ge t_0\}) \le 2\sqrt{R(\bar{f}^{\pm})}.$$

On the other hand, by Lemma 6.39, we have $R(\bar{f}^{\pm}) \le R(f)$, if we choose an appropriate sign. Hence, we have

$$h(\{(\bar{f}^{\pm})^2 \ge t_0\}) \le 2\sqrt{R(f)}.$$

This completes the proof. □

It follows from Theorem 6.36 and its proof that we obtain the following corollary:

Corollary 6.40 *Assume that the boundary of Y is empty. Then, Cheeger's inequality holds:*

$$\sqrt{\inf\{\ \mathrm{spec}(\Delta_0)\backslash\{0\}\ \}} \ \ge\ \frac{1}{2}\left(\inf_{S\subset Y, 0<\mathrm{Vol}(S)\le\frac{1}{2}\mathrm{Vol}(M)} h(S)\right).$$

Buser's inequality states the reverse inequality to Cheeger's [26, 107].

Theorem 6.41 *Let X be a Riemannian manifold whose Ricci curvature is bounded below by $-K$ for a positive constant $K > 0$. Then the inequality*

$$\inf\{\ \mathrm{spec}(\Delta_0)\backslash\{0\}\ \} \ \le\ C\left(\sqrt{K}h + h^2\right)$$

holds, where $C > 0$ is a constant that depends only on $\dim M$.

We will not present a proof of Theorem 6.41, because we will not use it for a proof of Theorem 6.31.

6.3.2 Quasi-Isometry Invariance

Definition 6.42 Let $\epsilon > 0$ be a positive number.

- A subset $P \subset X$ is called an ϵ-net, if for any element $x \in X$, there is $p \in P$ such that $d(x, p) \leq \epsilon$ holds.
- It is ϵ-separated, if for any two elements $p \neq q \in P$, $d(p, q) \geq \epsilon$ holds. An ϵ-separated net $P \subset X$ is maximal, if the union $P \cup \{x\}$ is no longer an ϵ-separated for any $x \in X \backslash P$.
- A discrete subset $P \subset X$ is an (ϵ, ϵ')-separated net, if it is ϵ-separated and is an ϵ'-net for $0 < \epsilon < \epsilon'$. Note that a maximal ϵ-separated subset is a 2ϵ-net.
- Let P be a discrete metric space, and $Q \subset P$ be a subset. We set

$$\partial Q := \{p \in Q^c : d(Q, p) = \min_{p' \in Q^c} d(Q, p')\}.$$

- Let (P, d_P) be an (ϵ, ϵ')-separated net. The path length $l(p, p') \geq 0$ on P is the minimum $l \geq 0$ such that there is a sequence $(p_0, p_1, \cdots, p_l)$ such that $d(p_i, p_{i+1}) \leq \epsilon'$ and $p_0 = p$, $p_l = p'$ holds.

Let (X, d_X) and (Y, d_Y) be two metric spaces. A map $f : X \to Y$ (not necessarily continuous) is said to be *quasi-isometric embedding*, if there are positive constants $A, B > 0$ such that the inequalities

$$A^{-1} \cdot d_X(x, x') - B \leq d_Y(f(x), f(x')) \leq A \cdot d_X(x, x') + B$$

hold for any $x, x' \in X$.

The map f is said to be quasi-isometry, if the image $f(X) \subset Y$ is a C-net for some $C \geq 0$.

Examples

- Let Γ be a finitely generated group and equip with a finite and symmetric ($A^{-1} = A$) generating set $A \subset \Gamma$. The word metric $l(g, g')$ is given by the minimum $l \geq 0$ such that there is a sequence $(a_0, \cdots, a_l)$ such that $a_0 = e$ and $g \cdot a_0 \cdot a_1 \cdots s_l = g'$ holds. The group Γ is the metric space with respect to the word metric. It is a $(1, 1)$-separated net.
- Let M be a compact manifold, and choose any base point $x_0 \in \tilde{M}$ in the universal covering space. The map

$$f : \pi_1(M) \to \tilde{M}$$

by $g \to x_0 \cdot g$ is the quasi-isometry, which we equip with a word metric on $\pi_1(M)$.

Define

$$h(X) := \inf_S h(S),$$

where $S \subset X$ is a compact submanifold with smooth boundary.

Let $Q \subset P$ be a finite subset. Then we set $h(Q) := \frac{\sharp(\partial Q)}{\sharp(Q)}$. Then we define $h(P) := \inf_Q h(Q)$.

Theorem 6.43 ([70, 84]) *Let X be a complete Riemannian manifold whose Ricci curvature is bounded from below by $-(n-1)K^2$ with $K > 0$. Let $P \subset X$ be a maximal $(\epsilon, 2\epsilon)$-separated net for $\epsilon \in (0, \mathrm{inj}(X)/2]$, where $\mathrm{inj}(X)$ is the injective radius of X. Then, $h(X) > 0$ holds, if and only if $h(P) > 0$ is positive.*

We follow [84]. The following are basic facts in Riemannian geometry. Let X be a complete Riemannian manifold whose Ricci curvature is bounded from below by $-(n-1)K^2$ with $K > 0$.

Lemma 6.44 *Then for any $x \in X$:*

1. *The uniform bounds* $\mathrm{vol}\, B_r(x) \leq V_K(r)$ *and* $\mathrm{area}\, \partial B_r(x) \leq A_K(r)$ *hold for any* $r > 0$.
2. *The function* $r \to \mathrm{vol}\, B_r(x)/V_K(r)$ *is monotone and non-increasing for any* $r > 0$.
3. *The local bound* $\mathrm{vol}\, B_r(x) \geq V_0 r^n$ *holds for any* $r \in (0, \mathrm{inj}(X)/2]$, *where* $n := \dim X$.
4. *For* $\epsilon \in (0, \mathrm{inj}(X)/2]$, *let* $H \subset B_\epsilon(p)$ *be a smooth hypersurface that separates* $B_\epsilon(p)$ *into two domains*

$$B_\epsilon(p) = D_1 \sqcup H \sqcup D_2.$$

Then, the following estimate holds:

$$\frac{\mathrm{area}\, H}{\min\{\mathrm{vol}\, D_1, \mathrm{vol}\, D_2\}} \geq c(\dim X, K, \epsilon).$$

Here $V_K(r)$ is the volume of the ball of radius r in the simply-connected complete Riemannian manifold of constant curvature $-K^2$, $A_K(r)$ is the area of the boundary of the ball of radius r, and $0 < V_0 = V_0(\dim X, \mathrm{inj}(X))$ is a positive constant.

Below we omit the notation K and write $V(r) := V_K(r)$ and $A(r) := A_K(r)$.

Proposition 6.45 *Let X be a complete Riemannian manifold whose Ricci curvature is bounded from below by $-(n-1)K^2$ with $K > 0$. Let $P \subset X$ be an $\epsilon > 0$-separated subset. Then, for $x \in X$ and $r > 0$, there exists a constant $\nu = \nu(\dim X, K, \epsilon, r) > 0$ such that*

$$\sharp\{p \in P : x \in B_r(p)\} \leq \nu.$$

Proof Note the inclusions

$$B_{\epsilon/2}(p) \subset B_{r+\epsilon/2}(x) \subset B_{2r+\epsilon/2}(p)$$

for $x \in B_r(p)$. By item 2 in Lemma 6.44, we have the estimate

$$\text{vol } B_{\epsilon/2}(p) \geq \frac{V(\epsilon/2)}{V(2r+\epsilon/2)} \cdot \text{vol } B_{2r+\epsilon/2}(p).$$

Note vol $B_{\epsilon/2}(p) \cap$ vol $B_{\epsilon/2}(p') = \phi$ for any $p \neq p' \in P$. Hence, we have

$$\begin{aligned}
\text{vol } B_{r+\epsilon/2}(x) &\geq \sum_{x \in B_r(p)} \text{vol } B_{\epsilon/2}(p) \\
&\geq \frac{V(\epsilon/2)}{V(2r+\epsilon/2)} \cdot \sum_{x \in B_r(p)} \text{vol } B_{2r+\epsilon/2}(p) \\
&\geq \frac{V(\epsilon/2)}{V(2r+\epsilon/2)} \cdot \text{vol } B_{r+\epsilon/2}(x) \cdot \sharp\{p \in P : x \in B_r(p)\}.
\end{aligned}$$

Hence, we have the estimate

$$\sharp\{p \in P : x \in B_r(p)\} \leq \frac{V(2r+\epsilon/2)}{V(\epsilon/2)}.$$

This completes the proof. □

Let us verify Theorem 6.43.

Proof First, let us verify that $h(X) > 0$ implies $h(P) > 0$. Let $P_S \subset P$ be any finite subset. Let $S := \cup_{p \in P_S} B_\epsilon(p) \subset X$. Then, we have $(\epsilon/2)^n \cdot V_0 \cdot \sharp P_S \leq \text{vol } S$ holds by item 3 in Lemma 6.44. Note the inclusion

$$\partial S \subset \cup_{p \in \partial' P_s} \partial B_\epsilon(p),$$

where $\partial' P_S := \partial(P \backslash P_S)$. Hence, we have the estimate area $\partial S \leq A(\epsilon) \cdot \sharp(\partial' P_S)$ by item 1 in Lemma 6.44.

Note that there is some $\lambda \geq 1$ such that $\sharp(B_r(P_S)) \leq \lambda^r \cdot \sharp P_S$ holds for any finite subset P_S, where $B_r(P_S)$ is the set of points whose distance from P_S is less than or equal to r with respect to the path norm on P. Hence, we have $\sharp(\partial' P_S) \leq \lambda \cdot \sharp(\partial P_S)$ holds. Then, we have the estimate area $\partial S \leq \lambda \cdot A(\epsilon) \cdot \sharp(\partial P_S)$. Then, we have

$$\frac{\text{area}(\partial S)}{\text{vol}(S)} \leq c' \cdot \frac{\sharp(\partial P_S)}{\sharp(P_S)},$$

where $c' := \lambda A(\epsilon)(\epsilon/2)^{-n} V_0^{-1}$. This implies the inequality

$$h(X) \leq c' \cdot h(P).$$

Let us verify the converse. Let $S \subset X$ be a bounded domain with smooth boundary. Denote this by

$$P_S := \{p \in P : \mathrm{vol}(S \cap B_\epsilon(p)) > \mathrm{vol}(B_\epsilon(p))/2\}.$$

Set $P_S^c(\epsilon) := (P \backslash P_S) \cap B_\epsilon(S)$, where $B_\epsilon(S) \supset S$ is ϵ neighbourhood of S. Note that $\{B_\epsilon(p) : p \in S \cup P_S^c(\epsilon)\}$ covers S. By item 4 in Lemma 6.44, the local isoperimetric inequality

$$\frac{\mathrm{area}(\partial S \cap B_\epsilon(p))}{\mathrm{vol}(S \cap B_\epsilon(p))} \geq c(\dim X, K, \epsilon)$$

holds for $p \in P_S^c(\epsilon)$. Then we have the estimates

$$\begin{aligned} \sum_{p \in P_S^c(\epsilon)} \mathrm{vol}(S \cap B_\epsilon(p)) &\leq c^{-1} \cdot \sum_{p \in P_S^c(\epsilon)} \mathrm{area}(\partial S \cap B_\epsilon(p)) \\ &\leq \nu \cdot c^{-1} \cdot \mathrm{area}(\partial S) \end{aligned}$$

by Proposition 6.45.

Suppose $P_S = \phi$. Then $P_S^c(\epsilon) = P \cap B_\epsilon(S)$, and the estimates hold:

$$\mathrm{vol}\, S \leq \sum_{p \in P_S^c(\epsilon)} \mathrm{vol}(S \cap B_\epsilon(p)) \leq \nu \cdot c^{-1} \cdot \mathrm{area}(\partial S).$$

Next suppose $P_S \neq \phi$. Then, we have the estimates

$$\begin{aligned} \mathrm{vol}\, S &\leq \sum_{p \in S \cup P_S^c(\epsilon)} \mathrm{vol}(S \cap B_\epsilon(p)) \\ &\leq \sum_{p \in P_S} \mathrm{vol}(S \cap B_\epsilon(p)) + \nu \cdot c^{-1} \cdot \mathrm{area}(\partial S) \\ &\leq V(\epsilon) \cdot \sharp P_S + \nu \cdot c^{-1} \cdot h(S) \cdot \mathrm{vol}(S). \end{aligned}$$

Hence, we have the estimate

$$\mathrm{vol}(S) \leq 2V(\epsilon) \cdot \sharp P_S,$$

if $h(S) \leq \frac{1}{2}\nu^{-1} \cdot c$ holds.

Set $H := \{p \in X : \mathrm{vol}(S \cap B_\epsilon(p)) = \mathrm{vol}(B_\epsilon(p))/2\}$. Then the inclusion

$$\partial P_S \subset B_{2\epsilon}(H) \tag{6.2}$$

holds (see Problem 6.66).

Let $Q \subset H$ be a maximal ϵ-separated subset of H. Then, we have the inclusions

$$\cup_{p\in\partial P_S} B_{\epsilon/2}(p) \subset B_{5/2\cdot\epsilon}(H) \subset \cup_{q\in Q} B_{7/2\cdot\epsilon}(q).$$

It follows from items 2, 3, and 4 in Lemma 6.44 that the following estimates hold:

$$\begin{aligned}
\left(\frac{\epsilon}{2}\right)^n \cdot V_0 \cdot \sharp P_S &\le \sum_{p\in\partial P_S} \mathrm{vol}(B_{\epsilon/2}(p)) \\
&\le \sum_{q\in\partial Q} \mathrm{vol}(B_{7/2\cdot\epsilon}(q)) \\
&\le \frac{V(7/2\cdot\epsilon)}{V(\epsilon)} \cdot \sum_{q\in\partial Q} \mathrm{vol}(B_\epsilon(q)) \\
&= 2\frac{V(7/2\cdot\epsilon)}{V(\epsilon)} \cdot \sum_{q\in\partial Q} \mathrm{vol}(S \cap B_\epsilon(q)) \\
&\le 2c^{-1} \cdot \frac{V(7/2\cdot\epsilon)}{V(\epsilon)} \cdot \sum_{q\in\partial Q} \mathrm{area}(\partial S \cap B_\epsilon(q)) \\
&\le 2\nu \cdot c^{-1} \cdot \frac{V(7/2\cdot\epsilon)}{V(\epsilon)} \cdot \sum_{q\in\partial Q} \mathrm{area}(\partial S),
\end{aligned}$$

where $n := \dim X$. Combining these estimates, we have the estimate

$$\frac{\mathrm{area}(\partial S)}{\mathrm{vol}(S)} \ge c' \cdot \frac{\mathrm{area}(\partial P_S)}{\mathrm{vol}(P_S)},$$

if $h(S) \le \frac{1}{2}\nu^{-1} \cdot c$ holds, where $c' := \frac{1}{2\nu} \cdot \left(\frac{\epsilon}{2}\right)^n \cdot V_0 V(\epsilon) V(7/2\epsilon)^{-1} c$. Hence, we have the estimate

$$h(S) \ge \min\left\{\frac{c}{2\nu}, c' \cdot h(P_S)\right\}.$$

This verifies the proof, because the estimate $h(P) \le h(P_s)$ holds. □

6.3.3 Proof of Theorem 6.31

We complete the proof of Theorem 6.31. It remains to verify the if part.

Proof Suppose $0 \in \mathrm{spec}(\Delta_0)$ holds. Then by Corollary 6.40,

$$0 \leq \inf_{S \subset Y} h(S) \leq \inf_{S \subset Y, 0 < \mathrm{Vol}(S) \leq \frac{1}{2} \mathrm{Vol}(M)} h(S) = 0$$

holds. Let $x_0 \cdot \pi_1(M) \subset \tilde{M}$ be the orbit of the fundamental group for a fixed point $x_0 \in \tilde{M}$. It induces a quasi-isometry $\pi_1(M) \cong \tilde{M}$.

Let $P \subset X$ be any maximal $\epsilon/2$-separated ϵ-net. Then there is a positive constant $c > 0$ independent of the choice of P such that the estimate

$$1 \leq \sharp(K \cap P) \leq c$$

holds, where $K \subset X$ is a fundamental domain of $\pi_1(M)$-action. Then, we have the equality

$$\inf_{S \subset \pi_1(M)} h(S) = \inf_{S \subset P} h(S).$$

Hence by Theorem 6.43,

$$\inf_{S \subset \pi_1(M)} h(S) = 0$$

holds, where $h(S) := \frac{\sharp(\partial S)}{\sharp S}$. Then by Proposition 6.34, the fundamental group $\pi_1(M)$ is amenable. □

6.4 Simplicial L^2-Cohomology Groups

Let X be a complete Riemannian manifold, and equip with a simplicial decomposition L of X. There is a L^2-cohomology group $\bar{H}^m_{(2)}(L)$ that is defined by using the simplicial structure on X.

Let us recall that a *simplicial complex* K is a set of simplices that satisfies:

- every face of a simplex in K is an element in K, and
- the non-empty intersection of two simplices in K is a face of them.

Any compact smooth manifold can admit a simplicial structure.

A *refinement* K' of a simplicial complex K is another simplicial complex such that a simplex in K is realized by a finite union of simplices in K'.

Let M be a compact manifold and equip with a simplicial structure K on M. Let $X := \tilde{M}$ be the universal covering space, and lift the simplicial structure to $\tilde{K}$ on X. We verify two key properties, where one is that the simplicial L^2-cohomology group is given independently of the choice of K. Secondly, $\bar{H}^m_{(2)}(\tilde{K})$ is isomorphic to the reduced de Rham L^2-cohomology group. In particular, we can just call it the simplicial L^2-cohomology group of $\tilde{M}$.

The covering transformation induces the unitary action of $\Gamma := \pi_1(M)$ on the L^2-Sobolev spaces $L^2_k(X; \Lambda^m)$. Then, it induces the unitary action on both the simplicial L^2-cohomology group and L^2-de Rham cohomology group.

The aim in this section is to verify the following:

Theorem 6.46 (**L^2-de Rham Theorem [33]**) *Suppose M is a compact manifold equipped with a simplicial decomposition K. Then, the two L^2-cohomology groups*

$$\bar{H}^m_{(2)}(\tilde{K}) \cong \bar{H}^m_{(2)}(X)$$

given by using the simplicial structure and the differential forms are mutually Γ-equivariantly isomorphic for all $0 \leq m \leq \dim M$.

6.4.1 Definition

Let K be a simplicial decomposition of M, and denote by K_i the subset consisting of all i-simplices. Let us lift it over the universal covering space, and denote the result by $\tilde{K}$ as the $\Gamma := \pi_1(M)$ invariant simplicial decomposition of $\tilde{M}$. In this section, we do not use the group action, but we do use it in Sect. 7.1.

Let $C^i_{(2)}(\tilde{K})$ be the space of real oriented L^2-cochains of degree i. Hence, an element $f \in C^i_{(2)}(\tilde{K})$ can be expressed as

$$f = \sum_{\sigma \in \tilde{K}_i} f_\sigma \cdot \sigma^*,$$

where $\sum_\sigma |f_\sigma|^2 < \infty$, and for $\tau \in \tilde{K}_i$,

$$\sigma^*(\tau) := \begin{cases} 1 & \tau = \sigma, \\ 0 & \tau \neq \sigma. \end{cases}$$

The linear space $C^i_{(2)}(\tilde{K})$ admits the structure of a Hilbert space with the inner product

$$\langle f, g\rangle := \sum_{\sigma} f_\sigma \cdot g_\sigma \in \mathbb{R}.$$

Because each simplex $\sigma \in C_i(\tilde{K})$ is oriented, there is the boundary operator $\partial : C_i(\tilde{K}) \to C_{i-1}(\tilde{K})$ given by taking the alternative sum of the faces. Hence, we have the coboundary operator

$$d_c : C^{i-1}_{(2)}(\tilde{K}) \to C^i_{(2)}(\tilde{K})$$

by $d_c(f)(\sigma) := f(\partial\sigma)$.

Definition 6.47 The reduced simplicial L^2-cohomology group is defined by

$$\begin{aligned}\bar{H}^i_{(2)}(\tilde{K}) :&= \mathrm{Ker}\{d_c : C^i_{(2)}(\tilde{K}) \to C^{i+1}_{(2)}(\tilde{K})\} \,/\, \overline{d_c(C^{i-1}_{(2)}(\tilde{K}))} \\ &=: Z^i_{(2)}(\tilde{K}) \,/\, \overline{B^i_{(2)}(\tilde{K})}\end{aligned}$$

where $-$ is the closure of the space.

Remark 6.48 Suppose that M is compact. Let $\{\sigma^1, \cdots, \sigma^l\}$ be the set of i-simplices of K, and denote by V^i the finite-dimensional vector space spanned by σ^j. Then the space of L^2-cochain $C^i_{(2)}(\tilde{K})$ is Γ-equivariantly isometric to the Hilbert space

$$C^i_{(2)}(\tilde{K}) \cong V^i \otimes l^2(\Gamma).$$

One merit to using the simplicial L^2-cohomology group is that V is finite dimensional. See Sect. 7.1.5. Note that the space of L^2-forms is Γ-equivariantly isometric to the Hilbert space

$$L^2(X; \Lambda^i) \cong L^2(K; \Lambda^i) \otimes l^2(\Gamma),$$

where $K \subset X$ is a fundamental domain, and the space $L^2(K; \Lambda^i)$ is infinite dimensional, if K is non-empty and has positive dimension.

Proposition 6.49 *Let $K' \supset K$ be a refinement of the simplicial complex over M. Then, the canonical map $\tilde{s}^* : C^i_{(2)}(\tilde{K}') \to C^i_{(2)}(\tilde{K})$ induces the isomorphism*

$$\tilde{s}^* : \bar{H}^i_{(2)}(\tilde{K}') \cong \bar{H}^i_{(2)}(\tilde{K})$$

between the reduced simplicial L^2-cohomology groups.

Corollary 6.50 *The reduced simplicial L^2-cohomology group is independent of the choice of a simplicial decomposition of M.*

Proof For two simplicial decompositions K and K' over M, there is a refinement K'' of both the simplicial decompositions K and K'. By Proposition 6.49, the isomorphisms hold:

$$\bar{H}^i_{(2)}(\tilde{K}') \cong \bar{H}^i_{(2)}(\tilde{K}'') \cong \bar{H}^i_{(2)}(\tilde{K}).$$

□

6.4.2 *Proof of Proposition 6.49*

Let K be a simplicial complex. We denote by K_* the subset consisting of all $*$-dimensional simplices in K.

Let K be a simplicial decomposition of a compact manifold M. Recall that the barycenter subdivision L is a refinement of K such that the set of vertices L_0 is given by the union of that of K with all the barycenters of all simplices of K. Hence the inclusion $K_0 \subset L_0$ holds.

Let $N := \dim K$. There is a hierarchy of refinements between K and L as

$$K, \; K^1, \; K^2, \cdots, \; K^N =: L,$$

where K^1 is a refinement of K such that the set of vertices K^1_0 is given by the union of that of K with all the barycenters of all N-dimensional simplices of K. Hence the inclusions $K_0 \subset K^1_0 \subset L^1_0$ hold. K^2 is a refinement of K^1 such that the set of vertices K^2_0 is given by the union of that of K with all the barycenters of all N- and $N-1$- dimensional simplices of K. Hence the inclusions $K_0 \subset K^1_0 \subset K^2_0 \subset L^1_0$ hold.

Successively, we define K^i as a refinement of K^{i-1} such that the set of vertices K^i_0 is given by the union of that of K with all the barycenters of all N-, $(N-1)$-, $\cdots$ and $(N-i+1)$- dimensional simplices of K. K^i is a refinement of K^{i-1}. We say that K^i is the i-th barycenter subdivision of K. See Problem 6.67.

Lemma 6.51 *Let $N = \dim K$. For each i, and an N-dimensional simplex σ in K^i, there is a unique $N - i$-dimensional subsimplex τ of σ that is also a simplex in K.*

Proof This can be checked quite easily. □

Let us reduce the proof of Proposition 6.49 as follows:

1. In order to verify Proposition 6.49, it is enough to assume that the refinement K' is the barycenter subdivision. It covers the case of any refinement by applying it successively for at most finitely many times.
2. In order to verify Proposition 6.49 in the case when K' is the barycenter subdivision, it is enough to verify the case when the pair $(K, K') = (K^i, K^{i+1})$ is the i-th and $i + 1$-th barycenter subdivisions for all i. It covers the case of the barycenter subdivision by applying it successively for at most finitely many times.
3. Below, we will verify the isomorphism in Proposition 6.49, when the refinement is the i-th barycenter subdivision (K^{i-1}, K^i).

Let us clarify some notations we use later:

- Recall that the chain group $C_*(K)$ is a vector space over $\mathbb{R}$ generated by the set of simplices modulo permutation of their vertices. For example, $\langle v_0, v_1, v_2\rangle = -\langle v_1, v_0, v_2\rangle$ in $C_2(K)$, and $\langle v_0, v_0, v_2\rangle = 0$.
- Let $N := \dim K$, and K^i be the i-th barycenter subdivision of K^{i-1}. For each simplex $\sigma = \langle v_0, \cdots, v_{N-i+1}\rangle \in K$, we denote the barycenter $\hat{\sigma}$ as a vertex in K^i. Note that each barycenter uniquely determines the simplex in K that contains the barycenter in its interior. Namely, the correspondence $\hat{\sigma} \leftrightarrow \sigma$ is one-to-one and onto.
- Let $\tau = \langle v_0, \cdots, v_k\rangle$ be a k-dimensional simplex. We let $\tau = \langle\tilde{\tau}, \tilde{\nu}\rangle$, if $\tilde{\tau} = \langle v'_0, \cdots, v'_l\rangle$ and $\tilde{\nu} = \langle v'_{l+1}, \cdots, v'_k\rangle$ are subsimplices of τ such that the equality $\langle v'_0, \cdots, v'_k\rangle = \langle v_0, \cdots, v_k\rangle$ holds.

There is a canonical chain map $s^1_* : C_*(K) \to C_*(K^1)$ by $s^1_*(\tau) = \tau$ for any $\tau \in K_*$ and $* \le N - 1$. For an N-dimensional simplex $\sigma = \langle v_0, \cdots, v_N\rangle$, we set

$$s^1_*(\sigma) = \sum_{j=0}^{N} \langle v_0, \cdots, \check{v}_j, \hat{\sigma}, v_{j+1}, \cdots v_N\rangle.$$

The operator s^1_* commutes with the boundary operator ∂.

Remark 6.52 Note that each component $\langle v_0, \cdots, \check{v}_j, \hat{\sigma}, v_{j+1}, \cdots v_N\rangle \neq 0$ is always non-zero. For example, $s^1_*(\sigma) = \langle\hat{\sigma}, v_1, v_2\rangle + \langle v_0, \hat{\sigma}, v_2\rangle + \langle v_0, v_1, \hat{\sigma}\rangle$ holds for $N = 2$ and $\sigma = \langle v_0, v_1, v_2\rangle$.

Now we define canonical chain maps $s^i_* : C_*(K^{i-1}) \to C_*(K^i)$ as follows. First, we set $s^i_*(\tau) = \tau$ for $* \le N - i$.

Next, let $\tau = \langle v_0, \cdots, v_k \rangle$ be a k-dimensional simplex for $k \geq N - i + 1$. Let σ_τ be the unique N-dimensional simplex in K^{i-1} that has τ as its subsimplex. By Lemma 6.51, there is a unique $N - i + 1$-dimensional subsimplex $\tilde{\tau} = \langle w_0, \cdots, w_{N-i+1} \rangle$ of σ that is also a simplex in K. Then, we set

$$s_*^i(\tau) = \begin{cases} \tau & \text{if } \tilde{\tau} \text{ is not a subsimplex of } \tau, \\ \sum_{j=0}^{N-i+1} \langle w_0, \cdots, \check{w}_j, \hat{\tilde{\tau}}, w_{j+1}, \cdots w_{N-i+1}, \tilde{\nu} \rangle & \text{if } \tau = \langle \tilde{\tau}, \tilde{\nu} \rangle. \end{cases}$$

The operator s_* still commutes with the boundary operator ∂.

Recall that a simplicial map between two simplicial decomposition is given by a map between their vertices that induces a map between their simplices. Let $t : K^i \to K^{i-1}$ be a simplicial map such that any vertex in K^{i-1} is fixed and the barycenter of a simplex is mapped to a vertex of the simplex.

Remark 6.53 The map t induces $t_* : C_*(K^i) \to C_*(K^{i-1})$, and some components may be crushed and vanish.

For example, let $v_0, v_3 \in K^{i-1}$ and $\tau := \langle v_0, v_3 \rangle$. Let $v_1 = \hat{\tau}$, and $\sigma := \langle v_0, v_1, v_2 \rangle \in K^i$. If $t(v_1) = v_0$, then,

$$t_*(\langle v_0, v_1, v_2 \rangle) = \langle v_0, t(v_1), v_2 \rangle = \langle v_0, v_0, v_2 \rangle = 0.$$

Lemma 6.54

- *The composition $t_* \circ s_* : C_*(K^{i-1}) \to C_*(K^{i-1})$ is the identity.*
- *There is a homotopy operator $H_* : C_*(K^i) \to C_{*+1}(K^i)$ of the composition $s_* \circ t_* : C_*(K^i) \to C_*(K^i)$ such that the equality holds:*

$$\partial \circ H_* + H_{*-1} \circ \partial = s_* \circ t_* - 1$$

Proof It is clear that the composition $t_* \circ s_* : C_*(K^{i-1}) \to C_*(K^{i-1})$ is the identity. Let us consider the rest case.

Let us construct an operator $H_* : C_*(K^i) \to C_{*+1}(K^i)$ as follows. We set $H_*(\tau) = 0$ for a simplex τ of K^i that does not contain the barycenter of any simplex in K^{i-1}. For a simplex of the form $\tau = \langle \hat{\sigma}, v_1, \cdots \rangle$, where $v_i \in K^{i-1}$ and σ is an $N - i + 1$-dimensional simplex in K, we set

$$H_*(\tau) := \langle \hat{\sigma}, t(\hat{\sigma}), v_1, \cdots \rangle \in C_{*+1}(K^i).$$

Note that:

- the image $H_*(\tau) = 0$ vanishes, if the subsimplex $\langle v_1, \cdots \rangle$ already contains $t(\hat{\sigma})$, or a subsimplex $\langle t(\hat{\sigma}), v_s \rangle$ contains $\hat{\sigma}$ for some s.
- $s_* \circ t_*(\langle \hat{\sigma}, v_1, \cdots \rangle) = \langle t(\hat{\sigma}), v_1, \cdots \rangle \in C_{*+1}(K^i)$ holds.

Then, we have the equalities

$$\begin{aligned}\partial \circ H_*(\tau) &= \partial\langle\hat{\sigma}, t(\hat{\sigma}), v_1, \cdots\rangle \\ &= \langle t(\hat{\sigma}), v_1, \cdots\rangle - \langle\hat{\sigma}, v_1, \cdots\rangle \\ &\quad + \sum_{j\geq 1} (-1)^{j+1} \langle\hat{\sigma}, t(\hat{\sigma}), v_1, \cdots, \hat{v}_j, \cdots\rangle.\end{aligned}$$

$$\begin{aligned}H_{*-1} \circ \partial(\tau) &= H_{*-1} \circ \partial\langle\hat{\sigma}, v_1, \cdots\rangle \\ &= H_{*-1}\left(\langle v_1, \cdots\rangle - \langle\hat{\sigma}, v_2, \cdots\rangle + \cdots\right) \\ &= H_{*-1}\left(\sum_{j\geq 1} (-1)^j \langle\hat{\sigma}, v_1, \cdots, \hat{v}_j, \cdots\rangle\right) \\ &= \sum_{j\geq 1} (-1)^j \langle\hat{\sigma}, t(\hat{\sigma}), v_1, \cdots, \hat{v}_j, \cdots\rangle.\end{aligned}$$

Hence, the equality

$$(H_{*-1} \circ \partial + \partial \circ H_*)(\tau) = (s_* \circ t_* - 1)(\tau)$$

holds. □

Remark 6.55 It follows from the proof of Lemma 6.54 that all the operators s_*, t, H_* and ∂ are all local, and hence they can lift to the bounded operators over the universal covering space.

Let us verify Proposition 6.49. Let K be a simplicial decomposition of M and K' be its barycenter subdivision. We lift them on the universal covering space and denote them by $\tilde{K}$ and $\tilde{K}'$.

Let us lift the operators s_*, t, H_* and ∂ over the universal covering space by Remark 6.55. The chain map $s : C_*(K) \to C_*(K')$ induces the cochain map $\tilde{s}^* : C^*_{(2)}(\tilde{K}') \to C^*_{(2)}(\tilde{K})$. The inclusions

$$\tilde{s}^*(Z^*_{(2)}(\tilde{K}')) \subset Z^*_{(2)}(\tilde{K}), \quad \tilde{s}^*(B^*_{(2)}(\tilde{K}')) \subset B^*_{(2)}(\tilde{K})$$

hold. Hence, it induces a linear map

$$\tilde{s}^* : \bar{H}^*_{(2)}(\tilde{K}') \to \bar{H}^*_{(2)}(\tilde{K}).$$

Consider the simplicial map t from K' to K. Because we use the simplicial decompositions of the universal covering space by taking their lifts, it follows that

t also induces a uniformly bounded simplicial map from $\tilde{K}'$ to $\tilde{K}$. Hence, it also induces a linear map

$$\tilde{t}^* : \bar{H}^*_{(2)}(\tilde{K}) \to \bar{H}^*_{(2)}(\tilde{K}').$$

The composition $\tilde{s}^* \circ \tilde{t}^* : C^*_{(2)}(\tilde{K}) \to C^*_{(2)}(\tilde{K})$ is the identity, because both the operators s and t are local and they satisfy the equality $t \circ s = id$.

Let us consider the converse composition. The linear map $H_* : C_*(K') \to C_*(K')$ also induces the bounded map

$$\tilde{H}^* : C^*_{(2)}(\tilde{K}') \to C^{*-1}_{(2)}(\tilde{K}')$$

such that they satisfy the equality

$$\tilde{H}^{*+1} \circ d_c + d_c \circ \tilde{H}^* = \tilde{t}^* \circ \tilde{s}^* - 1,$$

because these operators are defined by the lists of local operators.

Hence, $\tilde{s}^*$ induces the isomorphism

$$\tilde{s}^* : \bar{H}^*_{(2)}(\tilde{K}') \cong \bar{H}^*_{(2)}(\tilde{K}).$$

6.4.3 Proof of Theorem 6.46

Let us fix $k > 1$. Later we choose a large $k >> 1$. See (6.3). Recall the L^2_k-Sobolev space of the differential forms $L^2_k(X; \Lambda^i)$ and the L^2-cochain group $C^i_{(2)}(\tilde{K})$. There is a Γ-equivariant map

$$\int : L^2_k(X; \Lambda^i) \to C^i_{(2)}(\tilde{K}), \qquad \int \omega := \sum_{\sigma \in \tilde{K}_i} \left(\int_\sigma \omega \right) \cdot \sigma^*,$$

where $\Gamma = \pi_1(M)$.

This map induces the linear map

$$\int : \bar{H}^i_{(2)}(\tilde{M}) \to \bar{H}^i_{(2)}(\tilde{K}).$$

Let us verify that this map gives the isomorphism. Let us fix a simplicial decomposition K on M. For each vertex $v \in K_0$, let $U_v \subset M$ be the star closure of v that is an open subset of M. Recall that the star of an i-simplex $\sigma \in K_i$ is the union of $i+1$-simplices that have σ as their face. Define $S^l(\sigma)$ inductively by the star of $S^{l-1}(\sigma)$ with $S^0(\sigma) = \sigma$. The star closure U_v of $v \in K_0$ is defined by $U_v := S^N(v)$, where $N = \dim K$.

Examples

- Let K be a 1-dimensional simplicial complex with two 1-simplices $\{l_1, l_2\}$ that is attached at a vertex $v = \overline{l_1} \cap \overline{l_2}$, where $\bar{l}$ is the closure of l. Then, the star closure of v in K is $v \sqcup l_1 \sqcup l_2$.
- Let K be the 2-dimensional simplex that has three vertices v_1, v_2, v_3. Consider K^1 above. The set of 0-simplices $K_0^1 = \{v_0, v_1, v_2, v_3\}$, where v_0 is the barycenter of the disk. Any 1-simplices are parametrized as l_{ij}, where $0 \leq i \neq j \leq 3$. Any 2-simplices are parametrized as Δ_{ij0}, where $1 \leq i \neq j \leq 3$.

  The star closure of v_0 is given by the union

$$\begin{aligned} U_{v_0} &= v_0 \sqcup l_{01} \sqcup l_{02} \sqcup l_{03} \sqcup \Delta_{012} \sqcup \Delta_{012} \sqcup \Delta_{013} \\ &= K \setminus \{\, v_1 \sqcup v_2 \sqcup v_3 \sqcup l_{12} \sqcup l_{23} \sqcup l_{31} \,\}. \end{aligned}$$

Their union $\cup_{v \in K_0} U_v$ covers M. Let $\{\varphi_v\}_{v \in K_0}$ be a partition of unity subordinate to the covering $\{U_v\}_{v \in K_0}$.

Let us consider the lift $\tilde{K}$ on the universal covering space $X := \tilde{M}$. We can lift both the partition of unity $\{U_{\tilde{v}}\}_{\tilde{v} \in \tilde{K}_0}$ and $\{\varphi_{\tilde{v}}\}_{\tilde{v} \in \tilde{K}_0}$ respectively on X. Let us construct the map $W : C^p_{(2)}(\tilde{K}) \to L^2_k(X; \Lambda^p)$ by

$$W(\sigma^*) := \begin{cases} \varphi_0, & p = 0, \\ p! \sum_{i=0}^{p} (-1)^i \varphi_i d\varphi_0 \wedge \cdots \wedge \check{d\varphi_i} \wedge \cdots \wedge d\varphi_p, & p > 0 \end{cases}$$

for a simplex $\sigma = [v_0, \cdots, v_p]$ in $\tilde{K}$ and $\varphi_i := \varphi_{v_i}$. Then for $f = \sum_{\sigma \in \tilde{K}_p} f_\sigma \cdot \sigma^*$, we define

$$W(f) := \sum_{\sigma} f_\sigma \cdot W(\sigma^*).$$

Lemma 6.56 *$W : C^p_{(2)}(\tilde{K}) \to L^2_k(X; \Lambda^p)$ is a bounded and Γ-equivariant map. Moreover W is a cochain map, i.e., $d \circ W = W \circ d_c$.*

Proof Exercise (Problem 6.68). □

In particular W induces the linear map

$$W : \bar{H}^*_{(2)}(\tilde{K}) \to \bar{H}^*_{(2)}(X).$$

Lemma 6.57 *$\int \circ W$ is the identity.*

Proof Let us compute the 1-dimensional case. We may assume that M is connected. Then, for each 1-dimensional simplex l, there are two vertices v_0 and v_1 as its faces. Hence, $W(l^*) = \varphi_0 \cdot d\varphi_1 - \varphi_1 \cdot d\varphi_0$. Note the equality $\varphi_0 + \varphi_1 \equiv 1$ on l. Moreover

$\varphi_1(v_0) = 0$ and $\varphi_1(v_1) = 1$ holds. Then, the equalities hold:

$$\int_l \varphi_0 \cdot d\varphi_1 - \varphi_1 \cdot d\varphi_0 = \int_l (\varphi_0 + \varphi_1) \cdot d\varphi_1 = \int_l d\varphi_1 = 1.$$

Hence, the equality holds:

$$\left(\int \circ W\right)(l^*) = 1.$$

For other 1-simplex $l' \neq l$, at least one of v_0 or v_1 cannot be its face. This implies that $\varphi_0 \cdot d\varphi_1 = \varphi_1 \cdot d\varphi_0 = 0$ both vanish on l'. Hence, $(\int \circ W)(l'^*) = 0$ holds.

We examine the 2-dimensional case. Let $\Delta = \langle v_0, v_1, v_2 \rangle$ be a 2-simplex. Then

$$W(\Delta^*) = 2(\varphi_0 \cdot d\varphi_1 \wedge d\varphi_2 - \varphi_1 \cdot d\varphi_0 \wedge d\varphi_2 + \varphi_2 \cdot d\varphi_0 \wedge d\varphi_1).$$

Note the equality $\varphi_0 + \varphi_1 + \varphi_2 \equiv 1$ on Δ. Then, we have the equalities

$$\begin{aligned} W(\Delta^*) &= 2(\varphi_0 \cdot d\varphi_1 \wedge d(-\varphi_0) - \varphi_1 \cdot d\varphi_0 \wedge d(-\varphi_1) \\ &\quad + (1 - \varphi_0 - \varphi_1) \cdot d\varphi_0 \wedge d\varphi_1) \\ &= 2d\varphi_0 \wedge d\varphi_1 = d(\varphi_0 \cdot d\varphi_1 - \varphi_1 \cdot d\varphi_0) = dW(l_{01}^*). \end{aligned}$$

Note that $\varphi_0 + \varphi_1 \equiv 1$ holds on l_{01}. Moreover $d\varphi_0 \equiv 0$ on l_{12} and $d\varphi_1 \equiv 0$ on l_{02}. Then by Stokes' theorem, we have the equalities

$$\int_\Delta W(\Delta^*) = \int_\Delta dW(l_{01}^*) = \int_{l_{01}} W(l_{01}^*) = 1,$$

where the last equality follows from the case of a 1-simplex.

For the general case, we can proceed by induction. Let $\Delta_n = \langle v_0, v_1, \cdots, v_n \rangle$ be an n-simplex. We can easily compute and verify the equality

$$W(\Delta_n^*) = dW(\Delta_{n-1}^*).$$

This is left as an exercise to the reader (Problem 6.69).

Note that $\sum_{i=0}^{n-1} \varphi_i \equiv 1$ holds on Δ_{n-1}. Moreover $W(\Delta_{n-1}^*) \equiv 0$ vanishes on any face of Δ_n except for Δ_{n-1}. Then, we have the equalities:

$$\int_{\Delta_n} W(\Delta_n^*) = \int_{\Delta_n} dW(\Delta_{n-1}^*) = \int_{\Delta_{n-1}} W(\Delta_{n-1}^*) = \cdots = \int_l W(l^*) = 1.$$

This completes the proof. □

It follows from Lemma 6.57 that the surjectivity of the map $\int : \bar{H}^*_{(2)}(X) \to \bar{H}^*_{(2)}(\tilde{K})$ follows.

Note that an element of $\bar{H}^*_{(2)}(X)$ can be represented by an L^2-harmonic form by Corollary 6.7. Then, by the elliptic regularity, its Sobolev L^2_k-norms are all bounded, and hence its C^l-norms are also all bounded by the Sobolev embedding theorem.

Remark 6.58 One has to aware of non-compactness of the manifold, when the Sobolev embedding theorem is used, because the Sobolev constants depends on the C^0-norm of the metric and its derivatives. In this situation, we are using the lift of a Riemannian metric on the underlying compact manifold. Hence, they are uniformly bounded.

We claim that the composition $W \circ \int$ approximates a differential form by the image of W such that for any $\epsilon > 0$ and any $\omega \in L^2_k(X; \Lambda^p)$, there is a subdivision K' of K such that the corresponding map $W \circ \int$ with respect to K' satisfies the estimate holds:

$$\left\| \omega - \left(W \circ \int \right) (\omega) \right\|_{C^0} < \epsilon \cdot \|\omega\|_{L^2_k} \tag{6.3}$$

for all large $k >> 1$. In fact, it follows from the Sobolev embedding theorem that for any $\epsilon > 0$, there is a subdivision K' of K such that for any simplex $\sigma \in K'$, the uniform estimate $|\omega(p) - \omega(q)| \leq \epsilon \|\omega\|_{C^1(\sigma)} \leq C \cdot \epsilon \|\omega\|_{L^2_4(\sigma)}$ holds for any $p, q \in \sigma$. In particular, if ω is an L^2-harmonic form, then the right-hand side and hence the left-hand side can be arbitrarily small. Then, the estimate (6.3) follows from Lemma 6.57.

Let us verify that the map $\int : \bar{H}^*_{(2)}(X) \to \bar{H}^*_{(2)}(\tilde{K})$ is injective. Suppose $\int \omega = 0 \in \bar{H}^i_{(2)}(\tilde{K})$ holds. Then, $\int \omega = 0 \in \bar{H}^i_{(2)}(\tilde{K}')$ still vanishes for any subdivision K' by Proposition 6.49. Hence for any $\epsilon' > 0$, there is $f \in C^{p-1}_{(2)}(\tilde{K}')$ such that the estimate $\| \int \omega - d_c f \|_{L^2} < \epsilon'$ holds. Then, we have the estimate

$$\begin{aligned}
&\|\omega - dW(f)\|_{C^0} = \|\omega - W(d_c f)\|_{C^0} \\
&\leq \left\| \omega - \left(W \circ \int \right) (\omega) \right\|_{C^0} + \left\| W \circ \left(\int \omega - d_c f \right) \right\|_{C^0} \\
&\leq \left\| \omega - \left(W \circ \int \right) (\omega) \right\|_{C^0} + C\|W\| \cdot \left\| \int \omega - d_c f \right\|_{L^2} < \epsilon + C \cdot \epsilon'.
\end{aligned}$$

Because the right-hand side can be arbitrarily small, it follows that $[\omega] = 0 \in \bar{H}^i_{(2)}(X; \Lambda^p)$ holds.

Problems

6.59 (Spectrum on $\mathbb{R}$) Let $\mathbb{R}$ be the real line equipped with the standard metric.

(a) Verify that no non-zero L^2-harmonic function exists on $\mathbb{R}$.
(b) Verify that the unreduced L^2-cohomology group of degree 0 is non-zero on $\mathbb{R}$.

6.60 Let (X, g) be a complete Riemannian 4-manifold, and a be an element in $L_1^2(X; \Lambda^1)$ on X.

Verify that if the self dual part $d^+a = 0$ is zero, then $da = 0$ also vanishes.

6.61 Verify that P and Δ_m commutes in the proof of Lemma 6.9.

6.62 In the proof of Lemma 6.9, why the set $\{f_i\}_i$ can be assumed to be uniformly bounded?

6.63 Check that $\sim$ in Definition 6.11 is an equivalent relation.

6.64 (Amenable Group) Verify that the property of being amenable is closed under taking subgroups, quotient groups and directed unions.

6.65 In the proof of Proposition 6.24, describe how to decompose $f = \sum_{i=1}^n c_i \mathbf{1}_{F_i}$ for $c_i \geq 0$ and $F_1 \supset \cdots \supset F_n$ for a finitely supported and non-negative function $f \in l^1(\Gamma)$.

6.66 Verify the inclusion (6.2).

6.67 Draw pictures of K^1 and K^2 when K is a 3-dimensional simplex.

6.68 Verify Lemma 6.56.

6.69 Verify $W(\Delta_n^*) = dW(\Delta_{n-1}^*)$ in the proof of Lemma 6.57.

Chapter 7
L^2-Betti Number and von Neumann Trace

In this chapter, we present some basic subjects on functional analysis and operator algebras that are necessary to introduce the L^2-Betti numbers of the reduced L^2-cohomology groups of the universal covering space of a compact manifold. We give the following two basic subjects on the L^2-Betti numbers:

- We present a proof of Lück's theorem that states convergence of the normalized Betti numbers to the L^2-Betti numbers by finite coverings for a compact manifold with the residually finite fundamental group.
- We present a proof of Atiyah's Γ-index theorem that gives the coincidence of the L^2-Betti number of the space of the L^2-index of the lift on the universal covering space of an elliptic operator over a compact manifold with the Atiyah–Singer index.

7.1 L^2-Betti Numbers Over Universal Covering Spaces

7.1.1 Trace Class Operators

Let H be a separable Hilbert space and $\{e_i\}_i$ be an orthonormal basis. Let $A : H \to H$ be a bounded operator, and set $c_{ij}(A) := \langle Ae_i, e_j \rangle$. Then,

$$\|A\|_{HS} := \sqrt{\sum_{i,j} |c_{ij}(A)|^2} = \sqrt{\sum_i \|Ae_i\|^2}$$

is called the *Hilbert–Schmidt norm*. The value is independent of the choice of orthonormal basis (see Problem 7.52). A bounded operator is of *Hilbert–Schmidt class*, if the Hilbert–Schmidt norm is finite.

T. Kato, N. Nakamura, *Monopole Maps and L^2-cohomology*, Springer Monographs in Mathematics, https://doi.org/10.1007/978-981-96-8472-4_7

The Hilbert–Schmidt inner product is given by

$$\langle A, B\rangle_{HS} := \sum_{i,j} \overline{c_{ij}(A)} c_{ij}(B).$$

Definition 7.1 A bounded operator $T = AB$ is called a trace class operator, if it is given by a composition of Hilbert–Schmidt operators A and B.

We set the trace of T by

$$\begin{aligned}\operatorname{tr}(T) : &= \langle A^*, B\rangle_{HS} = \sum_{i,j} \overline{c_{ij}(A^*)} \cdot c_{ij}(B) \\ &= \sum_{i,j} c_{ji}(A) \cdot c_{ij}(B) = \sum_{i} c_{ii}(T).\end{aligned}$$

Remark 7.2

1. Note that $\operatorname{tr}(T)$ is given by the sum of eigenvalues, if it admits a discrete spectral decomposition by the eigenvectors.
2. The composition of a Hilbert–Schmidt operator and a bounded operator is Hilbert–Schmidt.
3. Both the compositions TT' and $T'T$ are of trace class, if either T is of trace class and T' is bounded, or both T and T' are of Hilbert–Schmidt class (Problem 7.53). Moreover, the trace is commutative:

$$\operatorname{tr}(TT') = \sum_{i,j} c_{ji}(T) \cdot c_{ij}(T') = \sum_{i,j} c_{ij}(T') \cdot c_{ji}(T) = \operatorname{tr}(T'T).$$

Lemma 7.3 *Suppose that a bounded operator $A : H \to H$ is given by a kernel function*

$$Au(x) := \int_M k(x, y)u(y) \operatorname{vol}_y,$$

where $H = L^2(M)$ and k is continuous. Then, the equality

$$\|A\|_{HS}^2 = \int_{M\times M} |k(x, y)|^2 \operatorname{vol}_x \operatorname{vol}_y$$

holds.

Proof

$$\begin{aligned}
\|A\|_{HS}^2 = \sum_j \|Ae_j\|^2 &= \sum_j \int_M \left| \int_M k(x,y)e_j(y)\,\mathrm{vol}_y \right|^2 \mathrm{vol}_x \\
&= \int_M \sum_j \left| \int_M k(x,y)e_j(y)\,\mathrm{vol}_y \right|^2 \mathrm{vol}_x \\
&= \int_M \int_M |k(x,y)|^2\,\mathrm{vol}_y\,\mathrm{vol}_x = \int_{M\times M} |k(x,y)|^2\,\mathrm{vol}_x\,\mathrm{vol}_y,
\end{aligned}$$

where Parseval's equality $\sum_j \left|\int_M k(x,y)e_j(y)\,\mathrm{vol}_y\right|^2 = \int_M |k(x,y)|^2\,\mathrm{vol}_y$ is used. This completes the proof. □

Corollary 7.4

- *If M is compact and k is continuous, then A is Hilbert–Schmidt.*
- *More generally, let φ and ψ be two compactly supported continuous functions on M. Then, the compositions $\psi \cdot A \cdot \varphi$ is Hilbert–Schmidt, if k is continuous.*
- *Suppose both A and B are given by their kernel functions k_A and k_B. Then, the formula*

$$\langle A, B\rangle_{HS} = \int_{M\times M} \overline{k_A(x,y)}k_B(x,y)\,\mathrm{vol}_x\,\mathrm{vol}_y$$

holds.

Theorem 7.5 *Suppose $T = AB$, and both A and B are given by their kernel functions k_A and k_B. Then, the formula*

$$\mathrm{tr}(T) = \int_M k_T(y,y)\,\mathrm{vol}_y$$

holds, where k_T is the kernel function of T.

Proof Note the equality

$$k_T(x,y) = \int_M k_A(x,z)k_B(z,y)\,\mathrm{vol}_z\,.$$

Then, we have

$$\begin{aligned}\operatorname{tr}(T) : &= \langle A^*, B\rangle_{HS} = \int_{M\times M} \overline{k_{A^*}(x,y)} k_B(x,y)\,\mathrm{vol}_x\,\mathrm{vol}_y \\ &= \int_{M\times M} k_A(y,x) k_B(x,y)\,\mathrm{vol}_x\,\mathrm{vol}_y \\ &= \int_M k_T(y,y)\,\mathrm{vol}_y\,.\end{aligned}$$

This completes the proof. □

Corollary 7.6 • *If M is compact and both k_A, k_B are continuous, then T is of trace class.*

- *More generally, let φ and ψ be two compactly supported continuous functions on M. Then, the compositions $\psi \cdot T \cdot \varphi$ is of trace class, if both k_A, k_B are continuous.*
- *Let D be a differential operator, and suppose that both k_A, k_B are smooth. Then, $D \circ T$ is of trace class, if M is compact.*

7.1.2 von Neumann Trace

Let Γ be a discrete group, and $\mathbb{C}\Gamma := \{\sum_{\text{finite}} a_\gamma \cdot \gamma\}$ be the group ring of Γ. Each element $\sum_\gamma a_\gamma \cdot \gamma$ is identified with a finitely supported function $f : \Gamma \to \mathbb{C}$ such that $f(\gamma) = a_\gamma$. Let δ_γ be the characteristic function. Then $f = \sum_\gamma a_\gamma \delta_\gamma$.

The convolution $*$ on $\mathbb{C}\Gamma$ is defined as

$$* : \mathbb{C}\Gamma \times \mathbb{C}\Gamma \to \mathbb{C}\Gamma$$

by $f * g(\gamma) = \sum_h f(h) \cdot g(h^{-1} \cdot \gamma)$.

Then, we have the equality

$$\delta_\gamma * \delta_h = \delta_{\gamma \cdot h}.$$

Let $l^2(\Gamma) := \{\sum_{g\in\Gamma} a_g \cdot \delta_g : a_g \in \mathbb{C},\ \sum_{g\in\Gamma} |a_g|^2 < \infty\}$ be the Hilbert space equipped with the natural inner product. Hence, we have the embedding $\mathbb{C}\Gamma \hookrightarrow l^2(\Gamma)$.

There is an isometric action of Γ on $l^2(\Gamma)$ by

$$\gamma^*(f)(h) := f(\gamma^{-1} \cdot h).$$

Let

$$\mathcal{N}(\Gamma) := \mathcal{B}(l^2(\Gamma), l^2(\Gamma))^{\Gamma}$$

be the algebra of the set of Γ-equivariant bounded operators.

Remark 7.7 The algebra $\mathcal{N}(\Gamma)$ is called the group von Neumann algebra. It has been deeply studied in the field of von Neumann algebras. See [117] and the references therein. Extensive topics in the field are out of scope of this book, and we shall restrict ourselves within the smallest subjects that are necessary for our theme.

Lemma 7.8 *An element $f \in \mathcal{N}(\Gamma)$ can be represented by the form*

$$f(v) = v * \left(\sum_{\gamma} f_{\gamma} \cdot \delta_{\gamma}\right),$$

where $f_{\gamma} \in \mathbb{C}$ and $v \in l^2(\Gamma)$.

Proof Let us express $f(\delta_{\gamma}) := \sum_h f_{h,\gamma} \cdot \delta_h$. Then, we have

$$f(\gamma^*(\delta_e)) = f(\delta_{\gamma}) = \sum_h f_{h,\gamma} \cdot \delta_h.$$

On the other hand, by Γ-equivariance, we have the equalities

$$\begin{aligned} f(\gamma^*(\delta_e)) &= \gamma^*(f(\delta_e)) = \gamma^*(\sum_h f_{h,e} \cdot \delta_h) \\ &= \sum_h f_{h,e} \cdot \delta_{\gamma \cdot h}. \end{aligned}$$

Hence, we have the equalities $f_{h,\gamma} = f_{\gamma^{-1} \cdot h, e}$ for any $\gamma, h \in \Gamma$. Then we have

$$\begin{aligned} f(\delta_{\gamma}) &= \sum_h f_{h,\gamma} \cdot \delta_h = \sum_h f_{\gamma^{-1} \cdot h, e} \cdot \delta_h \\ &= \sum_h f_{\gamma^{-1} \cdot h, e} \cdot \delta_{\gamma \cdot \gamma^{-1} \cdot h} = \sum_h f_{\gamma^{-1} \cdot h, e} \cdot \delta_{\gamma} * \delta_{\gamma^{-1} \cdot h} \\ &= \delta_{\gamma} * \left(\sum_h f_{h,e} \cdot \delta_h\right). \end{aligned}$$

□

Definition 7.9 The von Neumann Γ-trace $\mathrm{tr}_\Gamma : \mathcal{N}(\Gamma) \to \mathbb{C}$ is given by

$$\mathrm{tr}_\Gamma(f) := \langle f(\delta_e), \delta_e \rangle,$$

where $\delta_e \in l^2(\Gamma)$ is the characteristic function at the unit $e \in \Gamma$.

The trace satisfies three important properties (Problem 7.54):

Lemma 7.10

1. *Commutativity: $\mathrm{tr}_\Gamma(a \circ b) = \mathrm{tr}_\Gamma(b \circ a)$ for $a, b \in \mathcal{N}(\Gamma)$.*
2. *Faithfulness: $\mathrm{tr}_\Gamma(a^* \circ a) = 0$ holds, if and only if $a = 0$.*
3. *Positivity: $\mathrm{tr}_\Gamma(a^* \circ a) \geq 0$ holds.*

Example 1 Consider the circle $S^1 = \mathbb{R}/2\pi\mathbb{Z}$. Through the Fourier transform, we obtain the isometry

$$\widehat{\ } : L^2(S^1) \cong l^2(\mathbb{Z})$$

by $\widehat{e^{ik\cdot x}} = \delta_k$. Through the Fourier transform, we have the isomorphisms:

$$L^\infty(S^1) \cong \mathcal{N}(\mathbb{Z}) \cong l^\infty(\mathbb{Z}).$$

The von Neumann trace $\mathrm{tr}_\mathbb{Z} : L^\infty(S^1) \to \mathbb{C}$ is given by

$$\mathrm{tr}_\mathbb{Z}(f) := \frac{1}{2\pi} \cdot \int_{S^1} f.$$

Definition 7.11 A Hilbert $\mathcal{N}(\Gamma)$-module V is a Hilbert space V equipped with a linear isometric action of Γ, such that there is a Γ-equivariant isometric embedding $i : V \hookrightarrow H \otimes l^2(\Gamma)$, where H is a separable Hilbert space.

Example 2 Let $\mathcal{H}^m(X)$ be the space of L^2-harmonic m-forms. The closed embedding

$$\mathcal{H}^m(X) \otimes \mathbb{C} \subset L^2(X; \Lambda^m) \otimes \mathbb{C} \cong L^2(K; \Lambda^m) \otimes l^2(\Gamma)$$

gives the structure of a Hilbert $\mathcal{N}(\Gamma)$-module on $\mathcal{H}^m(X) \otimes \mathbb{C}$, where $K \subset X$ is a fundamental domain of Γ-action on X.

Hence, the reduced L^2-cohomology group $\bar{H}^m_{(2)}(X) \otimes \mathbb{C}$ admits the structure of a Hilbert $\mathcal{N}(\Gamma)$-module through the Γ-equivariant isomorphism in Corollaries 6.7 and 6.8.

Let $i : V \hookrightarrow H \otimes l^2(\Gamma)$ be a Γ-equivariant isometric embedding, where H is a separable Hilbert space. Then, the image $i(V) \subset H \otimes l^2(\Gamma)$ is a closed subspace, and hence there is an orthogonal decomposition

$$H \otimes l^2(\Gamma) \cong V \oplus V^{\perp}.$$

It is immediate to check that the closed linear subspace $V^{\perp}$ is Γ-invariant, because Γ acts on $H \otimes l^2(\Gamma)$ isometrically.

Let us take $B \in B(H) \otimes \mathcal{N}(\Gamma)$, where $B(H)$ is the space of bounded operators on H. Then, we can express $B = (B^{gh})_{g,h\in\Gamma}$ by

$$B(v \otimes \delta_g) := \sum_{h\in\Gamma} B^{gh}(v) \otimes \delta_h \ \in \ H \otimes l^2(\Gamma),$$

where $B^{gh} \in B(H)$.

Definition 7.12

- The von Neumann Γ-trace of B is given by

$$\mathrm{tr}_\Gamma(B) := \mathrm{tr}(B^{ee}).$$

- The von Neumann Γ-dimension $\dim_\Gamma(V)$ of V is given by

$$\dim_\Gamma(V) := \mathrm{tr}_\Gamma(P),$$

  where P is the orthogonal projection to V with respect to a Γ-equivariant isometric embedding $V \hookrightarrow H \otimes l^2(\Gamma)$.

7.1.3 *Smoothing Operators*

Let $\Gamma := \pi_1(M)$ and $X := \tilde{M}$. For an operator A on $L^2(X)$, denote the kernel function by $k_A(x, y)$. Let $C^m_b(X)$ be the Banach space of functions with uniformly bounded C^m-norm.

Lemma 7.13 *Let A be a Γ-equivariant boundedly self-adjoint operator on $L^2(X)$ with the kernel function. Suppose that A maps $L^2(X)$ boundedly to $C_b^m(X)$ for each $m \geq 0$. Then, A^2 satisfies*

$$\int_X |k_{A^2}(x, y)|^2 \operatorname{vol}_x, \int_X |k_{A^2}(x, y)|^2 \operatorname{vol}_y \leq C$$

for some constant C that is independent of the choice of $x, y \in X$.

The following proof is due to [147, Lemma 15.5].

Proof For simplicity of the notation, we verify only the scalar case $m = 0$.

Since the linear functional $s \mapsto As(p)$ for each $p \in X$ is continuous, there is $v_p \in L^2(X) \otimes \mathbb{C}$ such that $\langle s, v_p\rangle_{L^2} = As(p)$ holds for any $s \in L^2(X)$. By definition, $\sup_{p\in X} \|v_p\|_{L^2} < \infty$ holds. Moreover, the assignment $X \to L^2(X)$ by $p \to v_p$ is uniformly continuous map. In fact, the estimates

$$|\langle s, v_p - v_q\rangle_{L^2}| = |As(p) - As(q)| \leq \|As\|_{L^2} \cdot d(p, q) \leq \|s\|_{L^2} \cdot d(p, q)$$

hold.

Note $As(p) = \int_X s(q) \cdot v_p(q) \operatorname{vol}_q$. Because A is self-adjoint, we have the equality $\langle As_1, s_2\rangle_{L^2} = \langle s_1, As_2\rangle_{L^2}$. Hence,

$$\int_X s_1(q) \cdot \bar{s}_2(p) \cdot \bar{v}_p(q) \operatorname{vol}_p \operatorname{vol}_q = \int_X s_1(q) \cdot \bar{s}_2(p) \cdot v_q(p) \operatorname{vol}_p \operatorname{vol}_q .$$

In particular, we have the equality $v_p(q) = \bar{v}_q(p)$. Then, the equality $As(p) = \int_X s(q) \cdot v_q(p) \operatorname{vol}_q$ holds, and hence,

$$A^2 s(p) = \int_X k(p, q) \cdot s(q) \operatorname{vol}_q$$

holds, where $k(p, q) = Av_q(p)$. The function k is uniformly smooth in p-variable, because $q \to Av_q$ is uniformly continuous. By self-adjointness, it is also the case for the q-variable. Hence, k is uniformly smooth. □

Let $D := d \oplus d^*$ be the de Rham operator on X.

Corollary 7.14 *Let f be a nonnegative Schwartz class function. Then the kernel function $k(x, y)$ of $A := f(D)$ satisfies*

$$\int_X |k_A(x, y)|^2 \operatorname{vol}_x, \int_X |k_A(x, y)|^2 \operatorname{vol}_y \leq C$$

for some constant C that is independent of the choice of $x, y \in \tilde{M}$.

Proof By Proposition 4.15, A maps $L^2(X)$ boundedly to $C_b^m(X)$ for each $m \geq 0$. Let us express $f(D) = f^{1/2}(D) \circ f^{1/2}(D)$. Then we can apply Lemma 7.13 for $A = f^{1/2}(D)$. □

Lemma 7.15 *Suppose that the kernel functions k_1, k_2 of two bounded and Γ-equivariant operators A_1, A_2 satisfy*

$$\int_X |k_A(x, y)|^2 \, \mathrm{vol}_x, \int_X |k_A(x, y)|^2 \, \mathrm{vol}_y \leq C.$$

Then the Γ-trace commutes as

$$\mathrm{tr}_\Gamma(A_1 A_2) = \mathrm{tr}_\Gamma(A_2 A_1) \in \in \mathbb{R}.$$

Proof Let $K \subset X$ be a fundamental domain. By the assumption and Γ-equivariance, we have the equalities

$$\begin{aligned}
\mathrm{tr}_\Gamma(A_1 A_2 - A_2 A_1) &= \int\int_{K \times X} (\mathrm{tr}(k_1(x, y) k_2(y, x)) \\
&\quad - \mathrm{tr}(k_2(x, y) k_1(y, x))) \, \mathrm{vol}_x \, \mathrm{vol}_y \\
&= \sum_{\gamma \in \Gamma} \int\int_{K \times K} (\mathrm{tr}(k_1(x, y \cdot \gamma) k_2(y \cdot \gamma, x)) \\
&\quad - \mathrm{tr}(k_2(x, y \cdot \gamma) k_1(y \cdot \gamma, x))) \, \mathrm{vol}_x \, \mathrm{vol}_y \\
&= \sum_{\gamma \in \Gamma} \int\int_{K \times K} \Big(\mathrm{tr}(k_1(x \cdot \gamma^{-1}, y) k_2(y, x \cdot \gamma^{-1})) \\
&\quad - \mathrm{tr}(k_2(x \cdot \gamma^{-1}, y) k_1(y, x \cdot \gamma^{-1}))\Big) \, \mathrm{vol}_x \, \mathrm{vol}_y \\
&= \sum_{\gamma \in \Gamma} \int\int_{K \times K} \Big(\mathrm{tr}(k_2(y, x \cdot \gamma^{-1})) k_1(x \cdot \gamma^{-1}, y) \\
&\quad - \mathrm{tr}(k_1(y, x \cdot \gamma^{-1}) k_2(x \cdot \gamma^{-1}, y))\Big) \, \mathrm{vol}_x \, \mathrm{vol}_y \\
&= \mathrm{tr}_\Gamma(A_2 A_1 - A_1 A_2).
\end{aligned}$$

Hence $\mathrm{tr}_\Gamma(A_2 A_1 - A_1 A_2) = 0$ vanishes. This completes the proof. □

Definition 7.16 Let us say that a bounded operator $B : L^2 \to L^2$ is bounded on each L_k^2, if it extends to a bounded operator on L_k^2 for each $k \geq 0$.

Corollary 7.17 *Let A_1, A_2 be in Lemma 7.13. Then the Γ-traces*

$$\mathrm{tr}_\Gamma\left((B_1 A_1 B_1')(B_2 A_2 B_2')\right) = \mathrm{tr}_\Gamma\left((B_2 A_2 B_2')(B_1 A_1 B_1')\right)$$

mutually commute, where B_i' *are any bounded operators on* L^2*, and* B_i *are* D *or any bounded operators on each* L_k^2*, or their compositions.*

Proof Because both the operators $B_1 A_1 B_1'$ and $B_2 A_2 B_2'$ map $L^2(X)$ boundedly to $C_b^m(X)$ for each $m \geq 0$, their kernel functions have uniformly L^2-finite norms by Lemma 7.13. Then, the conclusion follows by Lemma 7.15. □

Consider the case $L^2(X; \tilde{E}) \cong L^2(M; E) \otimes l^2(\Gamma)$. Let V be a Hilbert space and take a Γ-equivariant embedding

$$i : V \hookrightarrow L^2(X; \tilde{E})$$

such that the corresponding orthogonal projection $P : L^2(X; \tilde{E}) \to i(V)$ is bounded on each L_k^2.

Lemma 7.18 *The value* $\mathrm{tr}_\Gamma(P)$ *is independent of the choice of the embedding* $i : V \hookrightarrow H \otimes l^2(\Gamma)$ *that satisfies the above conditions.*

Proof Let $i : V \hookrightarrow L^2(X; \tilde{E})$ and $j : V \hookrightarrow L^2(X; \tilde{F})$ be two Γ-equivariant isometric embeddings, and denote by P and Q as the orthogonal projections respectively. Let $U : L^2(X; \tilde{E}) \to L^2(X; \tilde{F})$ be the partial isometry, i.e., $U|_{\mathrm{im}(V)} = j \circ i^{-1}$ and $U|_{\mathrm{im}(V)^\perp} = 0$. Then the equalities

$$\mathrm{tr}_\Gamma(U \circ P \circ U^*) = \mathrm{tr}_\Gamma(U \circ P \circ P \circ U^*) = \mathrm{tr}_\Gamma(P \circ U^* \circ U \circ P) = \mathrm{tr}_\Gamma(P)$$

by Corollary 7.17. □

7.1.4 L^2-Betti Numbers of Simplicial L^2-Cohomology Group

Let M be a compact oriented smooth manifold with $\Gamma := \pi_1(M)$ and its universal covering space $X := \tilde{M}$. Consider a Γ-equivariant linear embedding

$$i : V \hookrightarrow L^2(X) \cong L^2(K) \otimes l^2(\Gamma),$$

where V is a closed linear subspace of $L^2(X)$, and $K \subset X$ is a fundamental domain.

Let $P : L^2(X) \to V$ be the orthogonal projection. Note that P is given by the kernel function $k(p, q)$ by

$$k(p, q) := \sum_i \varphi_i(p) \otimes \varphi_i(q)^*,$$

where $\{\varphi_i\}_i$ is the orthonormal basis of V and φ_i^* is the L^2-dual element. Then, the formula

$$Ps(p) = \int_X k(p, q) \cdot s(q) \, \mathrm{vol}_q \, .$$

holds. Note that $k(p, q)$ is Γ-invariant.

Lemma 7.19 *The equality*

$$\mathrm{tr}_\Gamma(P) = \int_K k(q, q) \, \mathrm{vol}_q$$

holds.

Proof Note that the kernel function of $P^{ee} : L^2(K) \to L^2(K)$ is given by $\chi_K(p) \cdot k(p, q) \cdot \chi_K(q)$, where χ_K is the characteristic function on K. Hence, $\mathrm{tr}(P^{ee}) = \int_K k(q, q) \, \mathrm{vol}_q$ by Theorem 7.5. □

Suppose M is equipped with a structure of a finite simplicial complex K, and denote its lift by $\tilde{K}$ on X. We consider the Hilbert Γ-module $\bar{H}^m_{(2)}(X) \otimes \mathbb{C}$.

In this subsection, we verify:

Proposition 7.20 *The number* $\dim_\Gamma \left(\bar{H}^m_{(2)}(X) \otimes \mathbb{C} \right) < \infty$ *is finite.*

Proof Let $\mathcal{H}^m(X)$ be the space of L^2-harmonic m-forms. Consider the closed embedding

$$\mathcal{H}^m(X) \otimes \mathbb{C} \subset L^2(X; \Lambda^m) \otimes \mathbb{C},$$

and $P \in B(L^2(X; \Lambda^m) \otimes \mathbb{C})$ be the projection to the image.

Note that the kernel function of $P = P^2$ is given by

$$k(x, y) := \sum_i \varphi_i(x) \otimes \varphi(y)^*$$

where $\{\varphi_i\}_i$ is the orthogonal basis of the image of P. Moreover the image of P satisfies the condition in Lemma 7.13, by Proposition 4.15. Hence, the conclusion follows by Lemma 7.13. □

Definition 7.21 The number

$$b^m_{(2)}(X) := \dim_\Gamma \left(\bar{H}^m_{(2)}(X) \otimes \mathbb{C} \right)$$

is called the L^2-Betti number of $X = \tilde{M}$.

7.1.5 Kazhdan's Equality on L^2-Cohomology Groups

Let (M, g) be a compact Riemannian manifold, and let $X := \tilde{M}$ be the universal covering space. The Riemannian metric g lifts on X and we denote the lift by $\tilde{g}$.

Suppose that the fundamental group $\Gamma := \pi_1(M)$ is residually finite such that there is a tower

$$\Gamma \supset \Gamma_1 \supset \Gamma_2 \supset \cdots$$

by normal subgroups of finite indices with $\cap_i \Gamma_i = 1$ (Definition 6.26). We denote their indices by $[\Gamma : \Gamma_m] \in \mathbb{N}$. Let X_m be the tower of normal coverings of finite indices corresponding to Γ_m.

Our main topic in this section is the following. The following equality is called the *Kazhdan's equality* and verified by Lück.

Theorem 7.22 ([116]) *The convergence*

$$\lim_{i\to\infty} \frac{b^m(X_i)}{[\Gamma : \Gamma_i]} = b^m_{(2)}(X) \tag{7.1}$$

holds, where $b^m(X_i)$ is the usual m-th Betti number of X_i.

Throughout this subsection, we assume that the cochains are $\mathbb{C}$-valued. Let K be a simplicial decomposition. Then, it induces the simplicial decompositions $K(i)$ on X_i. We denote the cochain complexes by

$$d^i_m : C^m(K(i)) \to C^{m+1}(K(i)).$$

Note the isomorphisms $C^m(\tilde{K}) \times_\Gamma \mathbb{C}[\pi_i] \cong C^m(K(i))$ for $\pi_i := \Gamma_i \backslash \Gamma$, where we regard $C^m(\tilde{K}) \cong \oplus^r \mathbb{C}\Gamma$ as a finite free $\mathbb{C}\Gamma$-module and r is the number of m-simplices in K.

7.1.6 Convergence of L^2-Betti Numbers

Let $\{\sigma^1_m, \cdots, \sigma^{s_m}_m\}$ be the set of m-simplices of K. Then each simplex lifts (non-canonically) as $\{\tilde{\sigma}^{i,g}_m\}_{g\in\Gamma}$ on $\tilde{K}$ that admits a free action of Γ. See Remark 6.48.

The L^2-cochain can be expressed by

$$C^m_{(2)}(\tilde{K}) := \{f := \sum_{i=1}^{s_m} \sum_{g \in \Gamma} a_{i,g} \cdot \left(\tilde{\sigma}_m^{i,g}\right)^* : \sum_{i,g} |a_{i,g}|^2 < \infty\},$$

where we regard an element as the bounded linear map $f : C_m^{(2)}(\tilde{K}) \to \mathbb{R}$ on the l^2-chain by $f(\tilde{\sigma}_m^{i,g}) = a_{i,g}$. The group Γ acts on $C^m_{(2)}(\tilde{K})$ by $h \cdot f(\tilde{\sigma}_m^{i,g}) := f(\tilde{\sigma}_m^{i,g} \cdot h) = f(\tilde{\sigma}_m^{i,g\cdot h}) = a_{i,g\cdot h}$.

There is the differential

$$d_m : C^m_{(2)}(\tilde{K}) \to C^{m+1}_{(2)}(\tilde{K})$$

as a bounded linear functional such that $d^2 = 0$ holds. Recall that the m-th simplicial L^2-cohomology group is defined by

$$\bar{H}^m_{(2)}(\tilde{K}) := \mathrm{Ker}\{d_m : C^m_{(2)}(\tilde{K}) \to C^{m+1}_{(2)}(\tilde{K})\} \,/\, \overline{d_{m-1}(C^{m-1}_{(2)}(\tilde{K}))}.$$

Consider a $\mathbb{C}\Gamma$-linear map

$$f : \oplus^r \mathbb{C}\Gamma \to \oplus^r \mathbb{C}\Gamma$$

between the left $\mathbb{C}\Gamma$-modules. It extends to a bounded Γ-equivariant operator

$$f^{(2)} : \oplus^r l^2(\Gamma) \to \oplus^r l^2(\Gamma).$$

Let $\{P(\lambda) \mid \lambda \in [0, \infty)\}$ be the spectral decomposition of $(f^{(2)})^* f^{(2)}$.

Let $\Gamma' \subset \Gamma$ be a normal subgroup. Then, f induces the linear map

$$f' : \oplus^r \mathbb{C}[\Gamma' \backslash \Gamma] \to \oplus^r \mathbb{C}[\Gamma' \backslash \Gamma].$$

Let Γ be a residually finite group such that there is a tower

$$\Gamma \supset \Gamma_1 \supset \Gamma_2 \supset \cdots$$

by normal subgroups of finite indices with $\cap_i \ \Gamma_i = 1$. Denote the induced family of linear maps by

$$f_i : \oplus^r \mathbb{C}[\Gamma_i \backslash \Gamma] \to \oplus^r \mathbb{C}[\Gamma_i \backslash \Gamma].$$

Let $E_i(\lambda)$ be the ordered set of eigenvalues μ of $(f_i)^* f_i$ satisfying $\mu \leq \lambda$ listed with multiplicity. Then, we set

$$F_i(\lambda) := \frac{|E_i(\lambda)|}{[\Gamma : \Gamma_i]},$$

where $|E_i(\lambda)|$ is the cardinality of $E_i(\lambda)$.

We set $F : [0, \infty) \to [0, a]$ as the spectral density function of $(f^{(2)})^* f^{(2)}$ given by

$$\lambda \to \dim_\Gamma (\mathrm{im}(P(\lambda)) = \mathrm{tr}_\Gamma(P(\lambda)).$$

Then, we set

$$\overline{F_\infty}^+(\lambda) := \lim_{\delta\to 0+} \overline{F_\infty}(\lambda+\delta), \quad \overline{F_\infty}(\lambda) := \limsup_{i\to\infty} F_i(\lambda),$$

$$\underline{F_\infty}^+(\lambda) := \lim_{\delta\to 0+} \underline{F_\infty}(\lambda+\delta), \quad \underline{F_\infty}(\lambda) := \liminf_{i\to\infty} F_i(\lambda).$$

Proposition 7.23 ([116]) *The following equalities hold:*

$$F(0) = \overline{F_\infty}^+(0) = \underline{F_\infty}^+(0) = \lim_{i\to\infty} F_i(0).$$

Let us verify Theorem 7.22 assuming Proposition 7.23.

Proof Let $X = \tilde{M}$ be the universal covering space. Let K be a simplicial decomposition of M, and N_m be the number of m-simplices in M.

For any finite-dimensional $\mathbb{C}$-representation V of Γ, let $C^m(K; V) := C^m(\tilde{K}) \times_\Gamma V$, where we regard $C^m(\tilde{K})$ as a finite free $\mathbb{C}\Gamma$-module. The differential d_* induces another differential d_*^V on $C^*_{(2)}(K; V)$. Then, we have the equalities

$$\begin{aligned} b^m(M; V) &= \dim_{\mathbb{C}} \left(\mathrm{Ker}(d_m^V)/\,\mathrm{im}(d_{m-1}^V)\right) \\ &= \dim_{\mathbb{C}} \left(\mathrm{Ker}(d_m^V)\right) - \dim_{\mathbb{C}} \left(\mathrm{im}(d_{m-1}^V)\right) \\ &= \dim_{\mathbb{C}} \left(\mathrm{Ker}(d_m^V)\right) + \dim_{\mathbb{C}} \left(\mathrm{Ker}(d_{m-1}^V)\right) - N_{m-1} \cdot \dim_{\mathbb{C}}(V) \\ &= \dim_{\mathbb{C}} \left(\mathrm{Ker}(d_m^V \oplus d_{m-1}^V)\right) - N_{m-1} \cdot \dim_{\mathbb{C}}(V). \end{aligned}$$

Then, we have the equalities

$$\frac{b^m(X_i)}{[\Gamma : \Gamma_i]} = \frac{b^m(M; \mathbb{C}[\Gamma_i \backslash \Gamma])}{[\Gamma : \Gamma_i]} = F_i(0) - N_{m-1},$$

where we applied $f_i := d_m^i \oplus d_{m-1}^i$.

We can consider the case on the universal covering space as follows. Note the following properties (Problem 7.55):

1. The isomorphism $C^m_{(2)}(\tilde{K}) \cong \oplus^{N_m} l^2(\Gamma)$ holds.
2. Let H and H' be Hilbert spaces equipped with isometric Γ-actions. Consider a Γ-equivariant bounded linear operator $d : H \to H'$. Then, the Γ-equivariant isomorphism

$$H \cong \mathrm{Ker}\, d \oplus \overline{d(H)}$$

holds. In fact, let $d = u \circ |d|$ be the Γ-equivariant polar decomposition of d. Then, the partial unitary operator u gives the isomorphism $u : H \,/\, \mathrm{Ker}\, d \cong \overline{d(H)}$. See [135].

We apply the second property above to d_{m-1}, and obtain the Γ-equivariant isomorphism

$$C^{m-1}_{(2)}(\tilde{K}) \cong \mathrm{Ker}\, d_{m-1} \oplus \overline{\mathrm{im}(d_{m-1})}.$$

Then, we have the isomorphism

$$\begin{aligned}\mathrm{Ker}(d_m \oplus d_{m-1}) &\cong \mathrm{Ker}(d_m)/\overline{\mathrm{im}(d_{m-1})} \ \oplus C^{m-1}_{(2)}(\tilde{K}) \\ &\cong \mathrm{Ker}(d_m)/\overline{\mathrm{im}(d_{m-1})} \ \oplus^{N_{m-1}} l^2(\Gamma).\end{aligned}$$

We take the Γ-dimensions of both sides. Then, we have the equality

$$b^m_{(2)}(X) = \dim_\Gamma\left(\mathrm{Ker}(d_m \oplus d_{m-1})\right) - N_{m-1}.$$

Let us apply Proposition 7.23 to $f^{(2)} := d_m \oplus d_{m-1}$. Then, we have

$$\begin{aligned}b^m_{(2)}(X) = F(0) - N_{m-1} &= \lim_{i\to\infty} F_i(0) - N_{m-1} \\ &= \lim_{i\to\infty} \frac{b^m(X_i)}{[\Gamma : \Gamma_i]}.\end{aligned}$$

This finishes the proof of Theorem 7.22. □

7.1.7 *Proof of Proposition 7.23*

For $\sum_g \lambda_g \cdot g \in \mathbb{C}\Gamma$, define $\mathrm{tr}_{\mathbb{C}\Gamma}\left(\sum_g \lambda_g \cdot g\right)$ by $\mathrm{tr}_{\mathbb{C}\Gamma}\left(\sum_g \lambda_g \cdot g\right) := \lambda_e$.

Lemma 7.24 $\mathrm{tr}_{\mathbb{C}\Gamma}(f^* f) = \mathrm{tr}_\Gamma\left((f^{(2)})^* f^{(2)}\right)$.

Proof This is easy to check. □

First, we present a Lück's finiteness lemma.

Lemma 7.25 *Let $p(x)$ be a polynomial. Then, there is some $i_0 = i_0(\{\Gamma_i\}_i, p)$ such that for any $i \geq i_0$, the equality holds:*

$$\mathrm{tr}_\Gamma \left(p((f^{(2)})^* f^{(2)}) \right) = \frac{1}{[\Gamma : \Gamma_i]} \cdot \mathrm{tr} \left(p(f_i^* f_i) \right).$$

Proof Let us represent f by an $r \times r$ matrix $B \in \mathrm{M}_r(\mathbb{C}\Gamma)$ with $\mathbb{C}\Gamma$-coefficient. Then, there are $g_0 = e, g_1, \cdots, g_a \in \Gamma$ and $\lambda_0, \cdots, \lambda_a \in \mathbb{C}$ such that $\lambda_i \neq 0$, $g_i \neq e$ for $i \geq 1$ and

$$\sum_{j=1}^{r} \left(p(BB^*) \right)_{j,j} = \sum_{i=0}^{a} \lambda_i \cdot g_i.$$

Hence, we have the equalities $\mathrm{tr}_\Gamma \left(p((f^{(2)})^* f^{(2)}) \right) = \lambda_0$, and

$$\mathrm{tr} \left(p(f_i^* f_i) \right) = \mathrm{tr} \left(\sum_{m=0}^{a} \lambda_m \cdot r(g_m) : \mathbb{C}[\Gamma_i \backslash \Gamma] \to \mathbb{C}[\Gamma_i \backslash \Gamma] \right),$$

where $r(g)$ is the right multiplication by the element $g \in \Gamma$.

Because the intersection of all the Γ_i is trivial, there is i_0 such that for all $i \geq i_0$, none of the elements g_m for $1 \leq m \leq a$ lies in Γ_i. Then,

$$0 = \mathrm{tr} \left(r(g_m) : \mathbb{C}[\Gamma_i \backslash \Gamma] \to \mathbb{C}[\Gamma_i \backslash \Gamma] \right)$$

vanishes, because $\Gamma_i \subset \Gamma$ is normal, and $r(g_i)$ acts by a permutation without fixed points.

Hence, we have the equality

$$\lambda_0 = \mathrm{tr}_{\mathbb{C}\Gamma} \left(p(f^* f) \right) = \mathrm{tr}_\Gamma \left(p((f^{(2)})^* f^{(2)}) \right) = \frac{1}{[\Gamma : \Gamma_i]} \cdot \mathrm{tr} \left(p(f_i^* f_i) \right)$$

for $i \geq i_0$ by Lemma 7.24. □

Note that then operator norms of $f^{(2)}$ and f_i are both bounded in terms of $K := r \cdot \sum_{j=1}^{r} \max \left(|B_{i,j}| : 1 \leq i \leq r \right)$.

Lemma 7.26 *For $\lambda \geq 0$, let $p_n(x)$ be a family of polynomials that approximates the characteristic function $\chi_{[0,\lambda]}(x)$ such that*

$$\lim_{n \to \infty} p_n(x) = \chi_{[0,\lambda]}(x), \quad |p_n(x)| \leq L$$

holds for some L and any $x \in [0, |f^{(2)}|^2]$. Then,

$$\lim_{n\to\infty} \mathrm{tr}_\Gamma(p_n((f^{(2)})^* f^{(2)})) = F(\lambda)$$

holds.

Proof Exercise (Problem 7.56). □

Let us verify the middle equality in Proposition 7.23.

Proof Let $g_n : \mathbb{R} \to \mathbb{R}$ be a piecewise linear functions such that

$$g_n(x) = \begin{cases} 1 + \dfrac{1}{n} & x \in (-\infty, \lambda] \\ 1 + \dfrac{1}{n} - n(x-\lambda) & x \in [\lambda, \lambda + n^{-1}] \\ \dfrac{1}{n} & x \in [\lambda + n^{-1}, \infty]. \end{cases}$$

The inequalities $\chi_{[0,\lambda]}(x) < g_{n+1}(x) < g_n(x)$ hold, and $g_n(x)$ converges to $\chi_{[0,\lambda]}(x)$ on $[0,\infty)$ as $n \to \infty$. Let $p_n(x)$ be a polynomial such that the inequalities $\chi_{[0,\lambda]}(x) < p_n(x) < g_n(x)$ hold on $[0, K^2]$. Recall the bound $K^2 \geq \| f_i^* f_i \|$. Then, we have the equalities

$$\begin{aligned}
\frac{1}{[\Gamma:\Gamma_i]} \cdot \mathrm{tr}(p_n(f_i^* f_i)) &= \frac{1}{[\Gamma:\Gamma_i]} \cdot \sum_{\mu \in E_i(K^2)} p_n(\mu) \\
&= \frac{|E_i(\lambda)|}{[\Gamma:\Gamma_i]} + \frac{1}{[\Gamma:\Gamma_i]} \cdot \\
&\left(\sum_{\mu \in E_i(\lambda)} (p_n(\mu) - 1) + \sum_{\mu \in E_i(\lambda + n^{-1}) - E_i(\lambda)} p_n(\mu) + \sum_{\mu \in E_i(K^2) - E_i(\lambda + n^{-1})} p_n(\mu) \right) \\
&\geq \frac{|E_i(\lambda)|}{[\Gamma:\Gamma_i]} = F_i(\lambda).
\end{aligned}$$

Conversely, we have

$$\begin{aligned}
&\frac{1}{[\Gamma:\Gamma_i]} \cdot \mathrm{tr}(p_n(f_i^* f_i)) \\
&\leq \frac{|E_i(\lambda)|}{[\Gamma:\Gamma_i]} \\
&+ \frac{1}{[\Gamma:\Gamma_i]} \cdot \Big(\sup\{p_n(\mu) - 1 | \mu \in [0,\lambda]\} \cdot |E_i(\lambda)| \\
&\quad + \sup\{p_n(\mu) | \mu \in [\lambda, \lambda + n^{-1}]\} \cdot |E_i(\lambda + n^{-1}) \setminus E_i(\lambda)| \\
&\quad\quad + \sup\{p_n(\mu) | \mu \in [\lambda + n^{-1}, K^2]]\} \cdot |E_i(K^2) \setminus E_i(\lambda + n^{-1})| \Big)
\end{aligned}$$

$$\le \frac{|E_i(\lambda)|}{[\Gamma:\Gamma_i]} + \frac{1}{[\Gamma:\Gamma_i]}\cdot$$
$$\Big(n^{-1}\cdot|E_i(\lambda)| + (1+n^{-1})\cdot|E_i(\lambda+n^{-1})\setminus E_i(\lambda)|$$
$$+n^{-1}\cdot|E_i(K^2)\setminus E_i(\lambda+n^{-1})|\Big)$$
$$\le \frac{|E_i(\lambda+n^{-1})|}{[\Gamma:\Gamma_i]} + \frac{1}{n}\cdot\frac{|E_i(K^2)|}{[\Gamma:\Gamma_i]}$$
$$\le F_i(\lambda+n^{-1}) + \frac{1}{n}\cdot r,$$

where r is the size of the matrix $BB^* \in \mathrm{M}_r(\mathbb{C}\Gamma)$.

By Lemma 7.25, there is i_0 such that for any $i \ge i_0$, the equality

$$\frac{1}{[\Gamma:\Gamma_i]}\cdot \mathrm{tr}(p_n(f_i^* f_i)) = \mathrm{tr}_\Gamma(p_n((f^{(2)})^* f^{(2)}))$$

holds. hence, we have the bounds

$$F_i(\lambda) \;\le\; \mathrm{tr}_\Gamma(p_n(f^{(2)})^* f^{(2)}) \;\le\; F_i(\lambda+n^{-1}) + \frac{r}{n}.$$

Then, we have the bounds $\overline{F_\infty}(\lambda) \;\le\; \mathrm{tr}_\Gamma(p_n(f^{(2)})^* f^{(2)}) \;\le\; \underline{F_\infty}(\lambda+n^{-1}) + \frac{r}{n}$, and by Lemma 7.26, we have the bounds

$$\overline{F_\infty}(\lambda) \;\le\; F(\lambda) \;\le\; \underline{F_\infty}^+(\lambda).$$

Then, we have

$$F(\lambda) \le \underline{F_\infty}^+(\lambda) \le \underline{F_\infty}(\lambda+\epsilon) \le \overline{F_\infty}(\lambda+\epsilon) \le F(\lambda+\epsilon).$$

Since $\lim_{\epsilon\to 0+} F(\lambda+\epsilon) = F(\lambda)$ holds, we obtain the equalities

$$F(\lambda) = \overline{F_\infty}^+(\lambda) = \underline{F_\infty}^+(\lambda) \quad \text{for } \lambda \ge 0.$$

□

Let $g : V \to W$ be a linear map between finite-dimensional Hilbert spaces. Let $p(t) := \det(t - g^*g)$ be the characteristic polynomial. Let $p(t) = t^k \cdot q(t)$ with $q(0) \ne 0$. Introduce two constants

$$K \ge \max(\|g\|, 1), \quad C \le |q(0)|.$$

Let $E(\lambda)$ be the ordered set of eigenvalues μ of g^*g satisfying $\mu \leq \lambda$ listed with multiplicity.

Lemma 7.27 *For* $0 < \lambda < 1$, *the inequality holds:*

$$\frac{|E(\lambda) \setminus E(0)|}{\dim_{\mathbb{C}}(V)} \leq \frac{-\log(C)}{\dim_{\mathbb{C}}(V) \cdot (-\log(\lambda))} + \frac{\log(K^2)}{-\log(\lambda)}.$$

Proof This is linear algebra, and is left the reader as an exercise (Problem 7.57). □

Let us verify the rest of the equalities in Proposition 7.23.

Proof Recall f_i and let p_i be the characteristic polynomials of f_i. Because f and its reductions f_i are given by the sums of differentials on the simplicial cochains, they have integral coefficients. Let $p_i(t) = t^{r_i} \cdot q_i(t)$ with $q_i(0) \neq 0$ as above. Because of integrability, $|q_i(0)| \geq 1$ must hold. In particular, we can choose $C = 1$. Apply Lemma 7.27 for $0 < \lambda < 1$, and we obtain the estimate

$$\begin{aligned} F_i(\lambda) - F_i(0) &= \frac{|E_i(\lambda)| - |E_i(0)|}{[\Gamma : \Gamma_i]} \\ &\leq \frac{-r \log(C)}{[\Gamma : \Gamma_i] \cdot (-\log(\lambda))} + \frac{r \log(K^2)}{-\log(\lambda)} \\ &= \frac{\log(K^2)}{-\log(\lambda)}. \end{aligned}$$

Hence,

$$F_i(\lambda) \ \leq \ F_i(0) + \frac{r \log(K^2)}{-\log(\lambda)}.$$

By taking inferior and superior limits, we obtain the estimates

$$\underline{F_\infty}(\lambda) \leq \underline{F_\infty}(0) + \frac{r \log(K^2)}{-\log(\lambda)},$$

$$\overline{F_\infty}(\lambda) \leq \overline{F_\infty}(0) + \frac{r \log(K^2)}{-\log(\lambda)}.$$

Then, letting $\lambda \to 0$, we obtain the equalities:

$$\underline{F_\infty}(0) = \underline{F_\infty}^+(0), \quad \overline{F_\infty}(0) = \overline{F_\infty}^+(0).$$

This completes the proof of Proposition 7.23. □

7.2 Atiyah–Singer Index Theorem on Covering Spaces

We present a survey on the Atiyah–Singer index theorem on covering spaces. Let D be an elliptic differential operator over a compact oriented smooth manifold M. Then, we have the Atiyah–Singer index theorem. See Sect. 2.3.

Theorem 7.28

$$\mathrm{ind}(D) = \int_M \hat{\mathcal{A}}(TM) \wedge ch(S').$$

Let $\tilde{M}$ be a finite cover of M with the finite group G, and lift all the components such as the elliptic operator over M and the Riemannian metric on M to the ones on $\tilde{M}$. We write them with a tilde to imply the lift. Let $|G|$ be the number of elements of G.

Corollary 7.29

$$\mathrm{ind}(\tilde{D}) = |G| \cdot \mathrm{ind}(D).$$

Proof By choosing a connection on the bundle with the Levi-Civita connection on TM, one can represent the cohomology class of $\hat{\mathcal{A}}(TM)$ by the differential forms of a polynomial of the curvature operator (see [147], Chapter 2). In particular, it can be given by a local computation of the curvature tensors.

It is immediate to see that the representation of the cohomology class of $\hat{\mathcal{A}}(T\tilde{M})$ is invariant by G-action. Hence, we have the equality

$$\int_{\tilde{M}} \hat{\mathcal{A}}(T\tilde{M}) \wedge ch(\tilde{S}') = |G| \cdot \int_M \hat{\mathcal{A}}(TM) \wedge ch(S').$$

This finishes the proof. □

Examples

- The signatures: $\sigma(\tilde{M}) = |G| \cdot \sigma(M)$.
- The Euler characteristics: $\chi(\tilde{M}) = |G| \cdot \chi(M)$.

Remark 7.30 The equality $\sigma(\tilde{M}) = |G| \cdot \sigma(M)$ is highly nontrivial, because the cohomology groups are not multiplicative by taking covering of M. It seems that there are no direct proof of this fact without using the Hirzebruch's signature theorem.

7.3 Atiyah's Γ-Index Theorem

The contents of Sect. 7.2 do not work when the group is infinite.

Let D be an elliptic differential operator over a compact oriented smooth Riemannian manifold (M, g), acting on vector bundles $D : \Gamma(E) \to \Gamma(F)$.

Let $P_0 \subset \mathcal{B}(L^2(M; E))$ be the orthogonal projection to the space of the kernel of D. Let D^* be the formal adjoint operator, and $P_1 \subset \mathcal{B}(L^2(M; F))$ be the projection to the space of the kernel of D^*. Then, we have the equalities

$$\mathrm{tr}(P_0) = \dim\ker D, \quad \mathrm{tr}(P_1) = \dim\ker D^* = \dim\operatorname{coker} D.$$

Hence, we have the equality

$$\begin{aligned} \operatorname{ind} D &= \dim\ker D - \dim\ker D^* \in \mathbb{Z} && (7.2)\\ &= \mathrm{tr}(P_0) - \mathrm{tr}(P_1). && (7.3) \end{aligned}$$

Let $X := \tilde{M}$ be the universal covering space with $\pi_1(M) =: \Gamma$, not necessarily a finite cover. We lift all the components such as $\tilde{g}, \tilde{D}, \tilde{E}$ and so on. They are Γ-invariant. Let $\ker\tilde{D} \subset L^2(X; \tilde{E})$ be the space of the kernel of $\tilde{D}$. The group Γ acts on $\ker\tilde{D}$, and $\ker\tilde{D}$ may be infinite dimensional, in general. Let $\tilde{P}_0 \subset L^2(X; \tilde{E})$ and $\tilde{P}_1 \subset L^2(X; \tilde{F})$ be the projections to the space of the kernel of $\tilde{D}$ and $\tilde{D}^*$ respectively.

Let $L \subset L^2(X; \tilde{E})$ be a Γ-invariant and closed linear subspace. Recall that the Γ-dimension of L is

$$\dim_\Gamma(L) := \mathrm{tr}_\Gamma(P_L),$$

where $P_L : L^2(X; \tilde{E}) \to L$ is the orthogonal projection (see Definition 7.12).

Definition 7.31 The Γ-index is defined by

$$\begin{aligned} \operatorname{ind}_\Gamma(\tilde{D}) : &= \dim_\Gamma \ker\tilde{D} - \dim_\Gamma \ker\tilde{D}^* \in \mathbb{R}\\ &= \mathrm{tr}_\Gamma(\tilde{P}_0) - \mathrm{tr}_\Gamma(\tilde{P}_1). \end{aligned}$$

Then, we have Atiyah's Γ-index theorem.

Theorem 7.32 ([8])

$$\mathrm{ind}_\Gamma(\tilde{D}) = \mathrm{ind}(D).$$

Corollary 7.33

- *If* $\mathrm{ind}(D) > 0$ *(resp.* < 0*), then there exist non-zero elements of* $\ker \tilde{D}$ *(resp.* $\ker \tilde{D}^*$*).*
  Moreover, they are infinite dimensional, if Γ *is infinite.*
- $\mathrm{ind}_\Gamma(\tilde{D})$ *is an integer.*

Remark 7.34 Theorem 7.32 also holds for any covering space. However, in this book, we mainly focus on the universal covering case. (See Problem 7.58.)

Example Let $\mathcal{H}^p(X) \subset L^2(X; \Lambda^p)$ be the space of L^2-harmonic p-forms. Then, the L^2-Betti number $b^p_\Gamma(X)$ is defined by

$$b^p_\Gamma(X) := \mathrm{tr}_\Gamma(P_{\mathcal{H}^p(X)}).$$

In particular, in the case of the de Rham operator $D = d + d^*$, the Γ-index of D coincides with the Euler characteristic

$$\mathrm{ind}_\Gamma(\tilde{D}) = \sum_{i=0}^{\dim M} (-1)^i \cdot b^p_\Gamma(\tilde{M}) = \chi(M).$$

Below we give a proof of Theorem 7.32. See [8] for details.

7.3.1 Almost Localized Indices Over Compact Manifolds

The following is a basic analytic aspect of theory of elliptic differential operators.

Lemma 7.35 *There is a bounded operator*

$$Q : L^2(M; F) \to L^2_1(M; E)$$

called the parametrix such that both of the operators S_0 *and* S_1 *given by*

$$S_0 := 1 - Q \circ D, \qquad S_1 := 1 - D \circ Q$$

have smooth kernels.

Moreover the operator Q can have its support which is almost localized close to their diagonals, as small as you choose.

In particular, both the kernel functions $s_0(x, y)$ and $s_1(x, y)$ of S_0 and S_1 can be also almost localized near the diagonal respectively. The operator Q is called an *almost local parametrix*. Note that the Green's operator is non-local.

Proof Let us set two unbounded operators by

$$\Delta_0 := D^*D : L^2(M; E) \to L^2(M; E),$$
$$\Delta_1 := DD^* : L^2(M; F) \to L^2(M; F).$$

Let $(\varphi_1, \lambda_1), (\varphi_2, \lambda_2), \cdots$ be the spectral decomposition of Δ_1. Note that $\{\varphi_i\}_i$ is an orthonormal basis of $L^2(M; E)$. Then the kernel function of $\exp(-\Delta_1)$ is given by

$$k_1(x, y) = \sum_{i=1}^{\infty} \exp(-\lambda_i) \cdot \varphi_i(x) \otimes \varphi_i^*(y), \tag{7.4}$$

where $\{\varphi_i^*\}_i$ is the dual basis of $\{\varphi_i\}_i$. (See Problem 7.59.)

Consider

$$\begin{aligned} 1 - \exp(-y) &= y - \frac{1}{2!} \cdot y^2 + \cdots \\ &= y \cdot (1 - \frac{1}{2!} \cdot y + \cdots) \\ &=: y \cdot f(y), \end{aligned}$$

Hence, we have that the equality

$$1 - \exp(-\Delta_1) = \Delta_1 f(\Delta_1) = DD^* f(\Delta_1)$$

holds. Note that, by the Schwartz kernel theorem (see [163]), there is the kernel distribution δ_1 of 1 which is supported only on the diagonal.

Let $\Delta := \{(x, x) : x \in M\} \subset M \times M$ be the diagonal, and let $\psi : M \times M \to [0, 1]$ be a smooth cut-off function such that $\psi|_\Delta \equiv 1$ and $\psi|_{N_\epsilon(\Delta)^c} \equiv 0$ for a sufficiently small and positive $\epsilon > 0$, where $N_\epsilon(\Delta)$ is ϵ-neighbourhood of the diagonal and $N_\epsilon(\Delta)^c$ is the complement.

By the Schwartz kernel theorem, there is a kernel distribution $\mu(x, y)$ of $D^* f(\Delta_1)$.

Then the equality

$$\delta_1(x, y) - k_1(x, y) = D_x \mu(x, y)$$

holds for $(x, y) \in M \times M$. By multiplying ψ on both sides, we have the equality

$$\begin{aligned}\delta_1(x, y) - \psi(x, y) \cdot k_1(x, y) &= \psi(x, y)\,(\delta_1(x, y) - k_1(x, y))\\ &= \psi(x, y) \cdot D_x\mu(x, y)\\ &= D_x\,(\psi(x, y) \cdot \mu(x, y)) - (D_x\psi(x, y)) \cdot \mu(x, y).\end{aligned}$$

We set

$$\begin{aligned}s_1(x, y) &:= \psi(x, y) \cdot k_1(x, y) - (D_x\psi(x, y)) \cdot \mu(x, y),\\ q(x, y) &:= \psi(x, y) \cdot \mu(x, y).\end{aligned}$$

Then, we have the equality

$$\delta_1(x, y) - s_1(x, y) = D_x q(x, y).$$

Let S_1 and Q be the corresponding operators to s_1 and q respectively. Then, the equality

$$1 - S_1 = DQ$$

holds.

Before proceeding with the proof, we need Lemma 7.36 and Corollary 7.37 below. □

Lemma 7.36 *$k_1 : M \times M \to \Gamma(F \otimes F^*)$ is smooth.*

Proof Note that the kernel function of $\Delta_1^m \exp(-\Delta_1)$ is given by $(\Delta_1)_x^m k_1(x, y)$ for any $m \geq 0$. Because $y^m \exp(-y)$ is Schwartz class on $[0, \infty)$, $\Delta_1^m \exp(-\Delta_1)$ is a bounded operator on $L^2(M; F)$. Hence, $(\Delta_1)_x^m k_1(x, y)$ is in L^2 on the x-variable for any $m \geq 0$. Then, by elliptic regularity, $k_1(x, y)$ is smooth on the x-variable. Because Δ_1 is self-adjoint, k_1 is also smooth on the y-variable. □

Corollary 7.37 *$\mu(x, y)$ is smooth on the complement of the diagonal.*

Proof It follows from the equality $\delta_1(x, y) - k_1(x, y) = D_x\mu(x, y)$ that the equality $k_1(x, y) = D_x\mu(x, y)$ holds outside the diagonal. Then, the conclusion holds by Lemma 7.36, because the image of $D^* f(\Delta_1)$ is L^2-orthogonal to the space of the kernel of D. □

Let us complete the proof of Lemma 7.35.

Proof The equality $s_1(x, y) = \psi(x, y) \cdot k_1(x, y)$ holds on $N_\epsilon(\Delta)$, because $(D_x\psi(x, y)) \cdot \mu(x, y) \equiv 0$ holds on $N_\epsilon(\Delta)$. Then, smoothness of s_1 follows by Lemma 7.36 and Corollary 7.37.

By replacing the role of D by D^*, we can also obtain the operator S_0' and the almost localized smooth kernel s_0', that satisfies the relation $S_0' = 1 - D^* Q'$.

In the case, we use the kernel function of the operator $Df(\Delta_0) = f(\Delta_1)D^*$ with the kernel function $\mu'(x, y)$. Note that the kernel function of the adjoint operator $(Df(\Delta_0))^* = Df(\Delta_1)$ coincides with $\mu(x, y)$. Hence if we choose $\psi(x, y)$ symmetrically in the variables (x, y), then we can obtain the equality

$$(S_0')^* = 1 - (Q')^* D = 1 - QD.$$

By setting $(S_0)' = S_0$, we complete the proof. □

Corollary 7.38 *The following equalities hold:*

$$\mathrm{tr}(\exp(-\Delta_0) = \mathrm{tr}(S_0), \quad \mathrm{tr}(\exp(-\Delta_1) = \mathrm{tr}(S_1).$$

Proof For $i = 0, 1$, let $k_i(x, y)$ be the kernel function of $\exp(-\Delta_i)$. This follows from the equalities

$$s_i(x, x) = \psi(x, x) \cdot k_i(x, x) - (D_x \psi(x, x)) \cdot \mu(x, x) = k_i(x, x)$$

for $i = 0, 1$ and $x \in M$. □

Let:

- P_0 and P_1 be the orthogonal projections to $\ker D$ and $\ker D^*$ respectively;
- $G : L^2(M; F) \to L^2(M; E)$ be a bounded operator given by $G \circ P_1 = 0$ and $G(v)$ be given by the unique element $D^{-1}(v) \in \ker D^\perp$ for any $v \in (\ker D^*)^\perp$.

Then we have the equalities $G \circ D = 1 - P_0$ and $D \circ G = 1 - P_1$. The bounded operator G is called *Green's operator.*

Remark 7.39

- Green's operator is bounded on each L^2_k (see Definition 7.16).
- The composition DG is bounded.

Lemma 7.40 *The operator $D\exp(-\Delta_0)G$ is of trace class, and the following holds:*

$$\mathrm{tr}(D\exp(-\Delta_0)G) = \mathrm{tr}(\exp(-\Delta_0)) - \mathrm{tr}(P_0).$$

Proof The operator $D\exp(-\Delta_0)G$ is of trace class, because it is a smoothing operator. Note the formulas $P_i = \exp(-\Delta_i)P_i$ for $i = 0, 1$. By using Corollary 7.17, we have the equalities (see Remark 7.39):

$$\begin{aligned}\mathrm{tr}(D\exp(-\Delta_0)G) &= \mathrm{tr}(D\exp(-\tfrac{1}{2}\Delta_0)\exp(-\tfrac{1}{2}\Delta_0)G) \\ &= \mathrm{tr}(\exp(-\tfrac{1}{2}\Delta_0)GD\exp(-\tfrac{1}{2}\Delta_0))\end{aligned}$$

$$
\begin{aligned}
&= \mathrm{tr}(\exp(-\frac{1}{2}\Delta_0)\exp(-\frac{1}{2}\Delta_0)GD) = \mathrm{tr}(\exp(-\Delta_0)GD) \\
&= \mathrm{tr}(\exp(-\Delta_0)) - \mathrm{tr}(P_0).
\end{aligned}
$$

This completes the proof. □

Corollary 7.41 *The equality holds:*

$$
\begin{aligned}
\mathrm{ind}(D) : &= \mathrm{tr}(P_0) - \mathrm{tr}(P_1) = \mathrm{tr}(\exp(-\Delta_0)) - \mathrm{tr}(\exp(-\Delta_1)) \\
&= \mathrm{tr}(S_0) - \mathrm{tr}(S_1).
\end{aligned}
$$

Proof Note the equalities

$$
D\exp(-\Delta_0)G = \exp(-\Delta_1)DG = \exp(-\Delta_1)(1 - P_1) = \exp(-\Delta_1) - P_1.
$$

Hence we have the equality

$$
\mathrm{tr}(\exp(-\Delta_0)) - \mathrm{tr}(P_0) = \mathrm{tr}(\exp(-\Delta_1)) - \mathrm{tr}(P_1).
$$

Then the conclusion follows by Corollary 7.38. □

7.3.2 Proof of Theorem 7.32

Let $X := \tilde{M}$ be the universal covering space with $\pi_1(M) =: \Gamma$, not necessarily a finite cover. We lift all the components g, D, E and so on to $X = \tilde{M}$ as $\tilde{g}$, $\tilde{D}$, $\tilde{E}$ and so on.

Then, they are Γ-invariant. Let $\ker \tilde{D} \subset L^2(X; \tilde{E})$ be the space of the kernel of $\tilde{D}$. The group Γ acts on $\ker \tilde{D}$, and $\ker \tilde{D}$ may be infinite dimensional, in general. Let $\tilde{P}_0 \subset L^2(X; \tilde{E})$ and $\tilde{P}_1 \subset L^2(X; \tilde{F})$ be the projections to the space of the kernel of $\tilde{D}$ and $\tilde{D}^*$ respectively. Note that all the operators $\tilde{P}_i$ and $1 - \tilde{P}_i$ are bounded on each L^2_k for $i = 0, 1$. (See Definition 7.16.)

Because Q and S_i are all almost local in Lemma 7.35, we can lift these operators over the universal covering space. We denote their lifts by $\tilde{Q}$, $\tilde{S}_0$ and $\tilde{S}_1$. Then, they satisfy the equalities

$$
\tilde{S}_0 = 1 - \tilde{Q} \circ \tilde{D}, \qquad \tilde{S}_1 = 1 - \tilde{D} \circ \tilde{Q}.
$$

Moreover their kernel functions are also given by the lifts.

Note the following:

Lemma 7.42 *The kernel functions of $\tilde{S}_i$ satisfy L^2-finiteness in Lemma 7.13.*

The following lemma holds by almost locality of the kernel functions:

Lemma 7.43 *The equalities* $\operatorname{tr}_\Gamma \tilde{S}_i = \operatorname{tr} S_i$ *hold for* $i = 0, 1$.

Proof Because both $\tilde{S}_0$ and $\tilde{S}_1$ are the lifts, their kernel functions satisfy the equalities $\operatorname{tr} \tilde{s}_0(\tilde{x}, \tilde{x}) = \operatorname{tr} s_0(\pi(\tilde{x}), \pi(\tilde{x}))$ and $\operatorname{tr} \tilde{s}_1(\tilde{x}, \tilde{x}) = \operatorname{tr} s_1(\pi(\tilde{x}), \pi(\tilde{x}))$, where $\pi : X = \tilde{M} \to M$ is the covering map. In particular, we have the equalities

$$\operatorname{tr}_\Gamma \tilde{S}_i = \int_K \operatorname{tr} \tilde{s}_i(\tilde{x}, \tilde{x}) \operatorname{vol}_{\tilde{x}} = \int_M \operatorname{tr} s_0(x, x) \operatorname{vol}_x = \operatorname{tr} S_i$$

for $i = 0, 1$, where $K \subset \tilde{M}$ is a fundamental domain. □

Our goal is to verify the following equalities:

Proposition 7.44 *The following equality holds:*

$$\operatorname{tr}_\Gamma \tilde{P}_0 - \operatorname{tr}_\Gamma \tilde{P}_1 = \operatorname{tr}_\Gamma \tilde{S}_0 - \operatorname{tr}_\Gamma \tilde{S}_1.$$

Proof Note the equalities

$$\tilde{P}_0 = \tilde{S}_0 \tilde{P}_0, \quad \tilde{P}_1 = \tilde{P}_1 \tilde{S}_1.$$

Then we have the equalities

$$\operatorname{tr}_\Gamma(\tilde{P}_0 \tilde{S}_0) = \operatorname{tr}_\Gamma(\tilde{S}_0 \tilde{P}_0 \tilde{S}_0) = \operatorname{tr}_\Gamma(\tilde{S}_0 \tilde{S}_0 \tilde{P}_0) = \operatorname{tr}_\Gamma(\tilde{S}_0 \tilde{P}_0) = \operatorname{tr}_\Gamma(\tilde{P}_0)$$

by Lemma 7.42 and Corollary 7.17. Similarly, we have the equality $\operatorname{tr}_\Gamma(\tilde{S}_1 \tilde{P}_1) = \operatorname{tr}_\Gamma(\tilde{P}_1)$. Let us set

$$T_i := (1 - \tilde{P}_i)\tilde{S}_i(1 - \tilde{P}_i)$$

for $i = 0, 1$. Then, the conclusion follows, if we verity the equality

$$\operatorname{tr}_\Gamma T_0 = \operatorname{tr}_\Gamma T_1.$$

Note that the equality $\tilde{D} T_0 = T_1 \tilde{D}$ holds by commutativity $\tilde{D} \tilde{S}_0 = \tilde{S}_1 \tilde{D}$. Below, we approximate the role of $\tilde{D}$ by an invertible operators. Then, taking the limit, we will be able to obtain the desired equality.

Let us consider the polar decomposition $\tilde{D} = \tilde{U} A$ (see [135]), where

$$\tilde{U}(v) := \begin{cases} 0 & v \in \ker \tilde{D}, \\ \frac{\tilde{D}}{\sqrt{\tilde{D}^* \tilde{D}}}(v) & v \in \ker \tilde{D}^\perp \end{cases}$$

is the partial unitary and $A = \tilde{U}^* \tilde{D}$ is self-adjoint. Note the equalities

$$\tilde{U}^* \tilde{U} = 1 - \tilde{P}_0, \quad \tilde{U} \tilde{U}^* = 1 - \tilde{P}_1.$$

Let us set $T_2 := \tilde{U}^* T_1 \tilde{U}$. Then, we have the equality $\mathrm{tr}_\Gamma\, T_2 = \mathrm{tr}_\Gamma\, T_1$ by Lemma 7.18. Moreover,

$$AT_0 = \tilde{U}^* \tilde{D} T_0 = \tilde{U}^* T_1 \tilde{D} = \tilde{U}^* T_1 \tilde{U} \tilde{U}^* \tilde{D} = T_2 A$$

holds by the equality $\tilde{D} T_0 = T_1 \tilde{D}$.

Let $P_n := E_A[\frac{1}{n}, n]$ be the spectral projections of the self-adjoint operator A, and set

$$T_{0,n} := P_n T_0 P_n, \qquad T_{2,n} := P_n T_2 P_n.$$

The operators $A_n := P_n A P_n + (1 - P_n)$ are all invertible. Moreover, by direct calculations, we can verify the equality

$$A_n T_{0,n} = T_{2,n} A_n.$$

(See Problem 7.60.) Hence, we have the equalities $A_n T_{0,n} A_n^{-1} = T_{2,n}$, and $\mathrm{tr}_\Gamma\, T_{0,n} = \mathrm{tr}_\Gamma\, T_{2,n}$. Then, $\mathrm{tr}_\Gamma\, T_{i,n} = \mathrm{tr}_\Gamma\, P_n T_i P_n = \mathrm{tr}_\Gamma\, T_i P_n$, and

$$\lim_{n\to\infty} \mathrm{tr}_\Gamma\, T_{i,n} = \mathrm{tr}_\Gamma\, T_i (1 - \tilde{P}_0) = \mathrm{tr}_\Gamma\, T_i$$

hold for $i = 0, 2$. Hence, we obtain the desired equality $\mathrm{tr}_\Gamma\, T_0 = \mathrm{tr}_\Gamma\, T_2 = \mathrm{tr}_\Gamma\, T_1$. □

Let us complete the proof of Theorem 7.32.

Proof By lemma 7.43, $\mathrm{tr}_\Gamma\, \tilde{S}_i = \mathrm{tr}\, S_i$ holds for $i = 0, 1$. By Proposition 7.44 and Corollary 7.41, we have

$$\mathrm{ind}_\Gamma\, \tilde{D} = \mathrm{tr}_\Gamma\, \tilde{P}_0 - \mathrm{tr}_\Gamma\, \tilde{P}_1 = \mathrm{tr}_\Gamma\, \tilde{S}_0 - \mathrm{tr}_\Gamma\, \tilde{S}_1 = \mathrm{tr}\, S_0 - \mathrm{tr}\, S_1 = \mathrm{ind}\, D.$$

This verifies the equality $\mathrm{ind}_\Gamma(\tilde{D}) = \mathrm{ind}(D)$. □

7.4 Zero-in-the-Spectrum Conjecture

The following is called the zero-in-the-spectrum conjecture:

Conjecture 7.45 The unreduced L^2-cohomology group is non-zero at some degree over the universal covering space of a compact aspherical manifold M with respect to any Riemannian metric.

Remark 7.46 Note the following:

- The condition $0 \in \operatorname{spec} \Delta_m$ is equivalent to $H^m_{(2)}(\tilde{M}) \neq 0$.
- The condition $0 \in \operatorname{spec} \Delta_0$ always holds, when M is compact.

Definition 7.47 Let (X, g) be a complete Riemannain manifold. It is called hyper-Euclidean, if there exists a proper Lipschitz map $F : X \to \mathbb{R}^m$ of non-zero degree, where $m = \dim X$.

Example It follows from the Cartan–Hadamard theorem [96, Theorem 8.1] that the exponential map

$$\exp_x : T_x X \cong X$$

is an expanding ($d_x(v, w) \leq d(\exp_x(v), \exp_x(w))$ hold for any $v, w \in T_x X$) diffeomorphism, if X is simply-connected and the sectional curvature $K_g \leq 0$ is non-positive everywhere. Hence,

$$\log_x := \exp_x^{-1} : X \cong T_x X$$

gives a map for the hyper-Euclidean data.

For example, $\mathbb{R}^m$ with the standard metric is hyper-Euclidean.

Theorem 7.48 *Let (M, g) be a compact Riemannian manifold such that the universal covering space is hyper-Euclidean with the lifted metric. Then, the Conjecture 7.45 holds.*

This is verified in Sects. 7.4.1 and 7.4.2.

7.4.1 Gromov–Lawson's Relative Index Theorem

To verify Theorem 7.48, we use the Gromov–Lawson relative index theorem [68].

Let (X, g) be a complete Riemannian manifold, and $E \to X$ and D be a vector bundle and an elliptic operator over $E \to X$ respectively.

Definition 7.49 It is said that D is *invertible at* ∞, if there exists a compact subset $K \subset X$ and a positive constant $c > 0$ such that for any $s \in L^2(X \backslash K; E|_{X \backslash K})$, the lower bound holds:

$$\|Ds\|_{L^2} \geq c \cdot \|s\|_{L^2}.$$

Remark 7.50 If $0 \neq \operatorname{spec} \Delta_m$ for all $0 \leq m \leq \dim X$, then $D := d + d^*$ is invertible.

For $i = 1, 2$, let $(E_i, \nabla_i) \to X$ be two vector bundles over X such that the isomorphism

$$(E_1|_{X \backslash K}, \nabla_1) \cong (E_2|_{X \backslash K}, \nabla_2) \tag{7.5}$$

holds. Then, the curvature forms $\mathbb{R}_i$ of ∇_i coincide on the complement of K for $i = 1, 2$. In particular, it gives the relative curvature form

$$ch(\nabla_1, \nabla_2) := ch(\nabla_1) - ch(\nabla_2) \in \Omega_c^{\text{even}}(X),$$

and the relative Chern character as a cohomology class

$$ch(E_1, E_2) := [ch(\nabla_1, \nabla_2)] \in H_c^{\text{even}}(X; \mathbb{Q}).$$

Suppose that $\dim X =: 2k$ is even. Then, the space of k-forms splits as

$$\Lambda^k \cong \Lambda^+ \oplus \Lambda^-$$

with respect to the involution $\iota := i^{k(k-1)}(-1)^{\frac{k(k-1)}{2}}$. It anti-commutes with $D = d + d^*$, and hence, it has the form

$$D = \begin{pmatrix} 0 & D^- \\ D^+ & 0 \end{pmatrix}$$

with respect to the decomposition. By twisting Λ^k with E_i, we have two covariant-derivatives

$$D_i^{\pm} : L_1^2(X; \Lambda^{\pm} \otimes E_i) \to L^2(X; \Lambda^{\mp} \otimes E_i)$$

for $i = 1, 2$. It also has the form

$$D_i = \begin{pmatrix} 0 & D_i^- \\ D_i^+ & 0 \end{pmatrix}$$

with respect to the decomposition.

Theorem 7.51 ([68]) *Assume that D is invertible at ∞. Then, $D^+ : L^2_1(X; \Lambda^+) \to L^2(X; \Lambda^-)$ is Fredholm.*

Moreover, assume the condition (7.5). Then, there is a positive constant $\delta > 0$ such that if the L^∞-norms of the curvatures at infinity are smaller than δ, then D_i are both Fredholm, and the following holds:

$$\text{ind}\, D_1^+ - \text{ind}\, D_2^+ = \int_X \mathcal{L}(X, g) \cdot ch(E_1, E_2),$$

where $\mathcal{L}(X, g) = 1 +$higher term, and the higher term is given by a universal polynomial of the components of the curvature form of (X, g).

This is a special case of the Gromov–Lawson relative index theorem. We will only give a proof of the Fredholm property of D^+ in Theorem 7.51.

Proof By the assumption, there is a positive constant $c > 0$ for any $s \in L^2(X \backslash K; E|_{X \backslash K})$, and the lower bound holds:

$$\langle D^2(s), s\rangle \geq c^2 \cdot \langle s, s\rangle.$$

It follows from the Weitzenböck formula that the difference

$$D_i^2 - D^2 \otimes \text{id} = R_i$$

holds. Hence, we have the estimates

$$\begin{aligned} \langle D_i^2(s), s\rangle &\geq \langle D^2(s) \otimes \text{id}, s\rangle + \|R_i\|_{L^\infty(X\backslash K)}\|s\|_{L^2} \\ &\geq (c^2 - \|R_i\|_{L^\infty(X\backslash K)}) \cdot \|s\|_{L^2} \geq (c^2 - \delta) \cdot \|s\|_{L^2} \end{aligned}$$

for $s \in L^2(X\backslash K; E_i)$.

Hence, it suffices to verify the Fredholm property for D. Let us consider the spectral decomposition $\{E_\lambda\}_{0 \leq \lambda}$ of D^2. Let us verify the finite dimensionality of $\text{im} E_\mu$ for a small $c^2 >> \mu > 0$. Let us take an element $s \in \text{im} E_\mu$, and let $\rho : X \to [0, 1]$ be a compactly supported cut-off function such that $\rho|_K \equiv 1$ and $\rho|K' \equiv 0$ for some compact subset $K \subset K'$. Then, we have the estimate

$$\begin{aligned} \mu \cdot \|s\|^2_{L^2(X)} \geq \|D(s)\|^2_{L^2(X)} &= \|D(\rho \cdot s)\|^2_{L^2(X)} + \|D((1-\rho) \cdot s)\|^2_{L^2(X)} \\ &\geq \|D(\rho \cdot s)\|^2_{L^2(K')} + c^2 \cdot \|(1-\rho) \cdot s\|^2_{L^2(X\backslash K)} \\ &\geq \|D(\rho \cdot s)\|^2_{L^2(K')} + c^2 \cdot \|s\|^2_{L^2(X\backslash K')}. \end{aligned}$$

Hence we have the estimate

$$(\mu + \|d\rho\|_{L^\infty}^2) \cdot \|s\|_{L^2(K')}^2 + \|\rho \cdot D(s)\|_{L^2(K')}^2 \geq (c^2 - \mu) \cdot \|s\|_{L^2(X\setminus K')}^2.$$

By the elliptic estimate, we have $\|\rho \cdot D(s)\|_{L^2(K')}^2 \leq \|D(s)\|_{L^2(K')}^2 \leq C \cdot \|s\|_{L^2(K')}^2$ for some positive constant $C > 0$. Hence, we have the estimate

$$(C + \mu + \|d\rho\|_{L^\infty}^2) \cdot \|s\|_{L^2(K')}^2 \geq (c^2 - \mu) \cdot \|s\|_{L^2(X\setminus K')}^2.$$

It follows from Rellich's lemma that any bounded sequence in $\operatorname{im} E_\mu$ has a convergent subsequence. This implies the finite dimensionality of $\operatorname{im} E_\mu$. □

7.4.2 *Proof of Theorem 7.48*

Consider the product $X' := X \times \mathbb{R}$, where we equip with the standard metric on $\mathbb{R}$. Then, the formula

$$\Delta_m^{X'} = \Delta_m^X + \Delta_m^{\mathbb{R}},$$

and hence,

$$\operatorname{spec} \Delta_m^{X'} = \{\, \lambda_1 + \lambda_2 : \lambda_1 \in \operatorname{spec} \Delta_m^X,\ \lambda_2 \in \operatorname{spec} \Delta_m^{\mathbb{R}} \,\}$$

hold. This implies that it is enough to verify Theorem 7.48, when $\dim X$ is even, because both $\lambda_1, \lambda_2 \geq 0$ are nonnegative.

Proof Let us verify Theorem 7.48, when $m := \dim X = 2k$ is even.

There is a complex vector bundle $E \to S^m$ with the m-th Chern class $c^k(E) \neq 0 \in H^m(S^m; \mathbb{Z})$ is non-zero. Take a connection ∇ on E that is trivial near the north pole $p_N \in S^m$. Then, by identifying $\mathbb{R}^m = S^m \setminus \{p_N\}$, the connection ∇ restricts $(E|_{\mathbb{R}^m}, \nabla)$ such that ∇ is trivial on the complement of a compact subset $K \subset \mathbb{R}^m$. By taking trivialization $i : E|_{\mathbb{R}^m} \cong \mathbb{R}^m \times \mathbb{C}^N$, the pull-back $i^*(\nabla)$ is regarded as a $u(N)$-valued 1-form ω on $\mathbb{R}^m$ such that the curvature form $\Omega = d\omega + \omega \wedge \omega$ satisfies $\Omega|_{\mathbb{R}^m\setminus K} \equiv 0$. Moreover $0 \neq \operatorname{tr}(\exp(-\frac{1}{2\pi i}\Omega) - N) \in H_c^m(\mathbb{R}^m; \mathbb{R})$ is a non-zero class.

Let $\epsilon > 0$ is a positive constant, and let $m_\epsilon : \mathbb{R}^m \to \mathbb{R}^m$ by $m_\epsilon(x) := \epsilon \cdot x$. Consider the renormalized connection 1-form $\omega_\epsilon := m_\epsilon^*(\omega)$ with $\|\omega_\epsilon\|_{L^\infty} = \epsilon \cdot \|\omega\|_{L^\infty}$. Its curvature is given by $\Omega_\epsilon := d\omega_\epsilon + \omega_\epsilon \wedge \omega_\epsilon$. By topological invariance of the compactly supported Chern class, we have

$$\int_{\mathbb{R}^m} \left(\operatorname{tr}\left(\exp\left(-\frac{1}{2\pi i}\Omega_\epsilon\right)\right) - N\right) \neq 0 \tag{7.6}$$

is independent of the choice of $\epsilon > 0$.

Now we proceed by a contradiction argument. Suppose $0 \neq \operatorname{spec} \Delta_i$ for all $0 \leq i \leq m$. Then, $D := d + d^*$ is invertible. See Remark 7.50.

By the assumption, there is a proper 1-Lipschitz map of non-zero degree $F : X \to \mathbb{R}^m$. Then, the following equalities hold:

$$\|F^*\omega_\epsilon\|_{L^\infty} \leq \|\omega_\epsilon\|_{L^\infty} = \epsilon \cdot \|\omega\|_{L^\infty}.$$

Let

$$\wedge (F^*(\omega_\epsilon))_x, i_g(F^*(\omega_\epsilon))_x : \Lambda^*(X) \otimes \mathbb{C}^N \to \Lambda^*(X) \otimes \mathbb{C}^N$$

be the algebraic operations given by the wedge product and the contraction on each $x \in X$. Then for $c(F^*(\omega_\epsilon)) := \wedge(F^*(\omega_\epsilon)) - i_g(F^*(\omega_\epsilon))$, we have the formulas

$$D_i = D \otimes \operatorname{id} + c(F^*(\omega_\epsilon)) : L_1^2(X; \Lambda^* \otimes \mathbb{C}^N) \to L_1^2(X; \Lambda^* \otimes \mathbb{C}^N)$$

for $i = 1, 2$. Both the operators are invertible, if we choose a sufficiently small $\epsilon > 0$. Then both $\operatorname{ind} D_1 = \operatorname{ind} D_2 = 0$ vanish. On the other hand, by Theorem 7.51, their difference $\operatorname{ind} D_1 - \operatorname{ind} D_2$ coincides with (7.6) which is non-zero. This is a contradiction. □

Problems

7.52 Verify that the value $\|A\|_{HS}$ is independent of the choice of any orthonormal basis.

7.53 Verify that both the compositions TT' and $T'T$ are of trace class, if either T is of trace class and T' is bounded, or both T and T' are of Hilbert–Schmidt class.

7.54 Verify Lemma 7.10.

7.55 Verify the following properties. (See Sect. 7.1.6.)

1. The isomorphism $C^m_{(2)}(\tilde{K}) \cong \oplus^{N_m} l^2(\Gamma)$ holds.
2. Let H and H' be Hilbert spaces equipped with isometric Γ-actions. Consider a Γ-equivariant bounded linear operator $d : H \to H'$. Then, the Γ-equivariant isomorphism

$$H \cong \operatorname{Ker} d \oplus \overline{d(H)}$$

holds.

7.56 Verify Lemma 7.26.

7.57 Verify Lemma 7.27.

7.58 Let M be a Riemann surface of genus $g \geq 2$. Let $\tilde{M}$ be an infinite cyclic covering of M. Verify that there exists a nowhere zero vector field on $\tilde{M}$. This suggests that in general, one needs some analytic set-up for a geometric object on the covering space, when one tries to relate it to Atiyah's Γ-index theorem. (See [90].)

7.59 Check the formula (7.4).

7.60 Verify the equalities $A_n T_{0,n} = T_{2,n} A_n$ in the proof of Proposition 7.44.

7.61 (Projection) Let P be a projection on a finite-dimensional $\mathbb{C}$-vector space V. Verify that $\mathrm{tr}(P)$ takes a value in $\mathbb{Z}$.

7.62 (Trace)

(1) Verify that tr_Γ takes a rational value $\mathbb{Q}\backslash\mathbb{Z}$, if Γ contains a torsion element.
(2) Verify that if tr_Γ takes an integer value for Γ and any projection in $\mathcal{N}(\Gamma)$, then it does not contain any nontrivial projection. (The Kaplanski conjecture states that the group von Neumann algebra does not have any nontrivial projection, if the group is torsion-free.)

7.63 (Multiplicativity) Let M be a compact oriented smooth manifold. Verify that the Betti numbers are not necessarily multiplicative under taking finite coverings.

7.64 (Existence of L^2-Harmonic Forms) Let M be a compact oriented smooth manifold, and equip with any Riemannian metric on M. Let $\tilde{M}$ be its universal covering space equipped with the lift of the metric.

Verify that there exist non-zero L^2-harmonic forms on $\tilde{M}$, if the Euler characteristic $\chi(M) \neq 0$ is non-zero.

Chapter 8
Aspherical $\frac{10}{8}$-Inequality and Singer's Conjecture

Singer's conjecture states that the reduced L^2-cohomology groups vanish except for the middle dimension over the universal covering spaces of aspherical manifolds. In this chapter, we combine Singer's conjecture and Furuta's $\frac{10}{8}$-inequality, and induce the aspherical $\frac{10}{8}$-inequality conjecture.

- We present variations of the $\frac{10}{8}$-inequality conjecture.
- We present various examples of 4-manifolds that satisfy the aspherical $\frac{10}{8}$-inequality.
- We induce a covering version of Furuta's $\frac{10}{8}$-inequality for a compact spin 4-manifold with residually finite group by applying Lück's convergence theorem.
- We introduce the L^2-Lefschetz theory and present a proof of Singer's conjecture for a class of Kähler hyperbolic manifolds by Gromov.
- We verify that Kähler hyperbolic 4-manifolds whose fundamental groups are residually finite satisfy the aspherical $\frac{10}{8}$-inequality conjecture.

8.1 Variations of the $\frac{10}{8}$-Inequalities

We have introduced Furuta's $\frac{10}{8}$-inequality (Theorem 1.1) and its proof (Sect. 5.5). Now we propose a covering version of the $\frac{10}{8}$-inequality:

Conjecture 8.1 Suppose M is a compact smooth spin four-manifold. Let $X = \tilde{M}$ be the universal covering space. Then, the covering $\frac{10}{8}$-inequality

(continued)

T. Kato, N. Nakamura, *Monopole Maps and L² -cohomology*, Springer Monographs in Mathematics, https://doi.org/10.1007/978-981-96-8472-4_8

Conjecture 8.1 (continued)

$$b^2_\Gamma(X) \geq \frac{10}{8}|\sigma(M)|$$

holds, where $b^2_\Gamma(X)$ is the second L^2-Betti number.

Note that the Γ-signature

$$\sigma_\Gamma(X) := \dim_\Gamma \mathcal{H}^+(X) - \dim_\Gamma \mathcal{H}^-(X)$$

is equal to the signature

$$\sigma_\Gamma(X) = \sigma(M)$$

of M by Atiyah's Γ-index theorem.

Our main theme in this chapter is to introduce the following $\frac{10}{8}$-inequality:

Conjecture 8.2 Suppose M is a compact aspherical smooth spin four-manifold. Then, the inequality

$$\chi(M) \geq \frac{10}{8}|\sigma(M)|$$

holds, where $\chi(M)$ is the Euler characteristic of M.

We call it the *aspherical* $\frac{10}{8}$-inequality conjecture [87].

Remark 8.3 Let $N \to M$ be a finite covering. If the aspherical $\frac{10}{8}$-inequality holds for M, then, it also holds for N, because both the signature and the Euler characteristic are multiplicative under a finite covering by Theorem 7.32.

Lemma 8.4 *The Singer conjecture plus the covering $\frac{10}{8}$-inequality implies the aspherical $\frac{10}{8}$-inequality.*

Proof The Singer conjecture states that if M is aspherical of even dimension $2m$, then the L^2-Betti numbers vanish:

$$b^i_\Gamma(X) = 0 \qquad i \neq m,$$

except for the middle dimension. Suppose the Singer conjecture is true for an aspherical four-manifold M. Then, we have the equalities

$$\begin{aligned}\chi(M) &= \chi_\Gamma(X) = b_\Gamma^0(X) - b_\Gamma^1(X) + b_\Gamma^2(X) - b_\Gamma^3(X) + b_\Gamma^4(X) \\ &= b_\Gamma^2(X),\end{aligned}$$

where the first equality comes from Atiyah's Γ-index theorem.

By the covering $\frac{10}{8}$-inequality combined with the above equality, we obtain the aspherical $\frac{10}{8}$-inequality:

$$\chi(M) = b_\Gamma^2(X) \geq \frac{10}{8}|\sigma(M)|.$$

□

Remark 8.5 Let us check that, in the case of non-positive signature values, the above inequality $\chi(M) \geq \frac{10}{8}|\sigma(M)|$ follows from another inequality

$$\chi_{\text{AHS}}(M) \geq -\frac{1}{8}\sigma(M),$$

where $\chi_{\text{AHS}}(M) = b^0(M) - b^1(M) + b^+(M)$. Assume that $\sigma(M) \leq 0$ is non-positive. Then, the equalities

$$\begin{aligned}&\chi(M) + \frac{10}{8}\sigma(M) \\ &\quad = b^0(M) - b^1(M) + b^2(M) - b^3(M) + b^4(M) + b^+(M) \\ &\qquad - b^-(M) + \frac{1}{4}\sigma(M) \\ &\quad = 2(b^0(M) - b^1(M) + b^+(M)) + \frac{1}{4}\sigma(M) \\ &\quad = 2\chi_{\text{AHS}}(M) + \frac{1}{4}\sigma(M)\end{aligned}$$

hold, according to the Poincaré duality.

8.2 Examples

Let us present examples that satisfy Conjecture 8.2.

8.2.1 Surface Bundles

Let $\pi : M \to B$ be a fiber bundle whose fiber and the base are both compact manifolds.

- Note that the estimate

$$b_1(M) \leq b_1(F) + b_1(B) \tag{8.1}$$

  holds. Consider the induced surjective homomorphism $\pi_* : \pi_1(M) \twoheadrightarrow \pi_1(B)$. By the homotopy extension theorem, $\ker \pi_* \subset \pi_1(F)$ holds. Hence, the inclusion $\pi_1(M) \subset \pi_1(F) \oplus \pi_1(B)$ holds. In general, the isomorphism $H^1(X;\mathbb{Z}) \cong \mathrm{Hom}(\pi_1(X), \mathbb{Z})$ holds.
- The Euler characteristics are multiplicative in the sense that the equality

$$\chi(M) = \chi(F) \cdot \chi(B) \tag{8.2}$$

  holds. (See Problem 8.40.)

Theorem 8.6 ([98]) *Let $M \to B$ is a surface bundle over a surface such that M is aspherical. Then the following hold:*

1. $\chi(M) \geq |\sigma(M)|$.
2. $\chi(M) \geq 2|\sigma(M)|$ *if B has the genus* ≥ 2.
3. $\chi(M) \geq 3|\sigma(M)|$ *if moreover M admits a complex structure.*

Note that M is aspherical if both the base and the fiber surfaces have their genus values ≥ 1.

Proof We prove only assertion 1. (The proof of assertions 2 and 3 requires the big theorems on symplectic 4-manifolds due to Taubes [162], which are outside the scope of this book. See the original paper [98].) Let g and h be the genera of B and F respectively.

$$\begin{aligned}
|\sigma(M)| &\leq b_2(M) = \chi(M) - 2 + 2b_1(M) \\
&\leq \chi(M) - 2 + 2b_1(B) + 2b_1(F) \\
&= (2 - b_1(B))(2 - b_1(F)) - 2 + 2b_1(B) + 2b_1(F) \\
&= 4gh + 2.
\end{aligned}$$

We refine the above inequality by taking asymptotic covering spaces.

- First, we take coverings of the base spaces. Note that $2g = b_1(B) = -\chi(B) + 2$ holds. Let $\rho : \tilde{B} \to B$ be a d-fold covering. Then, the pull-back $\rho^*(M) \to \tilde{B}$ gives another fiber bundle. Recall that both the signature and the Euler characteristic are multiplicative. Then, by applying the above inequality for the pull-back bundle, we have the inequalities

$$\begin{aligned} d|\sigma(M)| = |\sigma(\tilde{M})| \le 4\tilde{g}h + 2 &= 2(-\chi(\tilde{B}) + 2)h + 2 \\ &= 2(-d \cdot \chi(B) + 2)h + 2, \end{aligned}$$

where $\tilde{g}$ is the genus of $\tilde{B}$. By dividing both sides by d and letting $d \to \infty$, we obtain the inequality $|\sigma(M)| \le -2\chi(B) \cdot h = 4(g-1) \cdot h$.

- Next, we take another covering space fiberwisely. By Lemma 4.1 in [133], there exists a finite covering $B_1 \to B$ such that the pull-back bundle $M_1 \to M$ admits a fiber-wise m-fold covering space for each natural number m. Then, we can improve the inequality $|\sigma(M)| \le 4(g-1) \cdot (h-1) = \chi(B) \cdot \chi(F) = \chi(M)$.

□

Example Atiyah [7] constructed a complex algebraic surface Z_g for $g \ge 3$ with non-zero signature which admits the structure of fiber bundle whose base and fibers are both Riemann surfaces of genus ≥ 3. Since the base and fibers are both aspherical, we have $\pi_k(Z_g) = 0$ for $k \ge 2$ by the exact sequence of homotopy groups of fibrations, and therefore Z_g is also aspherical. The signature and Euler characteristic of Z_g are respectively given by

$$\sigma(Z_g) = (g-1)2^{2g+1}, \qquad \chi(Z_g) = (g-1)2^{2g+2}(2g-1).$$

Then, surely the inequality

$$\chi(Z_g) = 2(2g-1)|\sigma(Z_g)| > \frac{10}{8}|\sigma(Z_g)|$$

holds.

The construction of Z_g is as follows. Choose a Riemann surface C' with genus $g' \ge 2$, and its double covering $C \to C'$. Then the genus of C is $g = 2g' - 1$. Let $\tau: C \to C$ be the free involution associated with the double covering. Furthermore, we take the covering X of C associated with the homomorphism

$$\pi_1(C) \to H_1(C; \mathbb{Z}) \to H_1(C; \mathbb{Z}_2) \cong \mathbb{Z}_2^{2g}.$$

Then X has the property that any double covering of C becomes trivial when it is lifted to X. Let $f: X \to C$ be the covering map. Then the induced homomorphism $f_1^*: H^1(C; \mathbb{Z}_2) \to H^1(X; \mathbb{Z}_2)$ is zero.

Consider the graphs Γ_f and $\Gamma_{\tau f}$ of f and τf in $X \times C$.

Lemma 8.7 *The homology class $[\Gamma_f] + [\Gamma_{\tau f}]$ in $H^2(X \times C; \mathbb{Z})$ is divisible by 2.*

Proof Let $\gamma_f \in H^2(X \times C; \mathbb{Z}_2)$ be the mod 2 reduction of the class $[\Gamma_f]$. We need to prove that $\gamma_f + \gamma_{\tau f} = 0$. By the Künneth formula and Poincaré duality, we obtain

$$H^*(X \times C) \cong \mathrm{Hom}(H^*(C), H^*(X)),$$
$$\gamma_f = \sum_p f_p^*,$$

where f_p^* is the homomorphism induced by f on $H^p(\cdot; \mathbb{Z}_2)$. Then we have

$$f_1^* = 0$$
$$f_2^* = \deg f \pmod 2 = 0$$
$$f_0^* = 1 \text{ on } H^0(C) \cong H^0(X) \cong \mathbb{Z}_2.$$

Similarly we have

$$(\tau f)_1^* = f_1^* \tau_1^* = 0, \quad (\tau f)_2^* = 0, \quad (\tau f)_0^* = 1.$$

Thus we obtain $\gamma_f + \gamma_{\tau f} = \sum_p (f_p^* + (\tau f)_p^*) = 0$. □

Lemma 8.7 enables us to construct a double cover of $X \times C$ branched along $\Gamma_f + \Gamma_{\tau f}$ by a general procedure. Take a holomorphic line bundle L over $X \times C$ associated with the divisor $A = \Gamma_f + \Gamma_{\tau f}$. Then there is a holomorphic section s of L whose zero locus is A. Since the homology class of A is divisible by 2 by Lemma 8.7, there is a holomorphic line bundle $\tilde{L}$ such that $\tilde{L}^2 \cong L$. Let $Q\colon \tilde{L} \to L$ be the fiberwise squaring map. Then $Z_g := Q^{-1}(s(X \times C))$ is the required double cover. The composite projection

$$Z_g \to X \times C \to X$$

defines a fiber bundle over X such that the fiber over $x \in X$ is the double covering of C branched at $f(x)$ and $\tau f(x)$.

Since X is a 2^{2g}-fold covering of C and each fiber F of Z_g is a double branched covering of C, the Euler characteristics of X and F are obtained from the Riemann–Hurwitz formula (see [65]) as

$$\chi(X) = (2 - 2g)2^{2g}, \quad \chi(F) = 2(2 - 2g) - 2 = 2(1 - 2g).$$

Hence we have $\chi(Z_g) = \chi(X)\chi(F) = (g-1)2^{2g+2}(2g-1)$ by (8.2). The signature of Z_g can be computed by using the G-signature theorem [9, 75]. Let $\tilde{A} = \tilde{\Gamma}_f + \tilde{\Gamma}_{\tau f} \subset Z_g$ be the branched curve which is mapped isomorphically onto A. Then the self-intersection numbers of A and $\tilde{A}$ satisfy $\tilde{A}^2 = A^2/2$. It follows from the G-signature theorem that

$$\begin{aligned}\sigma(Z_g) &= 2\sigma(X \times C) - \tilde{A}^2 = 0 - A^2/2 = -(\Gamma_f^2 + \Gamma_{\tau f}^2)/2 = -\Gamma_f^2 \\ &= -\deg(f)\chi(C) = -2^{2g}(2 - 2g).\end{aligned}$$

8.2.2 Complex Surfaces of General Type

A complex surface is called of general type, if the Kodaira dimension is 2. For a complex surface of general type M, the inequality

$$c_1^2(M) \le 3c_2(M)$$

holds between the characteristic numbers, known as the *Miyaoka–Yau inequality*. (See [14].)

Theorem 8.8 ([171]) *A complex surface of general type with $c_1^2 = 3c_2$ is isomorphic to the unit ball in $\mathbb{C}^2$ divided by a discrete group action.*

Lemma 8.9 *Suppose that M is a complex surface of general type with $c_1^2 \geq 0$. Then, the inequality $\chi(M) \geq \frac{12}{8}|\sigma(M)|$ holds.*

The condition of $c_1^2 \geq 0$ holds if it is minimal, or $c_1^2 = 3c_2$ holds. The latter case is given by the ball quotient in Theorem 8.8.

Proof Recall the formulas $\sigma(M) = \frac{1}{3}(c_1^2 - 2c_2)$ with $\chi(M) = c_2(M)$. Moreover positivity $c_2 > 0$ holds.

We consider two cases, and suppose $\sigma > 0$ holds. Then the strict inequality

$$\frac{10}{8}|\sigma(M)| = \frac{10}{8}\frac{1}{3}(c_1^2 - 2c_2) \le \frac{10}{24}c_2 = \frac{10}{24}\chi(M)$$

holds according to the Miyaoka–Yau inequality $c_1^2 \le 3c_2$.

Suppose $\sigma(M) \le 0$ holds. Then

$$\frac{10}{8}|\sigma(M)| = \frac{10}{8}\frac{1}{3}(2c_2 - c_1^2) \le \frac{10}{12}c_2 = \frac{10}{12}\chi(M)$$

holds by non-negativity of the Chern number. □

There are examples of complex surfaces of general type with $c_1^2 = 3c_2$ [79, 134]. These complex surfaces are aspherical if they do not contain singularities. On the other hand, from the viewpoint of algebraic geometry, it would be more canonical if we can treat complex surfaces with singularities. There are studies of Seiberg–Witten theory on orbifold 4-manifolds. One might expect to obtain an orbifold version of the aspherical $\frac{10}{8}$-inequality conjecture

$$\chi_{\text{orb}}(M) \geq 10 \cdot |\operatorname{ind}_{\text{orb}} D|$$

for complex surfaces M of general type with $c_1^2 = 3c_2$ that (possibly) contain singularities. Here, $\text{ind}_{\text{orb}}\, D$ is the index of the Dirac operator for the orbifold M [56, 91, 92]

It would be also of particular interest for us to study complex hyperbolic surfaces [22].

8.3 Covering $\frac{10}{8}$-Inequality

We verify that the following class of 4-manifolds satisfies Conjecture 8.1.

Proposition 8.10 ([20], see [87]) *The covering $\frac{10}{8}$-inequality holds for a compact spin four-manifold with the residually finite fundamental group.*

Proof Let M be a compact spin four-manifold with the residually finite fundamental group. Let $\Gamma \supset \Gamma_1 \supset \Gamma_2 \supset \dots$ be the tower by normal subgroups of finite indices with $\cap_i\ \Gamma_i = 1$. Let M_i be Γ/Γ_i spin coverings of M with $\pi_1(M_i) = \Gamma_i$, and $X = \tilde{M}$ be the universal covering of M with $\pi_1(M) = \Gamma$.

Note that any covering space of a spin manifold is also spin. By Furuta, the inequalities hold:

$$b_2(M_i) \geq \frac{10}{8}|\sigma(M_i)| + 2.$$

Let $m_i = |\Gamma/\Gamma_i|$ and divide both sides by m_i as

$$\frac{b_2(M_i)}{m_i} \geq \frac{10}{8} \cdot \frac{|\sigma(M_i)|}{m_i} + \frac{2}{m_i}.$$

First, the equality $\frac{\sigma(M_i)}{m_i} = \sigma(M)$ holds because the signature is multiplicative under finite covering. It follows that

$$\lim_i \frac{b_2(M_i)}{m_i} = b_\Gamma^2(X)$$

converges to the L^2-Betti number (see Theorem 13.49 in [116]). Then, the conclusion holds because $\frac{2}{m_i} \to 0$. □

Bohr developed a very interesting argument that replies on some group-theoretic properties. In Bohr's case, M is not necessarily assumed to be aspherical.

Theorem 8.11 ([20]) *Suppose the intersection form of M is even, and its fundamental group is amenable or realized by π_1 of a closed hyperbolic manifold of* dim ≥ 3. *Then, the inequality $\chi(M) \geq \frac{10}{8}|\sigma(M)|$ holds.*

8.4 Ricci Nonnegative Case

By using the Bochner–Weitzenböck formula, we may expect the inequality

$$\chi(M) \geq \frac{10}{8} \cdot |\sigma(M)|,$$

when a compact spin smooth 4-manifold M has an infinite fundamental group and admits a Riemannian metric of nonnegative Ricci curvature.

It turns out that the conclusion is not so interesting, as we explain below.

Lemma 8.12 *Let M be a compact smooth 4-manifold with an infinite fundamental group. If M admits a Riemannian metric of nonnegative Ricci curvature, then the equality holds:*

$$\chi(M) = b_\Gamma^2(\tilde{M}).$$

Proof Suppose that M admits a Riemannian metric of nonnegative Ricci curvature. Let us lift it on the universal covering space $\tilde{M}$. It follows from the Bochner–Weitzenböck formula

$$\Delta_1 = \nabla^* \nabla + \mathrm{Ric}$$

on L^2 1-forms on $\tilde{M}$ that any L^2-harmonic 1-form α must satisfy the equality

$$\begin{aligned} 0 &= \langle \nabla^* \nabla \alpha, \alpha \rangle + \langle \mathrm{Ric}(\alpha), \alpha \rangle \\ &= \|\nabla \alpha\|^2 + \langle \mathrm{Ric}(\alpha), \alpha \rangle. \end{aligned}$$

Because both sides are nonnegative, $\nabla \alpha = 0$ must hold.

Hence, the L^2-harmonic 1-form α is parallel, and it is determined at the value $\alpha(p) \in T_p^* \tilde{M}$ of a point $p \in \tilde{M}$. Because $T_p^* \tilde{M}$ is 4-dimensional, it follows that the space of L^2-harmonic 1-forms is finite dimensional. This implies that the space is zero, because it is assumed that the fundamental group is infinite and hence, the space of L^2-harmonic forms must be infinite dimensional or zero. Hence, the L^2-Betti number $b_\Gamma^1(\tilde{M}) = 0$ vanishes. By the Poincaré duality, we also have $b_\Gamma^3(\tilde{M}) = 0$.

Note that $b^0_\Gamma(\tilde{M}) = 0$ vanishes, because $\tilde{M}$ is non-compact, and there are no non-zero L^2-harmonic functions on a connected and non-compact manifold. Again, by the Poincaré duality, we also have $b^4_\Gamma(\tilde{M}) = 0$.

It follows from Atiyah's Γ-index theorem that the equality holds:

$$\chi(M) = \sum_{i=0}^{4} (-1)^i \cdot b^i_\Gamma(\tilde{M}) = b^2_\Gamma(\tilde{M}).$$

This completes the proof. □

In particular, the Euler characteristic is nonnegative. Recently, the following has been verified:

Lemma 8.13 ([28]) *Let M be a compact spin smooth 4-manifold with an infinite fundamental group. Moreover, assume that M admits a Riemannian metric of non-negative Ricci curvature. Then, the signature $\sigma(M) = 0$ vanishes.*

Hence, the $\frac{10}{8}$-inequality in the first paragraph of this section trivially holds. See also [3].

Remark 8.14

- There is another vanishing theorem (Theorem 8.29).
- Note that Lemma 8.13 does not make sense for positive Ricci case, because in that case, the fundamental group is finite by Meyer's theorem.
- The universal covering space $\tilde{M}$ satisfies the conclusion of the Singer conjecture under the assumption of Lemma 8.12.

8.5 L^p-Exactness and Kähler Hyperbolicity

We discuss the Kähler hyperbolicity due to Gromov [66] in this section and the next. The relation between the Kähler hyperbolicity and the aspherical 10/8-conjecture will be discussed in Sect. 8.6.1.

Let α be a closed p-form $d\alpha = 0$.

Definition 8.15 α is L^p-exact, if there is $\beta \in L^p(X; \Lambda^{*-1})$, where $d\beta = \alpha$ holds for $1 \leq p \leq \infty$.

Example The 2-form $\alpha := dx \wedge dy$ is closed on $\mathbb{R}^2 \ni (x, y)$, and $\alpha = d\beta$ for $\beta = xdy$. The 1-form β is not in $L^p(\mathbb{R}^2; \Lambda^1)$. Any β with $d\beta = \alpha$ cannot be realized by an L^∞-form. In fact, suppose it could be. Then for r-ball $B(r) \subset \mathbb{R}^2$,

$$\text{Vol } B(r) = \int_{B(r)} \alpha = \int_{B(r)} d\beta = \int_{\partial B(r)} \beta,$$

where the left-hand side is order r^2 and the right is order r. This cannot happen.

Let $x_0 \in X$ be any base point, and for any subset $K \subset X$, we denote by CK the geodesic cone given by the union of geodesics from any points $x \in X$ to x_0.

Lemma 8.16 *Let (X, g) be a simply-connected complete Riemannian manifold with uniformly negative sectional curvature $-c' < K_g \leq -c < 0$.*

Then for any positive $\epsilon > 0$, there is $\delta > 0$ such that for any m-simplex Δ_m of diameter less than ϵ, the geodesic cone $C\Delta$, the area satisfies the uniform bound

$$\text{area}(C\Delta) \ \leq \ \delta \cdot \text{area}(\Delta).$$

Proposition 8.17 *Let (X, g) be a simply-connected complete Riemannian manifold with uniformly negative sectional curvature $-c' < K_g \leq -c < 0$.*

Then for any closed form $\alpha \in L^\infty(X; \Lambda^*)$, there is $\beta \in L^\infty(X; \Lambda^{*-1})$ such that $\alpha = d\beta$.

Proof Let us fix a base point $x_0 \in X$, and CK be as in Lemma 8.16.

Let $\varphi : [0, 1] \times X \to X$ be the smooth map such that $\varphi(\ \ , x) : X \to X$ is the unique geodesic with $\varphi(0, x) = x_0$ and $\varphi(1, x) = x$. Then, for any closed p-form α on X, consider the pull-backed p-form $\varphi^*(\alpha)$ on $[0, 1] \times X$. Let t be the coordinate on $[0, 1]$, and let us integrate as

$$K(\alpha) := \int_0^1 \varphi^*(\alpha)$$

with respect to the variable t. The result $K(\alpha)$ is a $p - 1$-form on X. It is easy to check that $d(K(\alpha)) = \alpha$ (exercise, see Problem 8.32).

We claim that $K(\alpha) \in L^\infty(X; \Lambda^{p-1})$, if $\alpha \in L^\infty(X; \Lambda^p)$. For a positive $\epsilon > 0$, let $\delta > 0$ as in Lemma 8.16. Let Δ_{p-1} be a simplex in X with diameter less than ϵ. Then, consider

$$\int_{\Delta_{p-1}} K(\alpha) = \int_{C\Delta_{p-1}} \alpha.$$

Then, we have the estimates

$$\left| \int_{\Delta_{p-1}} K(\alpha) \right| \leq \text{area}(C\Delta_{p-1}) \cdot \|\alpha\|_{L^\infty} \leq \delta \cdot \text{area}(\Delta_{p-1}) \cdot \|\alpha\|_{L^\infty}.$$

Hence, we obtain the estimate $\left(\mathrm{area}(\Delta_{p-1})\right)^{-1} \cdot \left|\int_{\Delta_{p-1}} K(\alpha)\right| \leq \delta \cdot \|\alpha\|_{L^\infty}$. This implies the estimate $\|K(\alpha)\|_{L^\infty} \leq \delta \cdot \|\alpha\|_{L^\infty}$. □

Example Let $\mathbf{H}_n$ be the hyperbolic space. Then for any closed form $\alpha \in L^\infty(\mathbf{H}_n; \Lambda^*)$, there is $\beta \in L^\infty(\mathbf{H}_n; \Lambda^{*-1})$ such that $\alpha = d\beta$.

Let us introduce a variant of the notion of L^p-exactness. Let (M, g) be a Riemannian manifold, and $X := \tilde{M}$ be the universal covering space.

Definition 8.18 A closed form $\alpha \in L^\infty(M; \Lambda^*)$ is $\tilde{L}^\infty$-exact, if there is $\tilde{\beta} \in L^\infty(X; \Lambda^{*-1})$ such that the lift $\tilde{\alpha}$ of α on the universal covering space satisfies the equality

$$\tilde{\alpha} = d\tilde{\beta}.$$

Let (M, ω) be a Kähler manifold. It is *Kähler hyperbolic*, if the Kähler form ω is $\tilde{L}^\infty$-exact [66].

Examples

- Let $\alpha = dx \wedge dy \in \Omega^2(T^2)$. Then $\tilde{\alpha} = d(xdy)$ holds, but xdy is not in L^∞.
- If a compact Kähler manifold admits a Riemannian metric of negative curvature, then it is Kähler hyperbolic by Proposition 8.17.
- An irreducible symmetric Hermitian space of non-compact type is Kähler hyperbolic. See Example 8.6.1.

8.6 L^2-Lefschetz Theory

Here is some linear algebra. Let T_0 be a real vector space of dimension $2m$, and take a nonsingular 2-form $\omega_0 \in \Lambda^2 T_0$. Define

$$L_0^k : \Lambda^l T_0 \to \Lambda^{l+2k} T_0$$

by $\alpha \to \omega_0^k \wedge \alpha$.

Lemma 8.19

$$L_0^k \text{ is } \begin{cases} \textit{injective} & \textit{for } 2l + 2k \leq 2m, \\ \textit{surjective} & \textit{for } 2l + 2k \geq 2m. \end{cases}$$

Proof First, consider the case $2l_0 + 2k = 2m$. The bilinear form

$$\Lambda^{l_0} T_0 \otimes \Lambda^{l_0} T_0 \to \mathbb{R}$$

given by $\xi \otimes \eta \to \xi \wedge \eta \wedge \omega_0^k \in \Lambda^{2m} T_0 \cong \mathbb{R}$. The bilinear paring is non-degenerate, because ω_0 is nonsingular. This implies that the map $\eta \hookrightarrow \eta \wedge \omega_0^k$ is injective. Because $\dim \Lambda^{l_0} T_0 = \dim \Lambda^{l_0+2k} T_0$ holds, it is an isomorphism.

Note the equality $L_0^k \circ L_0^{k'} = L_0^{k+k'}$. In particular, $L_0^k : \Lambda^l T_0 \to \Lambda^{l+2k} T_0$ is injective for $2l + 2k \leq 2m$. Next, choose $l_0 \leq l$ with $2l_0 + 2k = 2m$. Note that the exterior product $\wedge : \Lambda^{l-l_0} T_0 \otimes \Lambda^{l_0+2k} T_0 \to \Lambda^{l+2k} T_0$ is surjective. Because $L_0^k : \Lambda^{l_0} T_0 \cong \Lambda^{l_0+2k} T_0$ is isomorphic, the composition $\Lambda^{l-l_0} T_0 \otimes \Lambda^{l_0} T_0 \to \Lambda^l T_0 \to \Lambda^{l+2k} T_0$ is surjective. In particular, $L_0^k : \Lambda^l T_0 \to \Lambda^{l+2k} T_0$ is surjective for $2l+2k \geq 2m$. □

Corollary 8.20 *The algebraically dual operator* $\Lambda_0^k : \Lambda^{l+2k} T_0 \to \Lambda^l T_0$ *is injective for* $2l + 2k \geq 2m$, *where* $\Lambda_0^k := (L_0^k)^*$.

Let (M, ω) be a compact Kähler manifold of $\dim_{\mathbb{C}} = m$. There is the associated Kähler metric, and hence, the Laplace–de Rham operator Δ is given.

Denote

$$L^k : \Omega^l(M) \to \Omega^{l+2k}(M)$$

by $\varphi \to \omega^k \wedge \varphi$. Here is a local property of Kähler analysis.

Lemma 8.21 $[L^k, d] = [L^k, \Delta] = 0$ *hold.*

For a proof, see [65, p. 115].

Theorem 8.22 (The L^2-Lefschetz Theorem) *Let* $(X, \tilde{\omega})$ *be the complete Kähler manifold that is a lift of a compact Kähler manifold* (M, ω). *Then,* L^k *induces the map*

$$L^k : \mathcal{H}^l(X) \to \mathcal{H}^{l+2k}(X)$$

between L^2*-harmonic forms, such that they are*

$$\begin{cases} \textit{injective}, & \textit{if} \quad 2l + 2k \leq \dim_{\mathbb{R}} X, \\ \textit{surjective}, & \textit{if} \quad 2l + 2k \geq \dim_{\mathbb{R}} X. \end{cases}$$

Proof L^k induces the map $L^k : \mathcal{H}^l(X) \to \mathcal{H}^{l+2k}(X)$ by Lemma 8.21. Injectivity holds for $2l + 2k \leq \dim_{\mathbb{R}} X$ by Lemma 8.19.

Let $\Lambda^k := (L^k)^*$ be the algebraically dual operator. Then, Λ^k also induces the map

$$\Lambda^k : \mathcal{H}^{l+2k}(X) \to \mathcal{H}^l(X)$$

because the L^2-formal adjoint operator satisfies the equality $0 = [L^k, \Delta]^* = [\Lambda^k, \Delta]$. The algebraically dual operator $\Lambda_0^k : \Lambda^{l+2k}T^*X \to \Lambda^l T^*X$ is injective by Corollary 8.20 for $2l + 2k \geq \dim_{\mathbb{R}} X$. Hence, $\Lambda^k : \mathcal{H}^{l+2k}(X) \hookrightarrow \mathcal{H}^l(X)$ is injective for $2l+2k \geq \dim_{\mathbb{R}} X$. In particular $L^k : \mathcal{H}^l(X) \to \mathcal{H}^{l+2k}(X)$ has a dense image for $2l + 2k \geq \dim_{\mathbb{R}} X$. On the other hand, there is a positive constant $c > 1$ such that the estimates

$$c^{-1}\|\varphi\|_{L^2} \leq \|L^k\varphi\|_{L^2} \leq c\|\varphi\|_{L^2}$$

hold for any $\varphi \in L^2(X; \Lambda^{l_0})$ and $l_0 + k = m$. Hence, L^k is an isomorphism, when $l_0 + k = m$.

It remains to consider the case when $2l + 2k \geq 2m$ and $l + 2k \leq 2m$. There is $l_0 \leq l$ and $k_0 \geq k$ such that

$$2l_0 + 2k_0 = 2m, \quad l_0 + 2k_0 = l + 2k$$

hold, by letting

$$l_0 = 2m - (l + 2k), \quad k_0 = l + 2k - m.$$

Then, $L^{k_0} : \mathcal{H}^{l_0}(X) \cong \mathcal{H}^{l+2k}(X)$ factors through

$$\mathcal{H}^{l_0}(X) \xrightarrow{L^{l+k-m}} \mathcal{H}^l(X) \xrightarrow{L^k} \mathcal{H}^{l+2k}(X).$$

Hence, $L^k : \mathcal{H}^l(X) \twoheadrightarrow \mathcal{H}^{l+2k}(X)$ has closed range and is surjective. □

Corollary 8.23 *Let (M, ω) be a compact Kähler manifold such that ω is $\tilde{L}^\infty$-exact. Then $\mathcal{H}^l(X) = 0$ vanishes for $X := (\tilde{M}, \tilde{\omega})$ and $l \neq \dim_{\mathbb{C}} M$.*

Proof Let $\tilde{\omega} = d\eta$ for some $\eta \in L^\infty(X; \Lambda^1)$. Then, $L^k(\varphi) = \tilde{\omega}^k \wedge \varphi = d(\eta \wedge (d\eta)^{k-1} \wedge \varphi) =: d\psi$ for any L^2-closed form φ. Note that $\psi \in L^2(X; \Lambda^*)$ because both η and $d\eta$ are in L^∞. Then, $L^k(\varphi) = 0$ holds if $\varphi \in \mathcal{H}^l(X)$ by Corollary 6.6. Moreover, $\varphi = 0$ holds for $l \leq \dim_{\mathbb{C}} M - 1$ by Theorem 8.22.

For $l + k \geq \dim_{\mathbb{C}} M$, the map $L^k : \mathcal{H}^l(X) \twoheadrightarrow \mathcal{H}^{l+2k}(X)$ is a surjection. Hence, $\mathcal{H}^{l+2k}(X) = 0$ holds, because $\mathcal{H}^l(X) = 0$ holds for $0 \leq l \leq \dim_{\mathbb{C}} M - 1$. □

Corollary 8.24 *The Singer conjecture 2.9 holds for a Kähler hyperbolic manifold.*

We can induce the following, however we do not verify it because we will not use it later.

Theorem 8.25 ([66]) *Let $(X, \tilde{\omega})$ be a complete Kähler manifold such that $\tilde{\omega}$ is L^∞-exact. Then there is a positive constant $\lambda_0 > 0$ such that for any $\psi \in L^2(X; \Lambda^l)$ with $l \neq \dim_{\mathbb{C}} X$, the estimate holds:*

$$\langle \psi, \Delta \psi \rangle_{L^2} \geq \lambda_0^2 \langle \psi, \psi \rangle_{L^2}.$$

For $l = \dim_{\mathbb{C}} X$, the same estimate holds for any $\psi \in L^2(X; \Lambda^l) \cap (\mathcal{H}^l(X))^\perp$.

Moreover, $\mathcal{H}^m(X) \neq 0$ holds for $m = \dim_{\mathbb{C}} X$.

Corollary 8.26 *The strong Singer conjecture holds for a Kähler hyperbolic manifold.*

8.6.1 Kähler Hyperbolic Manifolds with Residually Finite Fundamental Groups

Let us combine previous results, and induce that the following class of 4-manifolds satisfies Conjecture 8.2.

Corollary 8.27 *Suppose a four-manifold M is aspherical and spin. Moreover, assume that $\pi_1(M)$ is residually finite and Kähler-hyperbolic.*

Then, the aspherical $\frac{10}{8}$ conjecture holds.

Proof By Corollary 8.24, $\tilde{M}$ satisfies the Singer conjecture. By Proposition 8.10, M satisfies the covering $\frac{10}{8}$-inequality conjecture 8.1. Then, the conclusion follows by Lemma 8.4. □

Examples

(1) A Kähler manifold is Kähler hyperbolic, if it is homotopy equivalent to a manifold which admits a metric of negative curvature.

A Kähler manifold M is Kähler hyperbolic, if $\pi_1(M)$ is word-hyperbolic and $\pi_2(M) = 0$ hold. In particular, an aspherical Kähler manifold M is Kähler hyperbolic, if $\pi_1(M)$ is word-hyperbolic. [66], see also [116].

(2) An irreducible symmetric Hermitian space of non-compact type G/K is Kähler hyperbolic, where G is a connected non-compact simple adjoint Lie group and K is a maximal connected and compact Lie subgroup of G with the center S^1. [170].

A complex manifold is Kähler hyperbolic, if it is biholomorphic to a bounded symmetric domain in the complex plane. [66].

(3) The product space of two Kähler hyperbolic manifolds is also Kähler hyperbolic.

In particular, a compact manifold is Kähler hyperbolic, if its universal covering space is a symmetric Hermitian space of non-compact type. In fact, it is a product of irreducible symmetric Hermitian spaces of non-compact type.

(4) A finitely generated linear group Γ is residually finite. Let G/K be the case of (2). If $M := \Gamma\backslash G/K$ is a compact Kähler manifold, then M or product spaces of M are all Kähler hyperbolic such that $\pi_1(M)$ is residually finite, where $\Gamma \subset G$ is a co-compact discrete subgroup.

It is one of approaches to atack the aspherical $\frac{10}{8}$-inequality, to seek for spin four-manifolds with two conditions of Kähler hyperbolicity and residually finiteness of the fundamental group. The aspherical $\frac{10}{8}$-inequality has been verified for such a class using a family of normal coverings of finite index. Our ultimate goal to is to eliminate the second condition of residual finiteness and to present a more straightforward method by developing the analysis on the universal covering spaces of compact spin four-manifolds [85, 87].

Remark 8.28 It is known that an oriented and definite four-manifold must have a diagonal and hence odd-type intersection form ([35], [38], see also [18]). An aspherical four-manifold with definite form is not expected to satisfy the aspherical $\frac{10}{8}$-inequality conjecture 1.5, if it exists. In fact in this case the inequality is given by

$$2(1 - b_1) + b_2 \geq \frac{10}{8} b_2,$$

which is equivalent to $8(1 - b_1) \geq b_2 \geq 0$. Then, we have a contradiction to the above inequality if $b_1 > 1$ holds. Thus, the $b_1 = 1$ (and hence $b_2 = 0$) and $b_1 = 0$ cases might survive. Both cases seem rare for aspherical four-manifolds.

Note that there are non-commutative and infinite groups whose abelianizations are trivial (perfect groups). If the fundamental group is perfect, then $b_1 = 0$ holds.

8.7 Infinite Amenable Groups

Here is a result by Cheeger–Gromov:

Theorem 8.29 ([27]) *Let X be the universal covering space of a compact aspherical smooth manifold M. Then the L^2-Betti numbers of all degrees vanish over X, if its fundamental group is infinite and amenable.*

Theorem 8.29 is based on the computational result of the L^2-cohomology of an amenable group. Hence, the aspherical condition is necessary.

In particular, the aspherical $\frac{10}{8}$-inequality holds in a trivial way:

Corollary 8.30 *Let M be an oriented compact aspherical smooth four-manifold with the infinite amenable fundamental group. Then, both the signature and the Euler characteristics vanish.*

Proof Apply Atiyah's Γ-index theorem to the cases of the signature and the Euler characteristics. □

Let us verify Theorem 8.29.

Proof of Theorem 8.29 Let $K \subset X$ be a fundamental domain of $\pi(M)$-action. By Proposition 6.34, there is a family of finite subsets $\{F_i\}_i \subset \pi_1(M)$ such that

$$\lim_{i\to\infty} \frac{\text{Area}(S_i)}{\text{Vol}(V_i)} = 0$$

holds, where

$$V_i := \cup_{g\in F_i} K \cdot g, \quad S_i := \cup_{g\in\partial F_i} K \cdot g.$$

Let us take a simplicial decomposition L of M, and let $\tilde{L}$ be the lift of L on X. It is enough to verify $\bar{H}^m_{(2)}(\tilde{L}) = 0$ for all $0 \le m \le \dim M$ by Theorem 6.46. Let $C^i_{(2)}(\tilde{L})$ be the space of L^2-cochains on the i-simplices, and denote the simplicial differentials by $d_c : C^i_{(2)}(\tilde{L}) \to C^{i+1}_{(2)}(\tilde{L})$.

It is easy to identify $\bar{H}^m_{(2)}(\tilde{L})$ with the space

$$\mathcal{H}^m_{(2)}(\tilde{L}) := \{ \alpha \in C^m_{(2)}(\tilde{L}) : d_c(\alpha) = d_c^*(\alpha) = 0 \}.$$

Let

$$P_m : C^m_{(2)}(\tilde{L}) \to \mathcal{H}^m_{(2)}(\tilde{L}),$$

$$Q_i : C^m_{(2)}(\tilde{L}) \to C^m_{(2)}(V_i)$$

be the orthogonal projection and the restriction respectively. Note that the restriction $Q_i : \mathcal{H}^m_{(2)}(\tilde{L}) \hookrightarrow C^m_{(2)}(V_i)$ is injective. Then, the following equality holds:

$$\dim_\Gamma \mathcal{H}^m_{(2)}(\tilde{L}) = \frac{1}{(\sharp L_m) \cdot (\sharp F_i)} \cdot \operatorname{tr}(Q_i P_m),$$

where $\sharp L_m$ is the number of m-simplices in L. The right-hand side is bounded as

$$\frac{1}{(\sharp L_m) \cdot (\sharp F_i)} \cdot \operatorname{tr}(Q_i P_m) \leq \frac{1}{(\sharp L_m) \cdot (\sharp F_i)} \cdot \operatorname{rank}(Q_i P_m).$$

Because the simplicial differential commutes with the restriction, we have that $d_c Q_i(h) = 0$ holds for any $h \in \mathcal{H}^m_{(2)}(\tilde{L})$. On the other hand, supp $d_c^* Q_i(h) \subset S_i$ holds, even though $d_c^* Q_i(h)$ may not vanish in general.

Because X is contractible, one can express $h = dg$, where g is an $m-1$-cochain (not necessarily in L^2). Then, for any $h, h' \in \mathcal{H}^m_{(2)}(\tilde{L})$, we have equalities of L^2-inner products:

$$\begin{aligned} \langle Q_i h, Q_i h' \rangle &= \langle Q_i d_c g, Q_i h' \rangle = \langle d_c Q_i g, Q_i h' \rangle \\ &= \langle Q_i g, d_c^* Q_i h' \rangle. \end{aligned}$$

In particular, the rank of the restriction $Q_i(\mathcal{H}^m_{(2)}(\tilde{L}))$ is bounded by $(\sharp L_m) \cdot (\sharp F_i)$.

$$\dim_\Gamma \mathcal{H}^m_{(2)}(\tilde{L}) \leq \frac{1}{(\sharp L_m) \cdot (\sharp F_i)} \cdot \operatorname{rank}(Q_i P_m) \leq \frac{\sharp \partial F_i}{\sharp F_i} \to 0$$

as $i \to \infty$. □

Examples

- The 4-torus T^4 is a compact aspherical smooth spin four-manifold with an amenable fundamental group.
- Let N be the 3-dimensional Heisenberg group. The fundamental groups are nilpotent, and hence amenable. Thus, the product $N \times S^1$ is a compact aspherical smooth spin four-manifold with an amenable fundamental group.

Question
Are there any group theoretic constraints of the fundamental group of a compact complex surface of general type?

Proposition 8.31 *Suppose that M is a compact complex surface of general type with $0 \neq c_1^2 = 3c_2$. Then the fundamental group cannot be amenable.*

Proof By Yau's ball quotient theorem, M is aspherical. If $\pi_1(M)$ could be amenable, then by Theorem 8.29, $\chi(M) = c_2(M) = 0$ must hold. □

There are various complex surfaces that are given by Yau's ball quotient, and they contain singularities.

Problems

8.32 Verify $d(K(\Omega)) = \omega$ in the proof of Proposition 8.17.

8.33 (Aspherical Fiber Bundles) Verify that the total space of a fiber bundle is aspherical, if both the base and the fiber spaces are aspherical.

8.34 Give more examples of aspherical 4-manifolds.

8.35 (Fundamental Groups of Aspherical Manifolds)
Verify that the fundamental group of a compact aspherical manifold must be infinite.

8.36 Let M be a compact spin 4-manifold with a finite fundamental group. Then, verify that the inequality $\chi(M) \geq \frac{10}{8}|\sigma(M)| + 2$ holds.

8.37 (Bohr [20])
Let M be a compact spin 4-manifold, and $\tilde{M}$ be the infinite cyclic covering corresponding to a surjective homomorphism $\pi_1(M) \twoheadrightarrow \mathbb{Z}$. Then, at least one of the following holds:

1. $\dim b_1(\tilde{M}) = \infty$, or
2. the inequality $\chi(M) \geq \frac{10}{8}|\sigma(M)|$ holds.

(Hint: Consider the induced family of finite normal coverings corresponding to $\pi_1(M) \to \mathbb{Z}_k$, and use Theorem [59].)

8.38 Give an example of an infinite group Γ whose commutator group vanishes.

8.39 Does there exist a compact 4-manifold with an infinite fundamental group and $b_1(M) = 0$?

8.40 Let $F \hookrightarrow M \xrightarrow{\pi} B$ be a fiber bundle, where M is compact. Then, verify that the Euler characteristics is multiplicative in the sense that the equality $\chi(M) = \chi(F) \cdot \chi(B)$ holds.

8.41 Verify that a compact aspherical 4-manifold with an amenable fundamental group cannot be equipped with a Kähler hyperbolic structure.

Solutions

Chapter 2

2.52 For example, consider the circle S^1.

2.53

(a) Describe $\mathbb{C}P^m = \{(z_0 : \cdots : z_m) \neq 0 | z_i \in \mathbb{C}\}/\sim$. Then we can give a finite group action of order n by

$$[z_0 : z_1 : \cdots : z_m] \mapsto [\exp(2\pi i/n) \cdot z_0 : z_1 : \cdots : z_m].$$

(b) Suppose a free action ρ of order ≥ 2 on $\mathbb{C}P^{2m}$ could exist, and consider the quotient space $\mathbb{C}P^{2m}/\rho$. Because the signature is multiplicative under taking a finite covering, the signature of $\mathbb{C}P^{2m}$ should be a multiple of n, in contradiction to $\sigma(\mathbb{C}P^{2m}) = 1$. (The same conclusion also holds for odd m. For a proof, use the Lefschetz fixed point theorem.)

2.54

(a) For a compact complex surface M, there exists some $h \in H^2(M;\mathbb{Z})$ that is an integral lift of $w_2(M)$ such that the formula $\langle h^2, [M]\rangle = 2e(M)+3\sigma(M)$ holds. (See [76, §4.6].) For the case of the sphere, $w_2(S^{2m}) = 0$ and $\sigma(S^{2m}) = 0$ hold: however, $e(S^{2m}) = 2$.
(b) Suppose a complex structure could exist on S^{2m} for $m \geq 4$. The Hirzebruch–Riemann–Roch theorem gives

$$\begin{aligned}\operatorname{ind} \bar{\partial} &= \int_{S^{2m}} \hat{A}(M) \wedge ch(TS^{2m}) \\ &= \int_{S^{2m}} \frac{1}{(m-1)!} \cdot c^m(S^{2m}) = \int_{S^{2m}} \frac{1}{(m-1)!} \cdot e(S^{2m}) = \frac{2}{(m-1)!}.\end{aligned}$$

T. Kato, N. Nakamura, *Monopole Maps and L^2-cohomology*, Springer Monographs in Mathematics, https://doi.org/10.1007/978-981-96-8472-4

However, the right-hand side cannot be an integer for $m \geq 4$.

2.55 Suppose X is aspherical. Let $\tilde{X}$ be the universal covering of X. Then $\pi_k(\tilde{X}) = 0$ for $k \geq 1$. By the long exact sequence of homotopy groups for fibration (see [123, §9.3]), we have $\pi_k(X) \cong \pi_k(\tilde{X}) = 0$ for $k \geq 2$.

Conversely, suppose $\pi_k(X)$ are trivial except for $k = 1$. Then $\pi_k(\tilde{X}) \cong \pi_k(X) = 0$ for $k \geq 2$ as above, and $\pi_1(\tilde{X})$ is trivial by definition. Therefore $\tilde{X}$ is weak homotopy equivalent to a point. By the Whitehead theorem (see [123, §10.3]), $\tilde{X}$ is contractible.

2.56 By the Poincaré–Hopf theorem, the Euler characteristic $\chi(M) = \int_M e(M) = 0$ must vanish. Note the equalities $c_2(M) = e(M)$. Then, the conclusion follows by the formula $c_1(M)^2 = 2e(M) + 3\sigma(M)$.

2.57 Let $v \in L^2(X; \Lambda^*)$ be a non-zero harmonic form. Because it satisfies an elliptic equation, v cannot vanish on any non-empty open subset of X. Let $K \subset X$ be a fundamental domain of the $\pi_1(M)$ action. If the space of the harmonic forms is finite-dimensional, then $\inf_v \{||v||_{L^2(K)} : v \text{ is harmonic with } ||v||_{L^2} = 1\} > 0$ must be positive. Let $g_i \in \pi_1(M)$ be $g_i \to \infty$. Then, $||g_i^*(v)||_{L^2(K)} \to 0$ must converge to zero, because v is in L^2.

2.58 First, assume that both manifolds are orientable. Then, the de Rham cohomology group is non-zero at the top degree. Because the de Rham cohomology group is homotopy invariant and the dimension coincides with the top degree, the conclusion holds.

If they are non-orientable, then take their double covering spaces and consider as above.

2.59 Let $f_t : M \to M'$ be a homotopy such that $f_0 = f$ and f_1 is diffeomorphic. Because the Pontryagin classes $p_i(M) \in H^{4i}(M; \mathbb{Z})$ are diffeomorphism invariant, the equalities $f_1^*(p_i(M')) = p_i(M)$ hold for any $i \geq 0$. Because the pull-back of cohomology groups is homotopy invariant, the same equalities $f^*(p_i(M')) = p_i(M)$ also hold for any $i \geq 0$.

2.60 Let $f : X \to Y$ be a smooth map between compact oriented smooth manifolds of the same dimension N. Take a regular value $y \in Y$. Then the inverse image $f^{-1}(y) =: \{x_1, \cdots, x_l\} \subset X$ consists of a finite set. Let $\omega \in \Omega^N(Y)$ be an N-form whose support is localized sufficiently near $y \in Y$ and $\int_Y \omega = 1$. Then the pull-back $f^*(\omega) = \mu_1 + \cdots + \mu_l$ splits with disjoint supports mutually and $x_i \in \text{supp}\, \omega_i$. Then, we have the equalities

$$\deg f = \int_X f^*(\omega) = \sum_{i=1}^{l} \mu_i = \sharp f^{-1}(y).$$

2.61 A compact spin smooth 4-manifold M with definite form must satisfy $b_2(M) = 0$, because a compact spin smooth manifold has even form (see [94]), and the definite form is odd (Theorem 2.36).

Chapter 3

3.90

(1) Let $\{e_1, \cdots, e_n\}$ be an orthonormal basis of V. Then for a subset $\{i_1, \cdots i_l\} \subset \{1, \cdots, n\}$, the assignment is given by

$$e_{i_1} \wedge \cdots \wedge e_{i_l} \to e_{i_1} \cdots e_{i_l}.$$

In particular, the basis of $\Lambda^* V$ is sent to the basis of $\mathrm{Cl}(V)$.

(2) This can be checked by computation of the orthonormal basis.

3.91 Check the equality $\nabla(f \cdot \omega \otimes \sigma) = \nabla(\omega \otimes f \cdot \sigma)$ for $f \in C^\infty(X)$.

3.92 Let the tensor $R := \nabla^2$. Then, the identities

$$\begin{aligned}\nabla^3(\sigma) = \nabla(\nabla^2(\sigma)) = \nabla(R(\sigma)) &= (\nabla R)(\sigma) + R(\nabla(\sigma)) \\ &= \nabla^3(\sigma) + \nabla^2(\nabla(\sigma)) = 2\nabla^3(\sigma)\end{aligned}$$

give the Bianchi identity.

3.93 (3.9) is obvious. For (3.8), It suffices to prove that $\omega_{pg}(g_* v) = g^{-1} \cdot \omega_p(v) \cdot g$ for $v = (p \cdot \exp(ta))'|_{t=0}$ for $a \in \mathfrak{g}$. Since $g_* v = (p \cdot \exp(ta) g)'|_{t=0} = (pg \cdot g^{-1} \exp(ta) g)'|_{t=0}$, we obtain $\omega_{pg}(g_* v) = g^{-1} a g = g^{-1} \cdot \omega_p(v) \cdot g$.

3.94 Choose two local trivializations φ_1 and φ_2 with the transition function $\Phi := \varphi_2^{-1} \circ \varphi_1$. Express the corresponding 1-form-valued matrices by

$$\tilde{\omega}_i := g_i^{-1} \cdot \omega_i \cdot g_i + g_i^{-1} \cdot dg_i$$

for $i = 1, 2$. Then, we have

$$\begin{aligned}\tilde{\omega}_2 &= g_2^{-1} \cdot \omega_2 \cdot g_2 + g_2^{-1} \cdot dg_2 \\ &= g_1^{-1} \Phi^{-1} \omega_2 \Phi g_1 + g_1^{-1} \Phi^{-1} d\Phi g_1 + g_1^{-1} dg_1 \\ &= g_1^{-1} \cdot \omega_1 \cdot g_1 + g_1^{-1} \cdot dg_1 = \tilde{\omega}_1.\end{aligned}$$

3.95 Recall the exact sequence $1 \to \mathrm{U}(1) \to \mathrm{Spin}^c(n) \to \mathrm{SO}(n) \to 1$ and the identification $\tilde{P}/\mathrm{U}(1) = P$. Since $\varphi \in \mathcal{G}$ induces the identity on $P = \tilde{P}/\mathrm{U}(1)$,

$$\mathcal{G} = \Gamma(\tilde{P} \times_{Ad} \mathrm{U}(1)) = C^\infty(X; \mathrm{U}(1)).$$

3.96 Check the equalities

$$\varphi_*(\nabla)(f \cdot \xi) = \varphi \nabla \varphi^{-1}(f \cdot \xi) = \varphi \nabla (f(\varphi^{-1}(\xi)))$$

$$= \varphi(df \otimes \varphi^{-1}(\xi)) + \varphi(f\nabla(\varphi^{-1}(\xi)))$$
$$= df \otimes \xi + f\varphi\nabla(\varphi^{-1}(\xi)) = df \otimes \xi + f\varphi_*(\nabla)(\xi)$$

for $f \in C^\infty(X)$ and $\xi \in \Gamma(E)$.

3.97 Take two orthonormal bases $\{e_i\}_{i=1}^n$ and $\{e'_i\}_{i=1}^n$ such that they satisfy the relation $e'_i = \sum_{j=1}^n a_{ij}e_j$ for some orthogonal-matrix-valued function (a_{ij}). Then, we have the equalities

$$\sum_{i=1}^n e'_i \cdot \tilde{\nabla}_{e'_i} = \sum_{i,j=1}^n a_{ij}e_j \cdot \tilde{\nabla}_{\sum_{k=1}^n a_{ik}\, e_k} = \sum_{i,j,k=1}^n a_{ij}a_{ik}e_j \cdot \tilde{\nabla}_{e_k}$$
$$= \sum_{j,k=1}^n \delta_{jk}e_j \cdot \tilde{\nabla}_{e_k} = \sum_{j=1}^n e_j \cdot \tilde{\nabla}_{e_j},$$

where δ_{jk} is the Kronecker delta.

3.98 The proof is easy. However, the idea is extensively developed in [68].

3.99 Let $(z_1, z_2) = (x_1 + iy_1, x_2 + iy_2)$ be the standard coordinate of $\mathbb{C}^2$. The adjoint of $\dfrac{\partial}{\partial \bar{z}_j}$ is $-\dfrac{\partial}{\partial z_j}$, and the adjoint of $d\bar{z}_j \wedge \cdot$ is $d\bar{z}_j \lrcorner \cdot$. Note that

$$\bar{\partial}\phi = \sum_j d\bar{z}_j \wedge \frac{\partial \phi}{\partial \bar{z}_j} = \sum_j \frac{\partial}{\partial \bar{z}_j}(d\bar{z}_j \wedge \phi).$$

Then we have

$$\bar{\partial}^*\phi = -d\bar{z}_j \lrcorner \frac{\partial \phi}{\partial z_j}.$$

Via the isomorphisms $TZ \cong T^{1,0} \cong (T^*)^{0,1}$, we have

$$\frac{\partial}{\partial x_j} \mapsto \frac{1}{\sqrt{2}}\left(\frac{\partial}{\partial x_j} - i\frac{\partial}{\partial y_j}\right) = \sqrt{2}\frac{\partial}{\partial z_j} \mapsto \frac{1}{\sqrt{2}}d\bar{z}_j,$$
$$\frac{\partial}{\partial y_j} \mapsto \frac{1}{\sqrt{2}}\left(\frac{\partial}{\partial y_j} + i\frac{\partial}{\partial x_j}\right) = i\sqrt{2}\frac{\partial}{\partial z_j} \mapsto \frac{i}{\sqrt{2}}d\bar{z}_j.$$

Therefore the Clifford multiplications of $\dfrac{\partial}{\partial x_j}$ and $\dfrac{\partial}{\partial y_j}$ are given by

$$\frac{\partial}{\partial x_j} \mapsto \frac{1}{\sqrt{2}}(d\bar{z}_j \wedge \cdot - d\bar{z}_j \lrcorner \cdot),$$

$$\frac{\partial}{\partial y_j} \mapsto \frac{1}{\sqrt{2}}(id\bar{z}_j \wedge \cdot - (id\bar{z}_j) \lrcorner \cdot) = \frac{i}{\sqrt{2}}(d\bar{z}_j \wedge \cdot + d\bar{z}_j \lrcorner \cdot).$$

(Here we used the relation in Remark 3.82.) Thus we obtain

$$\begin{aligned} D\phi &= \sum_j \frac{1}{\sqrt{2}} \left(d\bar{z}_j \wedge \frac{\partial \phi}{\partial x_j} - d\bar{z}_j \lrcorner \frac{\partial \phi}{\partial x_j} + i \left(d\bar{z}_j \wedge \frac{\partial \phi}{\partial y_j} \right) + i \left(d\bar{z}_j \lrcorner \frac{\partial \phi}{\partial y_j} \right) \right) \\ &= \sum_j \sqrt{2} \left(d\bar{z}_j \wedge \frac{\partial \phi}{\partial \bar{z}_j} - d\bar{z}_j \lrcorner \frac{\partial \phi}{\partial z_j} \right) = \sqrt{2}(\bar{\partial} + \bar{\partial}^*)\phi. \end{aligned}$$

($D = \sqrt{2}(\bar{\partial} + \bar{\partial}^*)$ holds for the canonical spin^c structures on Kähler manifolds by essentially the same computations. See [131, §3.4].)

Chapter 4

4.107 By Lemma 3.13, we may consider $\Lambda^2 V$ with Spin$^c(4)$-action to be embedded in $\mathrm{Cl}(V) \otimes \mathbb{C}$. Since the Spin$^c(4)$-action on V is given by conjugation $v \mapsto q_+ v q_-^{-1} + q_- v q_+^{-1}$ as in (3.5), the Spin$^c(4)$-action on $\Lambda^2 V$ is given by

$$v \wedge w \mapsto q_+(vw)_+ q_+^{-1} + q_-(vw)_- q_-^{-1},$$

where $(vw)_\pm \in (\mathrm{Cl}_0(V) \otimes \mathbb{C})^\pm$ correspond respectively to the self-dual and anti-self-dual parts of $v \wedge w$. Then the Spin$^c(4)$-action on $\Lambda^\pm V$ is identified with $\omega \in \mathrm{Im}\,\mathbb{H}^\pm \mapsto q_\pm \omega q_\pm^{-1}$ with respect to the bases (3.3).

4.108 Direct calculation.

4.109

$$\begin{aligned} D_{g\cdot A}(g\Phi) &= \left(D_A + \frac{1}{2}\rho(-2g^{-1}dg) \right)(g\Phi) \\ &= dg \cdot D_A\Phi + gD_A\Phi - dg \cdot D_A\Phi = gD_A\Phi. \end{aligned}$$

The curvature F_A^+ is fixed by the $\mathcal{G}$-action. Since the complex structure on $\mathbb{H}^+$ is given by right multiplication of complex numbers, we have

$$q_0(\phi u) = \phi u i \bar{u} \bar{\phi} = \phi i \bar{\phi} = q_0(\phi).$$

Therefore we obtain $q(g\Phi) = q(\Phi)$.

4.110 It is easy to see that there is a decomposition

$$L_k^2(X;\Lambda^1) = \overline{R(d) \oplus R(d^*)} \oplus H^1,$$

where $\overline{R}$ means the closure of R. Hence it is enough to verify that both $R(d)$ and $R(d^*)$ are closed linear subspaces in $L_k^2(X;\Lambda^1)$. This can be seen in the proof of Theorem 4.19.

4.111 The computation is straightforward.

4.112 Choose a point $(t_0, t_1) \in S^1 \times S^1$. Fix a trivialization of L along $\ell = \{t_0\} \times S^1$, and choose a nowhere vanishing section s_0 of $L|_\ell$. By parallel transport of s_0 along $S^1 \times \{t\}$, we obtain a nowhere vanishing section s on $L|_{S^1 \times S^1 \setminus \ell}$ and another section s_1 of $L|_\ell$ which can be different from s_0. Modifying s on a neighborhood of ℓ, we construct a section on the whole $S^1 \times S^1$. Modify s by using homotopy so that $s|_{\{t_0\}\times S^1} = s_0$, $s|_{\{t_0-\varepsilon\}\times S^1} = s_1$, and $s|_{S^1\times S^1\setminus(t_0-\varepsilon,t_0)\times S^1}$ is nowhere vanishing, and extend s across $(t_0 - \varepsilon, t_0) \times S^1$. Then the sum of zeros of the extended section counted with multiplicities is equal to the degree of the map of holonomy.

4.113 Note that the first Chern class $c_1(\mathbb{L})$ is in

$$H^1(X \times \mathcal{B}^*) = (H^2(X) \otimes H^0(\mathcal{B}^*)) \oplus (H^1(X) \otimes H^1(\mathcal{B}^*)) \oplus (H^0(X) \otimes H^2(\mathcal{B}^*)).$$

Since the restriction of $\mathbb{L}$ to $X \times \{pt\}$ is trivial, $c_1(\mathbb{L})$ has no component in $H^2(X) \otimes H^0(\mathcal{B}^*)$. Since $u = c_1(\mathbb{L})/[x_0]$, the component of $c_1(\mathbb{L})$ in $H^0(X) \otimes H^2(\mathcal{B}^*)$ is $1 \otimes u$. Similarly, since $\mu(\gamma_i) = c_1(\mathbb{L})/\gamma_i$, the component of $c_1(\mathbb{L})$ in $H^1(X) \otimes H^1(\mathcal{B}^*)$ is $\sum_i \gamma_i^\vee \otimes \mu(\gamma_i)$.

4.114 By Theorem 4.35, we have an isomorphism

$$H^*(\mathcal{B}^*) \cong H^*(\mathbb{C}P^\infty) \otimes H^*(T^{b_1}).$$

It suffices to prove that u is a generator of $H^2(\mathbb{C}P^\infty)$ and $(\mu(\gamma_1), \ldots \mu(\gamma_{b_1}))$ is a basis of $H^1(\mathcal{B}^*;\mathbb{Z}) = H^1(T^{b_1};\mathbb{Z})$ if $(\gamma_1, \ldots, \gamma_{b_1})$ is a basis of $H_1(X;\mathbb{Z})/\mathrm{Tor}$.

By Remark 4.38, $u = \mu([x_0])$ is a generator of $H^2(\mathbb{C}P^\infty)$.

Consider the b_1-dimensional torus

$$T^{b_1} := \{A_0 + a \mid a \in \sqrt{-1}\Omega^1(X), \text{harmonic}\}/\mathcal{G} \cong H^1(X;\mathbb{R})/H^1(X;\mathbb{Z}).$$

embedded in $\mathcal{B}^*$. Note that this embedding induces an isomorphism between $H^1(T^{b_1};\mathbb{Z})$ and $H^1(\mathcal{B}^*;\mathbb{Z})$. Let $\gamma_i^\vee \in H^1(X;\mathbb{Z})$ be the dual of γ_i. Consider the loops $\tilde{\beta}_i = \{t\gamma_i^\vee\}_{0\le t\le 1}$ $(i = 1, \ldots, b_1)$ in $T^{b_1} \subset \mathcal{B}^*$. Then their homology classes $\beta_i = [\tilde{\beta}_i]$ form a basis of $H_1(T^{b_1};\mathbb{Z})$. Let $\beta_i^\vee \in H^1(T^{b_1};\mathbb{Z}) = H^1(\mathcal{B}^*;\mathbb{Z})$ be the dual of β_i. Then it can be seen that $\beta_i^\vee = \mu(\gamma_i)$ from Proposition 4.39.

4.115 The map $\mathcal{M}(l) \to \mathcal{M}(k)$ is a homeomorphism because the map is continuous and bijective, and the moduli spaces $\mathcal{M}(l)$ and $\mathcal{M}(k)$ are compact and Hausdorff by Theorem 4.45 and Proposition 4.23.

4.116 The proof is parallel to the one of Proposition 4.48.

Chapter 5

5.52 Note that L_k^2 with $k \geq 4$ is an algebra by Lemma 4.12. Since ϕ and b is in L_k^2, $q(\phi)$ and $\rho(b)\phi$ are also in L_k^2. Since the inclusion $L_k^2 \hookrightarrow L_{k-1}^2$ is compact by Rellich's lemma (Lemma 4.11), the map c is also compact.

5.53 Note that E_U is the direct sum of a complex vector bundle E and a real vector bundle V, and $E_U^* / \mathrm{U}(1)$ is homotopic to $P(E) = S(E)/\mathrm{U}(1)$ where $S(E)$ is the unit sphere bundle of E. Note also that $P(E)$ is a fiber bundle with fiber $\mathbb{C}P^N$ satisfying the local triviality. Since the transition functions of $P(E)$ are induced from complex linear isomorphisms of the fibers of E, it can be seen from a Meyer–Vietoris argument that there exists global cohomology classes of $H^*(P(E))$ such that the restriction to each fiber freely generates the cohomology of the fiber. Then Lemma 5.20 follows from the Leray–Hirsch theorem (see [72, §4.D]).

5.54 Assume S^2 as $S^{\mathbb{R}^2}$, the one-point compactification of $\mathbb{R}^2$. We have a pinch map

$$\tau\colon S^2 = S^{\mathbb{R}^2} \to S^{\mathbb{R}^2}/S^{\mathbb{R}\oplus\{0\}} = S^2 \vee S^2.$$

For $[f], [g] \in [S^2 \wedge X, Y]$, the sum $[f] + [g]$ is represented by the map $h = (f \vee g) \circ \tau$. For $y \in Y$, let $M_y = f^{-1}(y)$ and $N_y = g^{-1}(y)$. The inverse image of h for a generic point is the disjoint union of M_y and N_y whose fundamental class is $[M_y] + [N_y]$.

5.55 Since $D(\mathbb{C}^n)$ is $\mathrm{U}(1)$-equivariantly homotopic to the origin O,

$$H_G^*(D(\mathbb{C}^n)) \cong H_G^*(O) = H^*(BG) = \mathbb{Z}[u].$$

Since $\mathrm{U}(1)$ acts on $S(\mathbb{C}^n)$ freely and $S(\mathbb{C}^n)/\mathrm{U}(1) = \mathbb{C}P^{n-1}$,

$$H_G^*(S(\mathbb{C}^n)) = H^*(\mathbb{C}P^{n-1}) = \mathbb{Z}[u]/(u^n).$$

The exact sequence

$$\to H_G^*(D(\mathbb{C}^n), S(\mathbb{C}^n)) \to H_G^*(D(\mathbb{C}^n)) \to H_G^*(S(\mathbb{C}^n)) \to$$

implies that $H_G^*(D(\mathbb{C}^n), S(\mathbb{C}^n)) \cong \mathbb{Z}[u]u^n$.

5.56 Since the cyclic group $C_2 = \{\pm 1\}$ acts freely on $S^\infty = S(\mathbb{R}^\infty)$ by multiplication, we have $BC_2 = \mathbb{R}P^\infty$. Similarly, since $\mathrm{SU}(2) = \mathrm{Sp}(1)$ acts freely on $S(\mathbb{H}^\infty)$, we have $B\mathrm{SU}(2) = \mathbb{H}P^\infty$.

5.57 The solution of this problem is parallel to that of Problem 5.55.

5.58 Consider the following commutative diagram:

$$\begin{array}{ccc} (D(\mathbb{R}^{m+1}), S(\mathbb{R}^{m+1})) & \xrightarrow{\bar{f}} & (D(\mathbb{R}^{n+1}), S(\mathbb{R}^{n+1})) \\ {\scriptstyle i_1}\uparrow & & \uparrow{\scriptstyle i_2} \\ (O, \emptyset) & \xrightarrow[\cong]{} & (O, \emptyset)\,. \end{array}$$

Applying $H_G^*(\cdot; \mathbb{F}_2)$ to the above diagram, we obtain the following commutative diagram:

$$\begin{array}{ccc} \mathbb{F}_2[q]q^{m+1} & \xleftarrow{\bar{f}^*} & \mathbb{F}_2[q]q^{n+1} \\ \downarrow & & \downarrow \\ \mathbb{F}_2[q] & = & \mathbb{F}_2[q], \end{array}$$

which implies that $m \le n$.

5.59 See [151, 165]. Instead, we present a proof for $k \le 0$ by using approximated moduli spaces here. Suppose k is positive. Take a finite-dimensional approximation $f \colon E_U \to U$. Since $b_1 = 0$ and $b^+ = 0$, the differential DF_0 at the origin has a form

$$DF_0 \colon \mathbb{C}^{x+k} \oplus \mathbb{R}^y \to \mathbb{C}^x \oplus \mathbb{R}^y.$$

As in the proof of Corollary 4.78 in Sect. 5.3, perturb f locally around the origin by choosing a U(1)-equivariant linear surjection $L \colon \mathbb{C}^{x+k} \to \mathbb{C}^x$ and a cut-off function. Then the zero locus of f around the origin is isomorphic to $\mathbb{C}^k$. Therefore a neighborhood of the origin of the quotient space $f^{-1}(0)/\mathrm{U}(1)$ has the form of a cone of $\mathbb{C}P^{k-1}$. By a further perturbation of f outside of the neighborhood of 0, we obtain an approximated moduli space M with a cone point which is a manifold outside the cone point. Removing a small neighborhood of 0 from M, we obtain a compact orientable manifold M' with boundary $\mathbb{C}P^{k-1}$ embedded in $\mathcal{B}^*$. Note that $\mathcal{B}^*$ is homotopic to $\mathbb{C}P^\infty$ if $b_1 = 0$. If we choose an orientation of M', then $\partial M' = \mathbb{C}P^{k-1}$ has a nontrivial fundamental class in $H^{2k-2}(\mathcal{B}^*)$. However, $[\partial M']$ should be 0 in $H^{2k-2}(\mathcal{B}^*)$ because $\partial M'$ is the boundary of the compact manifold M'. This is a contradiction.

5.60 Note that $H^*(\mathbb{R}P^2;\mathbb{F}_2) = \mathbb{F}_2[q]/(q^3)$ where the degree of q is 1, and $H^*(\mathbb{H}P^\infty;\mathbb{F}_2) = \mathbb{F}_2[v]$ where the degree of v is 4. Since all the differentials of the Leray–Serre spectral sequence of the fibration $\mathbb{R}P^2 \to BG \to \mathbb{H}P^\infty$ are trivial, we obtain

$$H^*(BG;\mathbb{F}_2) = H^*(\mathbb{R}P^2;\mathbb{F}_2) \otimes H^*(\mathbb{H}P^\infty;\mathbb{F}_2) = \mathbb{F}_2[q,v]/(q^3).$$

5.61 Since $D(\mathbb{H}^m \oplus \tilde{\mathbb{R}}^n)$ is G-equivariantly contractible to the origin O, we have

$$H^*_G(D(\mathbb{H}^m \oplus \tilde{\mathbb{R}}^n);\mathbb{F}_2) \cong H^*_G(O;\mathbb{F}_2) \cong H^*(BG;\mathbb{F}_2) \cong \mathbb{F}[q,v]/(q^3).$$

Since G acts on $S(\mathbb{H})$ freely, and $S(\mathbb{H})/G \cong \mathbb{R}P^2$, we obtain $H^*_G(S(\mathbb{H});\mathbb{F}_2) \cong \mathbb{F}_2[q]/(q^3)$. By the long exact sequence for the pair $(D(\mathbb{H}), S(\mathbb{H}))$, we can directly check the Thom isomorphism

$$H^*_G(D(\mathbb{H}), S(\mathbb{H});\mathbb{F}_2) = \left(\mathbb{F}[q,v]/(q^3)\right)\tau_G(\mathbb{H}),$$

where $\tau_G(\mathbb{H})$ is the G-equivariant Thom class which is the generator of degree 4. Note also that, by the restriction $H^*_G(D(\mathbb{H}), S(\mathbb{H});\mathbb{F}_2) \to H^*_G(D(\mathbb{H});\mathbb{F}_2) \cong H^*_G(O;\mathbb{F}_2)$, $\tau_G(\mathbb{H})$ is mapped to the G-equivariant Euler class $e_G(\mathbb{H}) = v$. Hence $e_G(\mathbb{H}^k) = v^k$.

We may assume $EG = S(\mathbb{H}^\infty)$ and $S(\tilde{\mathbb{R}}) = \{\pm 1\}$. Then

$$EG \times_G S(\tilde{\mathbb{R}}) = (S(\mathbb{H}^\infty) \times \{\pm 1\})/G = S(\mathbb{H})/\mathrm{U}(1) \cong \mathbb{C}P^\infty.$$

As above, we can also see that the Thom isomorphism,

$$H^*_G(D(\tilde{\mathbb{R}}), S(\tilde{\mathbb{R}});\mathbb{F}_2) = \left(\mathbb{F}[q,v]/(q^3)\right)\tau_G(\tilde{\mathbb{R}}),$$

holds, and $e_G(\tilde{\mathbb{R}}) = q$. Hence $e_G(\tilde{\mathbb{R}}^m) = q^m$.

5.62 The diagram of the problem induces the following commutative diagram:

$$\begin{array}{ccc}
(D(\mathbb{H}^{x+k} \oplus \tilde{\mathbb{R}}^y), S(\mathbb{H}^{x+k} \oplus \tilde{\mathbb{R}}^y)) & \xrightarrow{\bar{f}} & (D(\mathbb{H}^{x} \oplus \tilde{\mathbb{R}}^y), S(\mathbb{H}^{x} \oplus \tilde{\mathbb{R}}^{y+b})) \\
{\scriptstyle i_1}\big\uparrow & & \big\uparrow{\scriptstyle i_2} \\
(D(\{0\} \oplus \tilde{\mathbb{R}}^y), S(\{0\} \oplus \tilde{\mathbb{R}}^y)) & \xhookrightarrow{\bar{f}^{\mathrm{U}(1)}} & (D(\{0\} \oplus \tilde{\mathbb{R}}^y), S(\{0\} \oplus \tilde{\mathbb{R}}^{y+b})).
\end{array}$$

Applying $H^*_G(\cdot;\mathbb{F}_2)$ to the diagram and using the Thom isomorphism, we obtain

$$\begin{array}{ccc}
\left(\mathbb{F}[q,v]/(q^3)\right)\tau_G(\mathbb{H}^{x+k})\tau_G(\tilde{\mathbb{R}}^y) & \xleftarrow{\bar{f}^*} & \left(\mathbb{F}[q,v]/(q^3)\right)\tau_G(\mathbb{H}^{x})\tau_G(\tilde{\mathbb{R}}^{y+b}) \\
\downarrow i_1^* & & \downarrow i_2^* \\
\left(\mathbb{F}[q,v]/(q^3)\right)\tau_G(\tilde{\mathbb{R}}^y) & \xleftarrow{(\bar{f}^{\mathrm{U}(1)})^*} & \left(\mathbb{F}[q,v]/(q^3)\right)\tau_G(\tilde{\mathbb{R}}^{y+b}),
\end{array}$$

where i_1^*, i_2^* and $\left(\bar{f}^{\mathrm{U}(1)}\right)^*$ are given by multiplication of $e_G(\mathbb{H}^{x+k}) = v^{x+k}$, $e_G(\mathbb{H}^x) = v^x$ and $e_G(\tilde{\mathbb{R}}^b) = q^b$, respectively. By the commutativity of the diagram, $v^x q^b$ should be contained in the image of i_1^*. Since $q^b \neq 0$ if $b \leq 2$, we obtain $k = 0$.

Chapter 6

6.59

(a) A harmonic function is constant, and any non-zero constant is not in L^2 on $\mathbb{R}$.
(b) By *(a)* and Corollary 6.7, the reduced L^2-cohomology group vanishes at the degree 0. By Theorem 6.31, $\mathrm{spec}(\Delta_0)$ contains zero, because $\mathbb{R}$ is the universal covering space of S^1 and $\pi_1(S^1) \cong \mathbb{Z}$ is amenable. Then the conclusion follows from Lemma 6.9.

6.60 By the condition, the equality $da = d^- a$ holds. Then, the equalities

$$\int_M da \wedge da = \int_M d^- a \wedge d^- a = -||d^- a||^2_{L^2}$$

hold. On the other hand, $\int_M da \wedge da = \int_M d(a \wedge da) = 0$ vanishes by Stokes' theorem. Hence, $d^- a \equiv 0$ holds.

6.61 In fact, both the operators $P \circ \Delta_m = \Delta_m \circ P = 0$ vanish.

6.62 Because $d : L_1^2(X; \Lambda^{m-1}) \to L^2(X; \Lambda^m)$ has closed range, $d(L_1^2(X; \Lambda^{m-1})) \subset L^2(X; \Lambda^m)$ is a closed linear subspace. Hence it induces a Hilbert space structure on the right-hand side. Then, the map induces

$$d : L_1^2(X; \Lambda^{m-1})/\{\ker d\} \to d(L_1^2(X; \Lambda^{m-1})),$$

which is one-to-one onto. Because $\ker d \subset L_1^2(X; \Lambda^{m-1})$ is a closed linear subspace, one can take the orthogonal complement $(\ker d)^\perp \subset L_1^2(X; \Lambda^{m-1})$.

By the open mapping theorem, it has the bounded inverse

$$d^{-1} : d(L_1^2(X; \Lambda^{m-1})) \cong (\ker d)^\perp.$$

Then we can set $f_i := d^{-1}(v_i)$.

6.63 Immediate.

6.64 Let H be a subgroup of an amenable group G. Equip with a G-invariant finitely additive probability measure μ on G. Then, normalize as $\mu' := \frac{\mu}{\mu(H)}$. This gives the H-invariant finitely additive probability measure on H.

Suppose moreover $H \subset G$ is normal, and let $\pi : G \to G/H$ be the group homomorphism. Then, set the G/H-invariant finitely additive probability measure μ'' $\mu''([A]) := \mu(\pi^{-1}(A))$ on G/H.

Finally, for a directed set of groups $\{G_a\}_a$, set $G := \cup_a G_a$. Equip with a G_a-invariant finitely additive probability measure μ_a on each G_a. Then for $A \subset G$, set the G-invariant finitely additive probability measure μ on G by

$$\mu(A) := \limsup_a \mu_a(G \cap G_a).$$

Note $\mu(G) = \limsup_a \mu_a(G_a \cap G) = \limsup_a \mu_a(G_a) = 1$.

6.65 Set $c_1 = c_1' := \min\{f(g) > 0; g \in \Gamma\}$ and $F_1 := \{g \in \Gamma; f(g) \geq c_1'\}$. Next, set $c_2' := \min\{f(g) > c_1'; g \in \Gamma\}$, $c_2 := c_2' - c_1'$, and $F_2 := \{g \in \Gamma; f(g) \geq c_2'\}$. Inductively we set $c_3' := \min\{f(g) > c_2'; g \in \Gamma\}$, $c_3 := c_3' - c_2'$, and $F_3 := \{g \in \Gamma; f(g) \geq c_3'\}$, and so on.

6.66 Take $p \in \partial P_S$. By definition, $p \in P_S^c$ and there is $p' \in P_S$ such that $d(p, p') = 2\epsilon$. Let l be a shortest curve with $l(0) = p$ and $l(1) = p'$. Because both the inequalities

$$\mathrm{vol}(S \cap B_\epsilon(p')) > \mathrm{vol}(B_\epsilon(p'))/2),$$
$$\mathrm{vol}(S \cap B_\epsilon(p)) \leq \mathrm{vol}(B_\epsilon(p))/2)$$

hold, there is some $s \in [0, 1]$ such that the equality $\mathrm{vol}(S \cap B_\epsilon(l(s))) = \mathrm{vol}(B_\epsilon(l(s)))/2)$ holds. Hence, $l(s) \in H$, and $p \in B_{2\epsilon}(H)$.

6.68 For $\sigma = [v_0, \cdots, v_p]$, we have

$$\begin{aligned} dW(\sigma^*) &= d\left(p! \sum_{i=0}^{p} (-1)^i \varphi_i d\varphi_0 \wedge \cdots \wedge \check{d\varphi_i} \wedge \cdots \wedge d\varphi_p\right) \\ &= (p+1)! d\varphi_0 \wedge \cdots \wedge \cdots \wedge d\varphi_p. \end{aligned}$$

For a $(p+1)$-simplex $\tau := [v_0, \cdots, v_p, v_{p+1}]$ that contains σ, we have $d_c(\sigma^*)(\tau) = \sigma^*(\partial\tau) = (-1)^{p+1}$ holds. Note that $\sum_{i=1}^{p+1} \varphi_i \equiv 1$ holds on the support of $\varphi_i d\varphi_0 \wedge \cdots \wedge \cdots \wedge d\varphi_{p+1}$ that is contained in the interior of the maximal simplex containing τ. Then, we have the equalities

$$
\begin{aligned}
W(\tau^*) &= (p+1)!(-1)^{p+1}\varphi_{p+1}d\varphi_0 \wedge \cdots \wedge \cdots \wedge d\varphi_p \\
&\quad + (p+1)!\sum_{i=0}^{p}(-1)^i\varphi_i d\varphi_0 \wedge \cdots \wedge d\check{\varphi}_i \wedge \cdots \wedge d\varphi_{p+1} \\
&= (p+1)!(-1)^{p+1}(1-\sum_{i=0}^{p}\varphi_i)d\varphi_0 \wedge \cdots \wedge \cdots \wedge d\varphi_p \\
&\quad + (p+1)!\sum_{i=0}^{p}(-1)^i\varphi_i d\varphi_0 \wedge \cdots \wedge d\check{\varphi}_i \wedge \cdots \wedge d(1-\sum_{i=0}^{p}\varphi_i) \\
&= (p+1)!(-1)^{p+1}(1-\sum_{i=0}^{p}\varphi_i)d\varphi_0 \wedge \cdots \wedge \cdots \wedge d\varphi_p \\
&\quad + (p+1)!\sum_{i=0}^{p}(-1)^{p+1}\varphi_i d\varphi_0 \wedge \cdots \quad \wedge \cdots \wedge d\varphi_p \\
&= (p+1)!(-1)^{p+1}d\varphi_0 \wedge \cdots \wedge \cdots \wedge d\varphi_p.
\end{aligned}
$$

Hence by continuity, we have $W(d_c\sigma^*) = (p+1)!d\varphi_0 \wedge \cdots \wedge \cdots \wedge d\varphi_p$.

6.69 Similar to Problem 6.68.

Chapter 7

7.52 Direct computations.

7.53 The latter follows by definition. For the former, denote by $T = AB$ a product of Hilbert–Schmidt operators. Then check that both BT' and $T'A$ are also Hilbert–Schmidt operators.

7.54

(1) Let $a = \sum_{g\in\Gamma} a_g \cdot g$ and similarly for b. Then,

$$
\mathrm{tr}_\Gamma(a \circ b) = \sum_{gh=e} a_g \cdot b_h = \sum_{gh=e} b_h \cdot a_g = \mathrm{tr}_\Gamma(b \circ a).
$$

(2), (3) Note $(a^* \circ a)(e) = \sum_{gh=e} \overline{a_g} \cdot a_h$. Then,

$$
\mathrm{tr}_\Gamma(a^* \circ a) = \sum_{gh=e} \overline{a_{g^{-1}}} \cdot a_h = \sum_{gh=e} |a_h|^2 \ \geq\ 0.
$$

The left-hand side vanishes, if and only if $a_h = 0$ for any $h \in \Gamma$.

7.55

(1) Immediate.
(b) The orthogonal decomposition is compatible with the Γ-actions for all the spaces.

7.56

$$\lim_{n\to\infty} \mathrm{tr}_\Gamma(p_n((f^{(2)})^* f^{(2)})) = \lim_{n\to\infty} \mathrm{tr}_\Gamma \int_0^{||f||^2} p_n(\lambda) dP(\lambda)$$

$$= \lim_{n\to\infty} \int_0^{||f||^2} p_n(\lambda) dF(\lambda) = \int_0^{||f||^2} \lim_{n\to\infty} p_n(\lambda) dF(\lambda)$$

$$= \int_0^{||f||^2} \chi_{[0,\lambda]} dF(\lambda) = F(\lambda).$$

7.57 We only check the inequality

$$\frac{|E(\lambda) \setminus E(0)|}{\dim_{\mathbb{C}}(V)} \leq \frac{\log(K^2)}{-\log(\lambda)},$$

when $C = 1$. Let $0 \leq \mu_0 \leq \cdots \leq \mu_{\dim V}$ be the set of eigenvalues of $g^* g$. Set $0 = \mu_0 = \cdots = \mu_r < \mu_{r+1}$, and $\mu_s \leq \lambda < \mu_{s+1}$. Hence, we can express $q(t) = \prod_{i=r+1}^{\dim V} (t - \mu_i)$, and $|E(\lambda) \setminus E(0)| = s - r$. Then, we have

$$q(0) = \prod_{i=r+1}^{\dim V} (-\mu_i), \quad \prod_{i=r+1}^{s} \mu_i = |q(0)| \cdot \prod_{i=s+1}^{\dim V} \mu_i^{-1}.$$

hence, we have the inequality $\prod_{i=r+1}^{s} \lambda \geq \prod_{i=s+1}^{\dim V} K^{-2}$, and

$$\lambda^{|E(\lambda) \setminus E(0)|} \geq K^{-2 \dim V}.$$

7.58 Take a vector field V such that its zeros are isolated. The idea is to sweep-out all the zeros to infinity (see [90]). For each point $p \in \tilde{M}$ with $V(p) = 0$, choose a proper path $l_p : [0, 1) \to \tilde{M}$ such that $l_p(0) = p$, and $l_p([0, 1)) \cap l_{p'}([0, 1)) = \phi$ for any two distinct $p \neq p'$ with $V(p) = 0$ and $V(p') = 0$. Then each path l_p admits a non-zero vector field V_p along l_p, and the set $\{V_p\}_p$ generates a family of diffeomorphisms Φ_t on $\tilde{M}$ for $[0, 1)$ by extending the set of vector fields $\{V_p\}_p$. Then, one can obtain $(\Phi_t)_*(V)$ for $[0, 1)$, and it can be extended to a vector field V at $t = 1$. It has no zeros.

7.59 Check the formula

$$\int_M k_1(x, y) \varphi_i(y) \,\mathrm{vol}_y = \exp(-\lambda_i) \cdot \varphi_i(x).$$

7.60 Note that P_n and A commute each other. Then, the left-hand side is equal to $P_n A T_0 P_n$, and it follows from the equality $AT_0 = T_2 A$ that it coincides with $P_n T_2 A P_n$ that is the right-hand side.

7.61 Equip with an inner product on V, and take an orthonormal basis $\{e_1, \cdots, e_l\}$ of the image $P(V)$. Extend it to an orthonormal basis of V. Then P is expressed by the sum of id $\oplus 0$ with respect to the decomposition $V = P(V) \oplus P(V)^{\perp}$. Then $\mathrm{tr}(P) = \mathrm{rank}\, P(V)$ holds by Definition 7.1.

7.62

(1) Take $g \in \Gamma$ with $g^l = e$ for $l \geq 2$. Then,

$$P := \sum_{i=0}^{l-1} \frac{g^i}{l} \in \mathcal{N}(\Gamma)$$

is a projection, and $\mathrm{tr}_\Gamma(P) = l^{-1}$ holds.

(2) By Lemma 7.10 (3), $0 \leq \mathrm{tr}_\Gamma(P) \leq 1$ holds for any projection in $\mathcal{N}(\Gamma)$, because $\mathrm{tr}_\Gamma(\delta_e) = 1$ holds. The equalities hold only when $P = 0$ and $P = \delta_e$ for $\mathrm{tr}_\Gamma(P) = 0$ or 1 respectively. Hence, the conclusion holds.

7.63 For example, consider S^1.

7.64 This follows by Atiyah's Γ-index theorem.

Chapter 8

8.32 Because $\varphi(0, \quad)$ is constant, the pull-back $\varphi(0, \quad)^*(\alpha) \equiv 0$ vanishes. Follow a proof of Poincaré's Lemma and check the formula $dK + Kd$ is the identity.

8.33 Use the long exact sequence of the fibration of homotopy groups.

8.34 For example, see [69].

8.35 If the fundamental group of a compact manifold is finite, then the universal covering space is also compact. It cannot be contractible.

8.36 Note that $b_1(M) = 0$ and hence $b_3(M) = 0$ by Poincaré duality. Then, $\chi(M) = b_2(M) + 2 \geq \frac{10}{8}|\sigma(M)| + 2$ holds by Theorem 1.1.

8.37 Assume that $1 \leq \dim b_1(\tilde{M}) < \infty$ and $\chi(M) \leq \frac{10}{8}|\sigma(M)| + 1$. Let M_k be a finite cover corresponding to the surjective homomorphism $\pi_1(M) \to \mathbb{Z}_k$. Because the Euler characteristic and the signature are both multiplicative under taking finite coverings, we have the inequalities $\chi(M_k) + k \leq \frac{10}{8}|\sigma(M_k)|$. Then, by Theorem 1.1, we have

$$k \leq \frac{10}{8}|\sigma(M_k)| - \chi(M_k) = \frac{10}{8}|\sigma(M_k)| - b_2(M_k) - 2 + 2b_1(M_k) \leq 2b_1(M_k)$$

for any $k \geq 1$. This is a contradiction.

8.38 For example, consider $SL_3(\mathbb{Z})$ and its presentation.

8.39 It is known that any finitely presented group is isomorphic to the fundamental group of a compact (symplectic) 4-manifold [63], see also [64]. Combine this with Problem 8.38.

8.40 Take a Morse function $f : B \to \mathbb{R}$, and pull it back as $f \circ \pi : M \to \mathbb{R}$ which is a Morse–Bott function. Then, each critical set is the fiber of a critical point of f. Hence, we have the formula [23]

$$\chi(M) = \sum_{c \in B} \text{ind}(c) \cdot \chi(F) = \chi(B) \cdot \chi(F).$$

8.41 By Theorem 8.29, the L^2-Betti numbers of the universal covering space vanish on all degrees. On the other hand, the universal covering space of a compact Kähler hyperbolic manifold has a positive L^2-Betti number on the middle degree by Theorem 8.25.

References

1. Adams, J.F.: Prerequisites (on equivariant stable homotopy) for Carlsson's lecture. In: Algebraic Topology: Proceedings of the International Conference, Aarhus 1982. Lecture Notes in Mathematics, vol. 1051, pp. 483–532 (1984)
2. Akbulut, S.: 4-Manifolds. Oxford Graduate Texts in Mathematics, vol. 25. Oxford University Press, Oxford. xii, 262pp. (2016). ISBN 978-0-19-878486-9/hbk
3. Anderson, M.: Hausdorff perturbations of Ricci-flatmanifolds and the splitting theorem. Duke Math. J. **68**(1), 67–82 (1992)
4. Aronszajn, N.: A unique continuation theorem for solutions of elliptic partial differential equations or inequalities of second order. J Math. Pure Appl. IX. Sér. **36**, 235–249 (1957)
5. Atiyah, M.: K-Theory. Advanced Books Classics. Addison-Wesley, Boston (1967)
6. Atiyah, M.: Bott periodicity and the index of elliptic operators. Q. J. Math. Oxf. II. Ser. **19**, 113–140 (1968)
7. Atiyah, M.: The signature of fiber-bundles, In: Global Analysis in Honor of K. Kodaira, pp. 73–84. University of Tokyo Press, Tokyo (1969)
8. Atiyah, M.: Elliptic operators, discrete groups and von Neumann algebras. Astérisque **32–33**, 43–72 (1976)
9. Atiyah, F., Singer, I.: The index of elliptic operators. III. Ann. Math. **87**, 546–604 (1968)
10. Banach, S., Tarski, A.: Sur la écomposition des ensembles de points en parties respectivement congruentes. Fund Math. **6**, 244–277 (1924)
11. Baraglia, D.: Constraints on families of smooth 4-manifolds from Bauer-Furuta invariants. Algebr. Geom. Topol. **21**(1), 317–349 (2021)
12. Baraglia, D.: Tautological classes of definite 4-manifolds. Geom. Topol. **27**(2), 641–698 (2023)
13. Baraglia, D., Konno, H.: On the Bauer-Furuta and Seiberg-Witten invariants of families of 4-manifolds. J. Topol. **15**(2), 505–586 (2022)
14. Barth, W., Hulek, K., Peters, C., Van de Ven, A.: Compact complex surfaces. 2nd enlarged ed. Ergebnisse der Mathematik und ihrer Grenzgebiete. 3. Folge 4, xii, 436pp. Springer. Berlin (2004). ISBN 3-540-00832-2/hbk
15. Bauer, S.: A stable cohomotopy refinement of Seiberg-Witten invariants. II. Invent Math **155**(1), 21–40 (2004)
16. Bauer, S.: Refined Seiberg-Witten invariants. In: Donaldson, S. et al. (ed.) Different Faces of Geometry. International Mathematical Series, vol. 3, pp. 1–46. Kluwer Academic/Plenum Publishers, New York (2004). ISBN 0-306-48657-1/hbk
17. Bauer, S.: Almost complex 4-manifolds with vanishing first Chern class. J. Differ. Geom. **79**(1), 25–32 (2008)

T. Kato, N. Nakamura, *Monopole Maps and L^2-cohomology*, Springer Monographs in Mathematics, https://doi.org/10.1007/978-981-96-8472-4

18. Bauer, S., Furuta, M.: A stable cohomotopy refinement of Seiberg-Witten invariants: I. Invent. Math. **155**(1), 1–19 (2004)
19. Behzadan, A., Holst, M.: Multiplication in Sobolev spaces, revisited. Ark. Mat. **59**(2), 275–306 (2021)
20. Bohr, C.: On the signatures of even 4-manifolds. Math. Proc. Camb. Philos. Soc. **132**(3), 453–469 (2002)
21. Booss-Bavnbek, B., Wojciechowski, K.P.: Elliptic Boundary Problems for Dirac Operators. Birkhäuser, Boston (1993)
22. Borel, A.: Compact Clifford-Klein forms of symmetric spaces. Topology **2**(1–2), 111–122 (1963)
23. Bott, R.: Nondegenerate critical manifolds. Ann. Math. **60**(2), 248–261 (1954)
24. Brooks, R.: The fundamental group and the spectrum of the Laplacian. Comment. Math. Helv. **56**, 581–598 (1981)
25. Bryan, J.: Seiberg-Witten theory and $\mathbb{Z}/2^p$ actions on spin 4-manifolds. Math. Res. Lett. **5**(1–2), 165–183 (1998)
26. Buser, P.: A note on the isoperimetric constant. Ann. Sci. Éc. Norm. Supér. **15**, 213–230 (1982)
27. Cheeger, J., Gromov, M.: L^2-cohomology and group cohomology. Topology **25**(2), 189–215 (1986)
28. Chen, X., Ge, J., Han, F.: Almost nonnegative Ricci curvature and new vanishing theorems for genera. Math. Z. **306**, Art. 44, 1–11 (2024)
29. Chern, S.: On the curvature and characteristic classes of a Riemann manifold. Abh. Math. Sem. Univ. Hamburg **20**, 117–126 (1955)
30. Chodosh, O., Li, C.: Generalized soap bubbles and the topology of manifolds with positive scalar curvature. Ann. Math. **199**, 707–740 (2020)
31. Connes, A.: Noncommutative Geometry. Academic Press, Cambridge (1994)
32. de la Harpe, P.: Topics in Geometric Group Theory. Chicago Lectures in Mathematics. University of Chicago Press, Chicago (2000)
33. Dodziuk, J.: De Rham–Hodge theory for L^2-cohomology of infinite coverings. Topology **16**, 157–165 (1977)
34. Dodziuk, J.: L^2 harmonic forms on rotationally symmetric Riemannian manifolds. Proc. Am. Math. Soc. **77**(3):395–400 (1979)
35. Donaldson, S.: An application of gauge theory to four-dimensional topology. J. Differ. Geom. **18**, 279–315 (1983)
36. Donaldson, S.K.: Connections, cohomology and the intersection forms of 4-manifolds. J. Differ. Geom. **24**, 275–341 (1986)
37. Donaldson, S.: Irrationality and the h cobordism conjecture. J. Differ. Geom. **26**, 141–168 (1987)
38. Donaldson, S.: The orientation of Yang-Mills moduli spaces and 4-dimensional topology. J. Differ. Geom. **26**, 397–428 (1987)
39. Donaldson, S.: Polynomial invariants for smooth four-manifolds. Topology **29**(3), 257–315 (1990)
40. Donaldson, S., Kronheimer, P.: The Geometry of Four-Manifolds. Oxford University Press, Oxford (1990)
41. Donnelly, H.: The differential form spectrum of hyperbolic space. Manuscripta Math. **33**, 365–385 (1981)
42. Elkies, N.D.: A characterization of the $\mathbb{Z}^n$ lattice. Math. Res. Lett. **2**(3), 321–326 (1995)
43. Fang, F.: Smooth group actions on 4-manifolds and Seiberg-Witten theory. Differ. Geom. Appl. **14**(1), 1–14 (2001)
44. Farber, M., Weinberger, S.: On the zero-in-the-spectrum conjecture. Ann. Math. **154**, 139–154 (2001)
45. Feehan, P., Leness, T.: Witten's conjecture for many four-manifolds of simple type. J. Eur. Math. Soc. **17**(4), 899–923 (2015)
46. Fintushel, R., Stern, R.: Definite 4-manifolds. J. Differ. Geom. **28**(1), 133–141 (1988)

47. Freed, D., Uhlenbeck, K.: Instantons and Four-Manifolds, 2nd edn. Springer, Berlin (1991)
48. Freedman, M.: The topology of four-dimensional manifolds. J. Differ. Geom. **17**, 357–453 (1982)
49. Freedman, M., Quinn, F.: Topology of 4-Manifolds. Princeton University Press, Princeton (1990)
50. Frøyshov, K.: The Seiberg-Witten equations and four-manifolds with boundary. Math. Res. Lett. **3**(3), 373–390 (1996)
51. Frøyshov, K.: Equivariant aspects of Yang-Mills Floer theory. Topology **41**(3), 525–552 (2002)
52. Frøyshov, K.: An inequality for the h-invariant in instanton Floer theory. Topology **43**(2), 407–432 (2004)
53. Frøyshov, K.: Compactness and Gluing theory for Monopoles. Geometry and Topology Monographs, vol. 15, 198pp. Geometry & Topology Publications, Coventry (2008)
54. Frøyshov, K.: Monopole Floer homology for rational homology 3-spheres. Duke Math. J. **155**(3), 519–576 (2010)
55. Frøyshov, K.: 4-manifolds and intersection forms with local coefficients. J. Differ. Geom. **91**(2), 233–259 (2012)
56. Fukumoto, Y., Furuta, M.: Homology 3-spheres bounding acyclic 4-manifolds. Math. Res. Lett. **7**(6), 757–766 (2000)
57. Furuta, M.: Perturbation of moduli spaces of self-dual connections. J. Fac. Sci. Univ. Tokyo Sect I A **34**, 275–297 (1987)
58. Furuta, M.: Stable homotopy version of Seiberg-Witten invariant (1997, preprint). https://archive.mpim-bonn.mpg.de/id/eprint/838/
59. Furuta, M.: Monopole equation and the $\frac{11}{8}$-conjecture. Math. Res. Lett. **8**, 279–291 (2001)
60. Furuta, M., Kametani, Y.: Equivariant maps between sphere bundles over tori and KO^*-degree (2005, preprint). arXiv:math/0502511.
61. Furuta, M., Kametani, Y., Minami, N.: Stable-homotopy Seiberg-Witten invariants for rational cohomology $K3\#K3$'s. J. Math. Sci. Tokyo **8**(1), 157–176 (2001)
62. Garrido, A.: An introduction to amenable groups. https://www.math.uni-duesseldorf.de/~garrido/amenable.pdf
63. Gompf, R.: A new construction of symplectic manifolds. Ann. Math. **142**, 527–595 (1995)
64. Gompf, R., Stipsicz, A.: 4-Manifolds and Kirby Calculus. Graduate Studies in Mathematics, vol. 20, 558pp. American Mathematical Society, Providence (1999)
65. Griffiths, P., Harris, J.: Principles of Algebraic Geometry. Wiley Classics Library, 2nd edn., 813pp. Wiley, New York (1994)
66. Gromov, M.: Kähler hyperbolicity and L^2-Hodge theory. J. Differ. Geom. **33**(1), 263–292 (1991)
67. Gromov, M.: Asymptotic Invariants of Infinite Groups. London Mathematical Society Lecture Note Series, vol. 182, 295pp. Cambridge University Press, Cambridge (1993)
68. Gromov, M., Lawson, H.: Positive scalar curvature on complete Riemannian manifolds. I.H.E.S Publ. Math. **58**, 295–408 (1983)
69. Gromov, M., Thurston, W.: Pinching constants for hyperbolic manifolds. Inv. Math. **89**, 1–12 (1987)
70. Gromov, M., LaFontaine, J. (ed.), Pansu, P. (ed.) [Katz, M., Semmes, S.]: Metric Structures for Riemannian and Non-Riemannian Spaces. Translated from the French by Sean Michael Bates. With appendices by M. Katz, P. Pansu, and S. Semmes. Modern Birkhäuser Classics, 585pp. Birkhäuser, Basel (2007). ISBN 978-0-8176-4582-3/pbk
71. Guillemin, V., Pollack, A.: Differential topology. Reprint of the 1974 Original, 222pp. AMS Chelsea Publishing, Providence (2010). ISBN 978-0-8218-5193-7/hbk
72. Hatcher, A.: Algebraic Topology, 544pp. Cambridge University Press, Cambridge (2002). ISBN 0-521-79540-0/pbk
73. Hatcher, A.: Vector Bundles and K-Theory. https://pi.math.cornell.edu/~hatcher/VBKT/VBpage.html
74. Hempel, J.: Residual finiteness of surface groups. Proc. Am. Math. Soc. **32**, 323 (1972)

75. Hirzebruch, F.: The signature of ramified coverings. Global Analysis, Papers in Honor of K. Kodaira, 253–265 (1969)
76. Hirzebruch, F., Hopf, H.: Felder von Flächenelementen in 4-dimensionalen Mannigfaltigkeiten. Math. Ann. **136**, 156–172 (1958)
77. Hopkins, M., Lin, J., Shi, X.D., Xu, Z.: Intersection forms of spin 4-manifolds and the Pin(2)-equivariant Mahowald invariant. Commun. Am. Math. Soc. **2**, 22–132 (2022)
78. Iida, N.: A Bauer–Furuta-type refinement of Kronheimer and Mrowka's invariant for 4-manifolds with contact boundary. Algebr. Geom. Topol. **21**(7), 3303–3333 (2021)
79. Ishida, M.: An elliptic surface covered by Mumford's fake projective plane. Tôhoku Math. J. II. Ser. **40**(3), 367–396 (1988)
80. Ishida, M., Sasahira, H.: Stable cohomotopy Seiberg-Witten invariants of connected sums of four-manifolds with positive first Betti number. I: non-vanishing theorem. Int. J. Math. **26**(6), 1541004, 23pp. (2015)
81. Ishida, M., Sasahira, H.: Stable cohomotopy Seiberg-Witten invariants of connected sums of four-manifolds with positive first Betti number. II: applications. Commun. Anal. Geom. **25**(2), 373–393 (2017)
82. Jost, J., Peng, X., Wang, G.: Variational aspects of the Seiberg-Witten functional. Calc. Var. Partial Differ. Equ. **4**(3), 205–218 (1996)
83. Kametani, Y.: Spin structures and the divisibility of Euler classes (2018, preprint). arXiv:1809.04045
84. Kanai, M.: Rough isometries, and combinatorial approximations of geometries of non-compact Riemannian manifolds. J. Math. Soc. Japan **37**, 391–413 (1985)
85. Kato, T.: Induced map on K theoryfor certain $\Gamma-$equivariant maps between Hilbert spaces. J. Noncommut. Geom. **15**(4), 1433–1467 (2021)
86. Kato, Y.: Nonsmoothable actions of $\mathbb{Z}_2 \times \mathbb{Z}_2$ on spin four-manifolds. Topol. Appl. **307**, 107868, 13pp. (2022)
87. Kato, T.: Covering-monopole map and higher degree in non-commutative geometry. J. Noncommut. Geom. **15**(4), 995–1071 (2021)
88. Kato, T., Konno, H., Nakamura, N.: Rigidity of the mod 2 families Seiberg-Witten invariants and topology of families of spin 4-manifolds. Compos. Math. **157**(4), 770–808 (2021)
89. Kato, T., Nakamura, N., Yasui, K.: The simple type conjecture for mod 2 Seiberg-Witten invariants. J. Eur. Math. Soc. **25**, 4869–4877 (2023)
90. Kato, T., Kishimoto, D., Tsutaya, M.: Vector fields on non-compact manifolds. Algebr. Geom. Topol. **24**(7), 3985–3996 (2024)
91. Kawasaki, T.: The signature theorem for V-manifolds. Topology **17**(1), 75–83 (1978)
92. Kawasaki, T.: The index of elliptic operators over V-manifolds. Nagoya Math. J. **84**, 135–157 (1981)
93. Kervaire, M., Milnor. J.: Groups of homotopy shperes: I. Ann. Math. **77**(3), 504–537 (1963)
94. Kirby, R.: The Topology of 4-Manifolds. Lecture Notes in Mathematics, vol. 1374, 108pp. Springer, Berlin (1989)
95. Kiyono, K., Liu, X.: On spin alternating group actions on spin 4-manifolds. J. Korean Math. Soc. **43**(6), 1183–1197 (2006)
96. Kobayashi, S., Nomizu, K.: Foundations of Differential Geometry, vol. II. Wiley, Hoboken (1969)
97. Konno, H., Nakamura, N.: Constraints on families of smooth 4-manifolds from $\mathrm{Pin}^{-}(2)$-monopole. Algebr. Geom. Topol. **23**(1), 419–438 (2023)
98. Kotschick, D.: Signatures, monopoles and mapping class groups. Math. Res. Let. **5**, 227–234 (1998)
99. Kronheimer, P.B., Mrowka, T.S.: The genus of embedded surfaces in the projective plane. Math. Res. Lett. **1**(6), 797–808 (1994)
100. Kronheimer, P.B., Mrowka, T.S.: Embedded surfaces and the structure of Donaldson's polynomial invariants. J. Differ. Geom. **41**(3), 573–734 (1995)
101. Kronheimer, P.B., Mrowka, T.S.: Monopoles and contact structures. Invent. Math. **130**(2), 209–255 (1997)

102. Kronheimer, P.B., Mrowka, T.: Monopoles and Three-Manifolds. Paperback reprint of the 1st 2007 hardback edition. New Mathematical Monographs, vol. 10, 796pp. Cambridge University Press, Cambridge (2011). ISBN 978-0-521-18476-2/pbk
103. Kronheimer, P.B., Mrowka, T.: The Dehn twist on a sum of two $K3$ surfaces. Math. Res. Lett. **27**(6), 1767–1783 (2020)
104. Kuiper, N.H.: The homotopy type of the unitary group of Hilbert space. Topology **3**, 19–30 (1965)
105. Lang, S.: Real and functional analysis, Graduate Texts in Mathematics, 3rd edn., vol. 142, 580pp. Springer, New York (1993)
106. Lawson, H.B., Michelsohn, M.-L.: Spin Geometry. Princeton Mathematical Series, vol. 38, 427pp. Princeton University Press, Princeton (1989)
107. Ledoux, M.: A simple analytic proof of an inequality by P. Buser. Proc. Am. Math. Soc. **121**(3), 951–959 (1994)
108. Lee, R., Li, T.-J.: Intersection forms of non-spin four manifolds. Math. Ann. **319**(2), 311–318 (2001)
109. Li, T.-J.: Symplectic 4-manifolds with Kodaira dimension zero. J. Differ. Geom. **74**(2), 321–352 (2006)
110. Li, T.-J.: Quaternionic bundles and Betti numbers of symplectic 4-manifolds with Kodaira dimension zero. Int. Math. Res. Not. **2006**(20), 37385, 2 pp. (2006).
111. Li, T.J., Liu, A.: General wall crossing formula. Math. Res. Lett. **2**(6), 797–810 (1995)
112. Lin, J.: Pin(2)-equivariant KO-theory and intersection. Algebr. Geom. Topol. **15**(2), 863–902 (2015)
113. Lin, J.: Isotopy of the Dehn twist on $K3\#K3$ after a single stabilization. Geom. Topol. **27**(5), 1987–2012 (2023)
114. Liu, X.: On S_3-actions on spin 4-manifolds. Carpathian J. Math. **21**(1–2), 137–142 (2005)
115. Lott, J.: The zero-in-the-spectrum question, L'Ensignment Math. **42**, 341–376 (1996)
116. Lück, W.: Approximating L^2-invariants by their finite dimensional analogues. Geom. Funct. Anal. **4**(4), 455–481 (1994)
117. Lück, W.: L^2-Invariants: Theory and Applications to Geometry and K-Theory. Mathematics Monograph, vol. 44. Springer, Berlin (2004)
118. Manolescu, C.: Seiberg-Witten Floer stable homotopy type of three-manifolds with $b_1 = 0$. Geom. Topol. **7**, 889–932 (2003)
119. Manolescu, C.: A gluing theorem for the relative Bauer-Furuta invariants. J. Differ. Geom. **76**(1), 117–153 (2007)
120. Manolescu, C.: On the intersection forms of spin four-manifolds with boundary. Math. Ann. **359**(3–4), 695–728 (2014)
121. Manolescu, C.: Pin(2)-equivariant Seiberg–Witten floer homology and the triangulation conjecture. J. Am. Math. Soc. **29**(1), 147–176 (2016)
122. Matsumoto, Y.: On the bounding genus of homology 3-spheres. J. Fac. Sci., Univ. Tokyo Sect. I A **29**, 287–318 (1982)
123. May, J.P.: A Concise Course in Algebraic Topology. Chicago Lectures in Mathematics, 243pp. University of Chicago Press, Chicago (1999)
124. Milnor, J.: On manifolds homeomorphic to the 7-sphere, Ann f Maths. **64**, 399–405 (1956)
125. Milnor, J.: Differential Topology. Lectures at Princeton University, Fall Term. Notes by James Munkres (1958). https://www.maths.ed.ac.uk/~v1ranick/papers/difftop.pdf
126. Milnor, J.: Lectures on the h-Cobordism Theorem. Mathematical Notes 1. Princeton University Press, Princeton (1965)
127. Milnor, J., Husemoller, D.: Symmetric bilinear forms. Ergebnisse der Mathematik und ihrer Grenzgebiete. Band 73. Springer, Berlin (1973)
128. Milnor, J., Stasheff, J.: Characteristic Classes. Annals of Mathematics Studies, vol. 76. Princeton University Press, Princeton (1974)
129. Moise, E.: Affine structures in 3-manifolds. V: the triangulation theorem and Hauptvermutung. Ann. Math. **56**, 96–114 (1952)

130. Moore, J.: Lectures on Seiberg-Witten Invariants. 2nd edn. Lecture Notes in Mathematics, vol. 1629, 121pp. Springer, Berlin (2001). ISBN 3-540-41221-2/pbk
131. Morgan, J.: The Seiberg-Witten Equations and Applications to the Topology of Smooth Four-Manifolds. Princeton University Press, Princeton (1996)
132. Morgan, J., Szabó, Z.: Homotopy $K3$ surfaces and mod 2 Seiberg-Witten invariants. Math. Res. Lett. **4**(1), 17–21 (1997)
133. Morita, S.: Characteristic classes of surface bundles. Invent. Math. **90**, 551–577 (1987)
134. Mumford, D.: An algebraic surface with K ample, $(K^2) = 9$, $p_g = q = 0$. Am. J. Math. **101**, 233–244 (1979)
135. Murphy, G.J.: C^*-Algebras and Operator Theory. Academic Press, Cambridge (1990)
136. Nakamura, N.: Bauer-Furuta invariants under $\mathbb{Z}_2$-actions. Math. Z. **262**(1), 219–233 (2009)
137. Nakamura, N.: $\mathrm{Pin}^-(2)$-monopole equations and intersection forms with local coefficients of four-manifolds. Math. Ann. **357**(3), 915–939 (2013)
138. Namioka, I.: Folner's conditions for amenable semi-groups. Math. Scand. **15**, 18–28 (1964)
139. Nicolaescu, L.: Notes on Seiberg-Witten Theory. Graduate Studies in Mathematics, vol. 28, 484pp. American Mathematical Society (AMS), Providence (2000)
140. Novikov, S.P.: On the algorithmic unsolvability of the word problem in group theory. Trudy Mat. Inst. Steklov **44**, 1–143 (1955)
141. Novikov, S.P.: On manifolds with free abelian fundamental group and their application. Izv. Akad. Nauk SSSR Ser. Mat. **30**, 207–246 (1966)
142. Okonek, C., Teleman, A.: Seiberg-Witten invariants for manifolds with $b_+ = 1$ and the universal wall crossing formula. Int. J. Math. **7**(6), 811–832 (1996)
143. Ol'shanskii, A.: On the question of the existence of an invariant mean on a group. Uspekhi Mat. Nauk **35**(4), 199–200 (1980)
144. Ol'shanskii, A., Sapir, M.: Non-amenable finitely presented torsion-by-cyclic groups. Publ. Math. l'IHES **96**(1), 43–169 (2003)
145. Petersen, P.: Riemannian Geometry, 3rd edn. Graduate Texts in Mathematics, vol. 171. Springer, Berlin (2016)
146. Ranicki, A.: Algebraic and Geometric Surgery. Oxford University Press, Oxford (2002)
147. Roe, J.: Elliptic Operators, Topology and Asymptotic Methods. Pitman Research Notes in Mathematics Series, vol. 395, 2nd edn., 209pp. Longman, Harlow (1998)
148. Rokhlin, V.A.: New results in the theory of four dimensional manifolds (in Russian). Dokl. Akad. Nauk SSSR Ser. **84**, 221–224 (1952)
149. Ruberman, D., Strle, S.: Mod 2 Seiberg-Witten invariants of homology tori. Math. Res. Lett. **7**(5–6), 789–799 (2000)
150. Rudin, W.: Functional Analysis. International Series in Pure and Applied Mathematics, 2nd edn., 424pp. McGraw-Hill, New York (1991)
151. Salamon, D.: Spin Geometry and Seiberg-Witten Invariants. Book (unpublished) (1999)
152. Sanov, I.N.: A property of a representation of a free group. Dokl. Akad. Nauk SSSR **57**, 657–659 (1947)
153. Sasahira, H.: Spin structures on Seiberg-Witten moduli spaces. J. Math. Sci. Tokyo **13**(3), 347–363 (2006)
154. Segal, G.: Equivariant K-theory. Publ. Math. Inst. Hautes Étud. Sci. **34**, 129–151 (1968)
155. Smale, S.: Generalized Poincaré's conjecture in dimensions greater than four. Ann. Math. **74**(2), 391–406 (1961)
156. Smale, S.: An infinite dimensional version of Sard's theorem. Am. J. Math. **87**, 861–866 (1965)
157. Sung, C.: G-monopole classes, Ricci flow, and Yamabe invariants of 4-manifolds. Geom. Ded. **169**, 129–144 (2014)
158. Sung, C.: Equivariant Bauer-Furuta invariants on some connected sums of 4-manifolds. Tokyo J. Math. **40**(1), 53–63 (2017)
159. Szymik, M.: Characteristic cohomotopy classes for families of 4-manifolds. Forum Math. **22**(3), 509–523 (2010)

160. Szymik, M.: Bauer-Furuta invariants and Galois symmetries. Q. J. Math. **63**(4), 1033–1054 (2012)
161. Taubes, C.: Gauge theory on asymptotically periodic 4 manifolds. J. Differ. Geom. **25**, 363–430 (1987)
162. Taubes, C., Wentworth, R. (ed.): Seiberg-Witten and Gromov Invariants for Symplectic 4-Manifolds. Edited by Richard Wentworth. First International Press Lecture Series 2, vol. 2, 405pp. International Press, Somerville (2005). ISBN 1-57146-089-6/hbk
163. Taylor, M.: Partial Differential Equations. 1: Basic Theory. Applied Mathematical Sciences, vol. 115. Springer, Berlin (1996)
164. Teichner, P., Vogt, E.: All 4-manifolds have spin^c structures. https://math.berkeley.edu/~teichner/Papers/spin.pdf
165. Teleman, A.: Introduction à la théorie de jauge. Cours Spécialisés, vol. 18, 191pp. Société Mathématique de France (SMF), Paris (2012)
166. tom Dieck, T.: Transformation Groups. De Gruyter Studies in Mathematics, vol. 8, 312pp. Walter de Gruyter, Berlin (1987)
167. Wagon, S.: The Banach-Tarski Paradox. Encyclopedia of Mathematics and its Applications, vol. 24. Cambridge University Press, Cambridge (1985)
168. Warner, F.W.: Foundations of differentiable manifolds and Lie groups. Graduate Texts in Mathematics, vol. 94, 272pp. Springer, New York (1983)
169. Witten, E.: Monopoles and four-manifolds. Math. Res. Lett. **1**(6), 769–796 (1994)
170. Wolf, J.: On the classification of Hermitian symmetric spaces. J. Math. Mech. **13**, 489–495 (1964)
171. Yau, S.-T.: Calabi's conjecture and some new results in algebraic geometry. Proc. Natl. Acad. Sci. USA **5**(5), 1798–1799 (1977)

Index

T. Kato, N. Nakamura, *Monopole Maps and L^2-cohomology*, Springer Monographs in Mathematics, https://doi.org/10.1007/978-981-96-8472-4

The manufacturer's authorised representative in the EU is Springer Nature Customer Service Centre GmbH, Europaplatz 3, 69115 Heidelberg, Germany. If you have any concerns regarding our products, please contact ProductSafety@springernature.com

Printed and bound by CPI Group (UK) Ltd, Croydon, CR0 4YY
01/10/2026
02225912-0018